1403306931

CW01511871

Cranfield
UNIVERSITY

Environmental Law
and Ecological Responsibility

Reflexivity.
motivations

enablers   drivers.

Environmental Law and Ecological Responsibility

The Concept and Practice of Ecological Self-Organization

Edited by

Gunther Teubner
*European University Institute, Florence, Italy
and London School of Economics, UK*

Lindsay Farmer
Birkbeck College, London, UK

Declan Murphy
Trinity College, Dublin, Ireland

JOHN WILEY & SONS
Chichester · New York · Brisbane · Toronto · Singapore

Other Wiley Editorial Offices

John Wiley & Sons, Inc., 605 Third Avenue,
New York, NY 10158-0012, USA

Jacaranda Wiley Ltd, 33 Park Road,
Queensland 4064, Australia

John Wiley & Sons (Canada) Ltd, 22 Worcester Road,
Rexdale, Ontario M9W 1L1, Canada

John Wiley & Sons (SEA) Pte Ltd, 37 Jalan Pemimpin #05–04,
Block B, Union Industrial Building, Singapore 2057

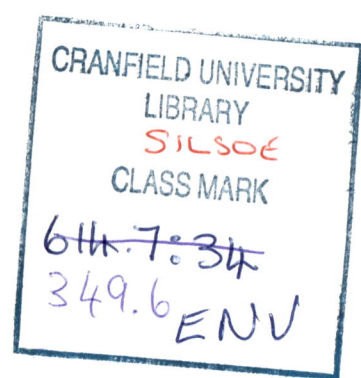

British Library Cataloguing in Publication Data

A catalogue record for this book is available from the British Library

ISBN 0 471 94986 8

Typeset in 10/12pt Sabon by Mathematical Composition Setters Ltd., Salisbury, Wiltshire
Printed and bound in Great Britain by Biddles Ltd, Guildford, Surrey

Contents

About the Contributors

MICHAEL BLECHER   Senior Counsel to the Albanian Government; born 1953; Referendar, University of Frankfurt, 1981; Assessor, Wiesbaden 1984; Dr. jur., European University Institute, Florence (1990); Researcher, Centre for European Legal Policy, Bremen (1991—); Author: *Zu einer Ethik der Selbstreferenz oder Theorie als Compassion—Möglichkeiten einer Kritischen Theorie der Selbstreferenz von Gesellschaft und Recht* (1991).

NATHALIE BOUCQUEY   Researcher at the Law Department of the European University Institute, Florence; born 1966; Licenciée en droit (1990) and Licenciée en philosophie (1991) of the Catholic University of Louvain. Translator (with Gaby Maier): *Le droit: un système autopoïétique* (1993); *Droit et réflexivité: l'auto-référence en droit et dans l'organisation* (forthcoming).

ERIC BREGMAN   Adjunct Professor of Law and Partner, Sive Paget & Riesel, PC; born 1944; AB, Harvard College (1965); LLB, University of Pennsylvania Law School (1968); Law Clerk to Hon. Marvin E. Frankel, US District Judge, Southern District of New York (1968–1970); Adjunct Professor of Law, Advanced Civil Procedure and Pre-trial Litigation, Cardozo School of Law; Yeshiva University (1986—); Adjunct Professor of Law, Environmental Law and Litigation, Pace University School of Law (1993—).

LEONIE BREUNUNG   Sociologist; born 1946; Dipl.-Soz., University of Göttingen (1971); Research Collaborator, University of Freiburg (1972–74); Assistant, University of Hannover (1975–89); Dr. jur., Vrije Universiteit Amsterdam (1982); Research Collaborator, University of Hannover (1990—); Author (with Pieper and Stahlmann): *Der Sachverständige im Zivilprozeß (1982); articles in professional journals and collective publications.*

GERT BRÜGGEMEIER   Professor of Private Law; born 1944; Referendar, Saabrücken (1969); Dr. jur., University of Frankfurt/M (1973); Assessor, Wiesbaden (1973); Professor of Law, University of Bremen (1978—); Visiting Professor, University of California at Berkeley (1980/81); Judge of the Court of Appeal, Bremen (1988—); Director of the Centre for European Law and

Policy, University of Bremen (1991—). Author: *Vorstudien zu einer Wettbewerbsrechtstheorie* (1973); *Zur Entwicklung des Rechts im organisierten Kapitalismus*, 2 vols. (1977/79); *Deliktsrecht* (1986); (with Eike Schmidt) *Zivilrechtlicher Grundkurs*, 4th ed. (1991); various articles in professional journals and collective publications.

LINDSAY FARMER Lecturer in Law, Birkbeck College, London; born 1963; LLB Edinburgh University (1985); M.Phil. in Criminology, Cambridge University (1986); Ph.D., European University Institute, Florence (1993); Lecturer in Law, Strathclyde University (1987–89); Lecturer, Birkbeck College (1993—); various articles in journals and collective publications.

MAARTEN HAJER Assistant Professor of Sociology; born 1962, MA in Urban and Regional Planning, University of Amsterdam (1987); MA in Political Science, University of Amsterdam (1988); Graduate in Politics, University College, Oxford (1988–90); Research Fellow at the Leyden Institute for Law and Public Policy, University of Leyden (1989–91); he recently completed a D.Phil. thesis on environmental discourse in Great Britain and the Netherlands for the University of Oxford; Assistant Professor of Sociology, University of Munich (1993—); Author: *City Politics: Hegemonic Projects and Discourse* (1989); *De Stad als Publiek Domein* (1989); various articles in professional journals and collective publications.

TORU HIJIKATA Associate Professor of Sociology; born 1956; LLB, Chuo University, Tokyo (1981); MA in Sociology, Chuo University (1984); Lecturer, Tokiwa University (1987–88); Assistant Professor, Seigakuin University, Saitama (1988–92); Associate Professor, Seigakuin University (1992—); Visiting Professor, Forschungsinstitut für Philosophie, Hannover (1992—); Author (and editor with Luhmann); *Kitarubeki chi (Knowledge to come)* (1990); various articles in professional journals and collective publications; some translations of Luhmann's work.

KARL HOFSTETTER Vice President and General Counsel, Schindler Management AG; born 1956; Lic.iur., Zürich University (1981); JSM, Stanford Law School (1984); LLM, UCLA School of Law (1985); Dr.jur., Zürich University (1986); Zürich Bar (1987); New York Bar (1989); Habilitation, Zürich University (1989—); General Counsel, Schindler Management AG (1991—); Member of State Parliament, Lucerne (1993—); Author: *Tariffähigkeit und Tarifzuständigkeit von Arbeitnehmerorganisationen: Ein Rechtsvergleich zwischen dem deutschen, amerikanischen und schweizerischen kollektiven Arbeitsrecht* (1986); various articles in professional journals.

ARTHUR JACOBSON Max Freund Professor of Litigation and Advocacy; born 1948; BA Harvard College (1969); JD Harvard Law School (1974); PhD

(Government) Harvard University (1978); Associate, Cleary, Gottlieb, Steen & Hamilton, New York (1975–77); Assistant Professor of Law (1977–80); Associate Professor of Law (1980–81); Professor of Law (1981–88); Max Freund Professor of Litigation and Advocacy. Cardozo School of Law, Yeshiva University (1988—); Counsel, Sive, Paget & Riesel, PC, New York (1989—); Author (with Anthony D'Amato): *Justice and the Legal System* (1992); various articles in professional journals and collective publications.

IDA KOPPEN   Executive Co-ordinator of the European Policy Unit, European University Institute, Florence; born 1957; BA, Honors College, Environmental Studies, University of Oregon (1981); LLM, University of Amsterdam (1985); Fulbright Visiting Scholar, MIT-Harvard Public Disputes Program (1988–1989); Salzburg Seminar (1990); trainer at the Dutch Ministry for the Environment of the course "The Implementation Challenge" (1992—); Author: "The European Court of Justice", in J.D. Liefferink *et al.* (Eds.) *European Integration and Environmental Policy* (1993); "Environmental Mediation: an Example of Applied Autopoiesis?" in R.J. in 't Veld *et al.* (Eds.) *Autopoiesis and configuration theory: new approaches to societal steering* (1991); (with Karl-Heinz Ladeur) "Environmental Rights", in A. Cassese *et al.* (Eds.) *Human Rights and the European Community: the Substantive Law* (1991).

KARL-HEINZ LADEUR   Professor of Law; born 1943; Referendar, University of Bonn (1967); Assessor, Düsseldorf (1971); Assistant, Universities of Giessen, Bielefeld; Dr. jur., University of Bremen (1976); Ass. Professor, University of Bremen (1977–1983); Professor of Law (1983—); Visiting Professor, Amiens (1981); Visiting Scholar, Stanford Law School (1987); Author: (with Ridder) *Zum politischen Mandat von Universität und Studenschaft* (1973); *Rechtssubjekt und Rechtsstruktur* (1978); (with Hase) *Verfassungsgerichtsbarkeit und politisches System* (1980); (Ed. *et al.*) *Ordnungsmacht* (1981); *"Abwägung"—ein neues Paradigma des Verwaltungsrechts?* (1984); *Postmoderne Rechtstheorie* (1992); various articles in professional journals and collective publications.

DECLAN MURPHY   Lecturer, University College Dublin; born 1965; BCL, National University of Ireland (1986); Barrister at Law, of King's Inns, Dublin (1988); Researcher on the political economy of corporate insolvency law, European University Institute, Florence (1988—); Lecturer in commercial law and financial regulation, University College, Dublin (1992—).

JOACHIM NOCKE   Akademischer Oberrat; born 1942; Referendar, University of Hamburg (1967); Assistant, University of Hamburg (1974–79); Dr. jur., University of Hamburg (1979); Akademisher Oberrat, University of

Hannover (1980—); Author: *Wissen in der Organisation* (1980); articles in professional journals and collective publications.

FRANÇOIS OST   Professor of Law and Jurisprudence; born 1952; Professor at the Facultés Universitaires Saint-Louis, Bruxelles (1980—); Dean of the Law Faculty (1982–1993); Director of CEDRE (Centre d'étude du droit et de l'environnement) Bruxelles; Co-Director of the European Academy of Legal Theory; Editor-in-chief of the *Revue Interdisciplinaire d'Études Juridiques*. Author (with M. van de Kerchove): *Le Système Juridique entre Ordre et Désordre, Jalons pour une théorie critique du droit* (1988); *Le Droit ou Les Paradoxes du Jeu* (1992); (co-ed.) *Images et Usages de la Nature du Droit*; various articles in professional journals and collective publications.

MICHAEL POWER   Lecturer in Accounting and Finance; born 1957; BA Philosophy, Politics and Economics, St. Edmund Hall, University of Oxford (1979); MPhil. History and Philosophy of Science (1980); PhD, Girton College, University of Cambridge (1983); Deloitte Haskins & Sells (Chartered Accountants) (1983–1987); Lecturer and Coopers & Lybrand Fellow, London School of Economics and Political Science; Associate Member of the Institute of Chartered Accountants in England and Wales (1987—); Associate Member of the Institute of Taxation (1989—); Research Fellow for the International Accounting Standards Committee (1990–91); Associate Scholar, European Institute for Advanced Studies in Management (1992—); Editorial Board, *Accounting, Organizations and Society*. Author (and ed. *et al.*): "Law and Accountancy", a special issue of the *Modern Law Review* (1991); various articles in professional journals and collective publications.

ECKARD REHBINDER   Professor of Law; Co-Director of the Institute for Foreign and International Law, Frankfurt/M; born 1936; Referendar, University of Frankfurt (1960); Dr. jur., University of Frankfurt (1964); Assessor, Wiesbaden (1964); Assistant Professor, University of Frankfurt (1965–1969); Research Fellow, Institute for Foreign and International Trade Law, Georgetown University Law Center, Washington, DC (1966–67); Privatdozent, University of Frankfurt (1968); Professor of Law, University of Bielefeld (1969–72); Professor of Law, University of Frankfurt (1972—); Dean of the Law Faculty, University of Frankfurt (1981–82); Member of the German Council of Environmental Advisors, Wiesbaden (1987–90, 1992—); Visiting Professor of Law, University of Michigan, Ann Arbor (1980); University of California, Berkeley (1984); European University Institute, Florence (1981 and 1991); Author: *Extraterritoriale Wirkungen des deutschen Kartellrechts* (1965); *Konzernaußenrecht und allgemeines Privatrecht* (1969); (with Burgbacher and Knieper) *Bürgerklage im Umweltrecht* (1972); *Vertragsgestaltung* (1982, 2nd edn 1993); (with Kayser and Klein) *Kommentar zum Chemikaliengesetz* (1985); (with Stewart) "Environmental Protection Policy" in Cappelletti, Seccombe & Weiler *Integration Through Law*, vol. 2 (1982); *Das Vorsorgeprinzip im internationalen Vergleich* (1991); various articles in professional journals and collective publications.

PHILIP SELZNICK   Professor Emeritus of Law and Sociology; born 1919; MA in Sociology, Columbia University (1942); PhD in Sociology, Columbia University (1947); Instructor, Sociology, University of Minnesota (1946–47); Instructor/Assistant Professor, University of California, Los Angeles (1947–52); Assistant Professor, University of California, Berkeley (1952–53); Associate Professor (1953–57); Professor of Sociology (1957–77); Professor of Law and Sociology (1977–83); Emeritus (1 January 1984); Founding Chairman, Center for the Study of Law and Society, UC Berkeley (1963–67); Founding Chairman, Jurisprudence and Social Policy Program, School of Law, UC Berkeley (1978–82); Fellow of the American Academy of Arts and Sciences (elected in 1961); Doctor of Jurisprudence, *honoris causa*, University of Utrecht (1986). Author (major publications include): *TVA and the Grass Roots* (1949); *Law, Society and Industrial Justice* (1969); (with P. Nonet) *Law and Society in Transition* (1978); *The Moral Commonwealth: Social Theory and the Promise of Community* (1992).

SEAN SMITH   Fellow and Lecturer, Emmanuel College, Cambridge; born 1961; LLB Glasgow University (1984); Research student at the European University Institute, Florence (1987—); Fellow, Emmanuel College, Cambridge (1991—); various articles in professional journals and collective publications.

GUNTHER TEUBNER   Otto-Kahn-Freund-Professor for Comparative Law and Legal Theory, London School of Economics; born 1944; studies of law and legal sociology in Göttingen, Tübingen and Berkeley, California (1970); Dr. jur. Tübingen (1971); Assessor, Habilitation Tübingen (1977); Professor of private law and legal sociology, Frankfurt and Bremen (1977); European University Institute, Florence (1981); London School of Economics (1992—); Visiting Professor, Berkeley, Ann Arbor, Stanford, Berlin, Wien. Author: *Standards und Direktiven in Generalklauseln* (1971); *Public Status of Private Associations* (1974); *Gegenseitige Vertragsuntreue* (1975); *Organisationsdemokratie und Verbandsverfassung* (1978); (co-author) *Alternativkommentar zum bürgerlichen Recht* (1979/80); *Recht als autopoietisches System* (1989) (English translation: *Law as an Autopoietic System* (1993)), *Droit et réflexivité* (1994). (Ed.) *Dilemmas of Law in the Welfare State* (1985, 1988); *Contract and Organisation* (1986); *Juridification* (1987); *Autopoietic Law* (1988); *Regulating Corporate Groups* (1989); *State, Law, Economy as Autopoietic Systems* (1992); *Paradoxes of Self-Reference in the Humanities, Law and the Social Sciences* (1992); *Entscheidungsfolgen als Rechtsgründe* (1994).

SECTION I

INTRODUCTION

CHAPTER 1

Ecological Self-Organization

Lindsay Farmer and Gunther Teubner
London/Florence

THE END OF ENVIRONMENTAL LAW

This volume deals with the self-organization of enterprises. In the last 10 years theories have been developed, between the social sciences and law, that have suggested a rather sceptical view of the potential of direct outside intervention as a means of control of economic organizations. Instead, externally induced, internal self-organizing processes have come to be seen as the means of rendering organizations more sensitive to the demands of their environment. This shift has coincided with the demand for a more responsive, and effective, form of ecological regulation, as the extent of the damage to the natural environment has become clear in the same period. The fruits of the collision between these two movements are represented in this collection of essays.

All of the contributors accept that the old-style regulation of enterprise based on a combination of state "command" and enterprise liability for damage caused to the environment has reached its limits. The reasons for its ineffectiveness can be seen to be a combination of those factors relating to its inability to be responsive to the demands of enterprise, those factors relating to the impossibility of generating sufficient knowledge to provide an effective form of regulation in any case, and finally, the unprecedented scale of the threat posed by damage to the natural environment. Indeed, while there is a general acceptance of these conditions, debate has raged over the most effective means of response to the challenge posed by the destruction of the natural environment. Many governments have taken the environmentalist "turn" when confronted by popular concern with green issues. They have been happy to dismantle the existing structures of regulation, only to be confronted by the accusation that behind the façade of fashionable theories of deregulation they are simply shirking their governmental responsibilities: the market or economy will never provide an effective safeguard for the environment.

Environmental Law and Ecological Responsibility: The Concept and Practice of Ecological Self-Organization
Edited by G. Teubner, L. Farmer and D. Murphy
© 1994 John Wiley & Sons Ltd

Equally, attempts to legislate for the environment have met the charge of having constrained enterprise and hampered development in the pursuit of an impossible ideal.

Some relief from this conflict has been found in strategies of "sustainable development" seeking to harness enterprise to the protection of the environment. Roughly speaking, there are three legislative strategies of ecological sustainability and, although the contributors to this volume partially integrate these, they also attempt to develop an approach which differs significantly from all of them. The first is an optimization strategy. Its guiding idea is that an economic cost-benefit analysis should be drastically enlarged to also include non-economic considerations. It argues that political decisions must analyse and take into account the legal, political and social consequences of ecological legislation, as well as the economic costs. Ecological sustainability requires a comprehensive comparison of the advantages and disadvantages of the envisaged legislation. The second may be termed a method of empirical decision. The goal is the rationalization of divergent value preferences, to enable their representation in a decision tree. It is hoped that this approach will allow a greater potential number of values to be taken into account. The third may be called a participatory procedure. Legal procedures are designed to make it possible for a greater number of people to participate in ecological decisions. Again, it is hoped that this method might incorporate more, and more diverse, values into the decisions.

However, from the standpoint of the contributors to this volume, all three strategies suffer from a common deficit: they do not take uncertainty seriously enough. All the above strategies presuppose the availability of natural and social science knowledge which is presumed to be able to cope successfully with the ecological problems at hand. In addition, they are only concerned with the question of how best to organize the existing knowledge and develop it in the short term. This, however, is to beg the question. The starting point for the essays in this volume is this: that uncertainty about ecological risks is irreducible. From this it follows that those strategies which begin from the assumption that it is possible to successfully reduce this fundamental uncertainty are highly questionable. Such strategies place too much confidence in science and do not seriously consider how frequently it is that catastrophes occur before the scientific knowledge necessary to handle them has been developed. Alternatively, they rely too heavily on technology and the economy without taking the negative feedback loops sufficiently into account. Finally, they have a too ready belief in political solutions, without thematizing the dangers of incrementalism or the even greater risks of those fashionable "zero-risk" policies.

A possible alternative to these strategies may be self-organization—understood in a rather broad sense. Its message is this: try to avoid catastrophes and learn by trial and error. Self-organization in these terms does not favour the blind self-reproduction of closed systems. The guiding idea is that of setting up a network of self-organizing forms of co-operation between political institutions, the setting of legal norms, economic and technological action, all of which is subsumed under the ecological imperative.

The particular challenge that this poses for law is the theme that is explicitly addressed by the contributors to this volume. The problem is that of experimenting with novel forms of environmental regulation, assessing existing initiatives, and responding to the challenges of both government and enterprise. The point is summarized very neatly by Selznick when he points out that the central question for a system of *self*-regulation is that of the form and limits of *external* regulation. Beginning from the remarkable resistance by economic organizations to political attempts to influence them from the outside, we must look at whether and how external regulation can learn from this frustrating experience. If one starts from the idea that resistance to regulation is not primarily a question of political reforms being sabotaged by economic power, but is instead a problem of the survival mechanism of self-reference, then it becomes conceivable that regulatory strategies cannot and should not try to break this resistance. It becomes more plausible to think in terms of trying to understand the mechanisms of this resistance and exploiting it for different purposes. In the context of family therapy this is expressed in the idea of "difference therapy". Do not try to teach from the outside. Instead, the system must be induced to produce more knowledge about itself and to reflect upon this knowledge.

Further questions follow from this. How do enterprises react to environmental law? How can the law influence their internal self-regulation in order to make them more responsive to the needs of the environment? And, how can the law encourage enterprises to organize their efforts *collectively* in order to confront the type of complex environmental questions that face contemporary society? These rarely confronted questions provide the basis for all the contributions to this volume, revealing the shared concerns of scholars of different nationalities and from a number of intellectual disciplines. The essays debate the potential of self-organization as a response to the environmental crisis, covering positions that range from an extreme cynicism about the motives of enterprise, to (one might say) an extreme optimism. This range, and the unity of concerns that give the collection its cohesion and richness, can be seen if we briefly consider some of the principle themes that link the essays.

A dominant concern of all the participants was the need to link the theoretical and practical levels. Although the volume has been divided into three sections, moving towards a more detailed discussion of the theoretical concerns, this is in some respects an arbitrary division. Throughout, the more theoretical essays dealing with concepts of self-organization and autopoiesis are anchored to concrete issues, and the more technical and legal essays are informed by theories of self-regulation. Nor is this a trivial point. It reflects the attempt both to discern trends in the contemporary regulation of ecological responsibility, and to subject these movements to the sort of considered and detailed examination that is lacking from current studies in environmental law. Hijikata makes this point in a rather different context in his essay on Japanese enterprise, when he points out that "amidst all the hard work, we have barely begun to reflect on our life". One might paraphrase this to say that in the rush to respond to the crisis confronting the environment, we lawyers have barely begun to reflect on the purposes of environmental law.

The importance of this linkage is best illustrated by looking briefly at the type of discussion that it provokes in relation to two of the other themes of the volume. The first of these is the recognition that the organizational form of enterprise is changing. This is recognized in rather diverse ways by the contributors. For some, drawing on management theory, this is the change to "total quality management", the move to styles of management based on "quality circles" and "competition" between different parts of the enterprise. The modern enterprise, it is argued, cannot survive without introducing flexibility and the revisability of decision-making into its organizational structure (Blecher; Brüggemeier; Hofstetter). Others consider significant the attempt to introduce this sort of organization into the government of the environment (Hajer; Koppen), while others still take up Teubner's writing on the growth of the "polycorporate network", a new form of organization between contract and organization. What is perhaps striking about all these metaphors is their *organic* quality. We are being asked to drop those conceptions that viewed the enterprise mechanistically and which invited the imposition of controls from the outside. The new enterprise is seen rather as a delicate and complex organism, interference with which may have unforeseen and unforeseeable effects. The regulator must recognize its ecology and acknowledge that the natural environment may benefit more from the encouragement of its natural growth, rather than a rude disturbance of the natural order. In both theoretical and practical terms, the problem of the environment is seen as a question of the environment of enterprise; ecology, as Hijikata puts it so well, is the fate of enterprise. The conflict thus arises in relation to the implications of this for the law. At one extreme there is the position argued by Hofstetter, that equates the organic quality of the enterprise with the organic quality of the market, in its classical formulations, and concludes that while the law may have some role in stimulating the self-organization of the enterprise, it must beware upsetting the equilibrium of enterprises by increasing their liability. Ost challenges this position to argue that the voluntary character of much contemporary regulation threatens its efficacy, for the motives of enterprise cannot be trusted. A similar tone of scepticism is raised by Breunung and Nocke who conclude that, at the end of the day, enterprise is enterprise and will continue to serve the profit motive before all else, making irrelevant all attempts to introduce green considerations within the enterprise. Crudely, one might suggest that "Ohnism" (Blecher—the new Japanese theory of management) is "Ohnanism" for enterprise! In Ost's formulation, law must play the role of the "excluded third", setting clear moral and practical goals for enterprise, and working to ensure that these are reached. A number of essays take up the challenge of working to bring these positions together. The changing form of the enterprise is recognized, and it is argued that this must be met with a greater flexibility in the law. However, it is also argued that this will be ineffective unless some sense of the purpose of environmental law is also retained—flexibility and revisability of decisions can no more be ends in themselves for the law than they can be for the enterprise. In the theoretical formulation of Ladeur, this is an argument for the "second-order proceduralization" of environmental law, aiming at the regeneration of the self-stabilizing potential of nature by stimulating mutual "irritation" between enterprise and administration. This type of aim achieves a rather more practical formulation in the proposal developed by Boucquey and Teubner to stimulate enterprises to organize themselves into *collective* risk management by innovations in the legal form of the allocation of liability.

The second theme relates to the importance of information. There is an acceptance that there is necessarily an imbalance in knowledge, both between government and enterprise and between enterprise and public. If this is a game, as Ost's metaphor suggests, it is one in which poker-faced enterprise holds most of the cards because of its greater resources. The challenge, therefore, is to compensate for this by increasing levels of visibility, in such a way as to render information less a private good (for enterprise) than a public one. Once again current proposals for environmental contracts (Rehbinder), environmental performance review (Bregman and Jacobson; Hajer), environmental officers (Blecher; Breunung and Nocke) and so on, which aim to make the enterprise account for environmental decisions, are subjected to detailed theoretical scrutiny. This is clearest in relation to the issue of the "eco-audit", discussed by several of the contributors, which may be described as the attempt to quantify the ecological profit and loss of the individual enterprise by means of accounting techniques. This initiative is given a generally cautious welcome, as something that fits with the changing forms of management, but the benefits of an extended theoretical analysis become clear in the essay by Power. In this essay he lays down a more general challenge to those who simply regard the eco-audit as a means of increasing the visibility of enterprise. On the one hand, this is because it is accounting itself that creates the categories of visibility. Hence, those which are accepted as institutionally legitimate may not in fact reveal anything at all, or may even deter enquiry. On the other hand, he suggests that in a certain sense the initiatives do not go far enough, since they do not take up accountancy's potential as a mode of *transforming* organizational rationality. If accountancy is not an economic rationality, but a "hybrid body of expertise" that facilitates the flow of information within and between organizations, then it contains potential, not only as a means of representing external interests within the organization, but also as a means of reconstructing the interests of the organization. Once again we have returned to the theme of how to stimulate modes of self-organization within the firm.

In short, self-organization means something other than a *laissez faire* approach as managerialism would suggest. There are three factors in particular that would inform a law of ecological self-organization using strong external pressures for internal self-regulation:

(a) A critique of the optimistic assumption that enough technological information can be generated to reduce ecological dangers to the merely "residual risk" of a not-yet-complete knowledge. This implies scepticism towards the prevailing approach of basing ecological law on a quantification of ecological risks.

(b) The reflexivity of ecological law. The primary role of environmental law is not that of prescribing ecologically correct behaviour on the basis of accepted technological knowledge. Instead, environmental law should "irritate" economic and technological practices and induce them to produce new ecological knowledge. Law should favour reversibility and institutionally guarantee that options and alternatives in the economic and technological sectors are kept open.

(c) Legal institutions of self-organization. Environmental law should systematically favour those legal institutions that try to combine external constraints with the internal potential of producing ecological knowledge. Of special interest here are those institutions—ecological accounting, environmental officers, ecological contracts and covenants, liability rules—which have the potential to induce methods for coping with ecological risks.

FORMS OF ECOLOGICAL SELF-ORGANIZATION

The contributions have been divided into three sections, confronting three aspects of this problem. The first addresses the question of the way in which existing legal mechanisms for allocating environmental liability influence self-regulation within the enterprise, and how effective they can be in responding to environmental problems. The second turns to look at existing alternative legal strategies in an attempt to assess whether they might better correspond to the self-regulatory capacities of enterprises. Finally, in the third, a more general theoretical question is raised. Here it is asked under what conditions the regulation of self-organization can make enterprise more responsive to ecological issues.

The opening essays of the volume look at trends in current liability law. Teubner looks at the recent reductions of the causal requirement in liability law, showing how this creates forms of collective liability—even if lawyers are reluctant to recognize it as such. His central concern is the question of how firms react to this collective liability, and the way in which its creation has had perverse effects. He argues that collective liability should be interpreted as an organizational opportunity to deal collectively with ecological issues for the firms that fall within this wider construction. He confronts the question of how the law can support these opportunities in order to facilitate collective forms of ecological risk management, arguing in particular that a collective approach to risk management can be developed if the law provides mechanisms through which the collective liability can be reinternalized within the firm. This argument is developed by Boucquey in relation to the operation of markets for pollution rights, the so-called "bubble", an initiative that has proved popular in the USA. Pointing out that these have given rise to particular concentrations of pollution (the "hot spots"), that run contrary to the purpose of environmental law, she argues that the ecological purpose should be reintroduced through law, to create an ecological institution equivalent to the market that would help, through ecological self-organization, to solve the liability problems to which the bubbles have given rise.

Brüggemeier and Hofstetter look at different aspects of enterprise liability under the current law. Brüggemeier looks at the specific characteristics of both environmental risks and the post-Fordist organization of industrial production in order to see how various liability strategies fare in coping with the ecological problem. Accepting that civil liability will continue to play only a minor role in relation to administrative control and regulation, he nonetheless argues for the existence of a "production liability", analogous to product liability. He argues that through the evolution of this concept, the tort liability of enterprise can develop into a liability for management defects. This would create incentives for firms to introduce the best available quality and environmental protection systems, thus remedying some of the traditional deficiencies of tort law. Hofstetter, by contrast, looks at the law relating to the ecological liability of multinational groups with limited liability shareholders. The question that he faces is one of how liability law can deal with the potential distortions caused by limited liability corporate structures. Arguing that the central issue is that of how risk should be allocated in order to sustain the development of enterprise, he discusses problems in the economic analysis of law. Drawing on systems theory he concludes that liability law should be reconceived as "an instrument offering certainty within uncertainty to private enterprise" (p. 111), in such a way as to encourage investment. Parent liability can be a means of stimulating self-organization, developing the risk-learning capacities of enterprises, but he concludes with the warning that it is important that the law should not create too much liability. The final essay of the section, that of Smith, lays down a more fundamental challenge to the views expressed by Brüggemeier and Hofstetter as to the potential of liability law to handle environmental risks. He argues that, far from being an instrument by means of which society effectively responds to risk, tort law is actually too ill-fashioned and insensitive to register either the degree or scale of actual environmental risk. The law, he claims, hides risk from society. The attraction of tort law is, of course, clear: complex problems, particularly those involving environmental pollution are broken down and presented in a manageable form. The usefulness of the law should then be a result of its ability to immunize action from the uncertainty of the future. In practice, Smith argues, tort law seems to make the legality of an action depend on an uncertain future, forever postponing judgment about the legality of activities until harm has occurred. Businesses no longer know which precautions to take, and the risk is ultimately passed on to the insurance economy. The challenge posed to law by environmental risks, then, is precisely that of developing new forms of tort to deal with novel risks. It is rather that of beginning to *unthink* what it is that tort law actually does in our society in order to understand what it can do.

In the first essay of the second section, Rehbinder deals in a comparative way with the theme of environmental contracts in the USA, Germany and Japan. These are agreements between enterprises and bodies interested in environmental quality—governmental and local authorities, environmental groups and local communities, with a particular emphasis placed upon the latter. He argues that in contrast to the normal perception of these agreements as a response to over-regulation on the part of government, they are perhaps better looked on as a response to under-regulation—a means of dealing with the deficiencies of the system. He argues that, in spite of wide national differences in their use, their advantages might be seen in their ability to overcome acceptance conflicts, the establishment of a relationship between the community and the enterprise, that they provide a means of compensating for the disadvantages of those affected, and that broadly they offer a mechanism for the fine-tuning of the law.

The issue of agreements, this time between government and enterprise, are raised again in the essays of Hajer and Koppen on Dutch environmental policy. Hajer chronicles the history of *"verinnerlijking"* or partnership, as a central part of Dutch policy in the last decade. This was based on the idea of sustainable development—that enterprise and environment were not incompatible, but that firms could indeed be central actors in the process of environmental renewal if they worked with government and internalized ecological aims. Characterizing the movement to *verinnerlijking* as a discourse coalition, he argues for the limits of such a positive management approach because it proceeded on the basis of assumptions that were never questioned: for the enterprise continuity was always put before sustainability and, in the rush for the novel idea, there was never any serious examination of the conflicting interests at stake. He concludes by examining the possibilities of enhancing ecological responsibility through communicative means of public ecological discourse, such as environmental performance review. Koppen looks on the Dutch initiatives in a rather more favourable light, analysing the importance of informal negotiations in the conception and adoption of environmental provisions. Taking the case study of Dutch waste reduction policy, she analyses negotiations in terms of their capacity to bridge different systemic languages. She argues that recent negotiations can be characterized as "third generation policy instruments", which are distinctive for their communicative informality. Through these instruments common goals and common perceptions of problems can be defined between government and enterprise, creating a shared notion of self-regulation supported by the quasi-contractual nature of the agreement which is formed.

The issue of the efficacy and desirability of agreements is raised once again in the essay by Bregman and Jacobson, which additionally examines a wider range of instruments in American environmental law in terms of their capacity to promote self-regulation. These are building codes, marketable pollution rights, private enforcement rights, and environmental performance review. On the basis of their definition of self-regulation as a system which allows the participation of the regulated in the setting of the content, the procedures for enforcement and, in a broader sense, the definition of the public interest, they argue that only environmental performance review can be considered to meet the criteria of adequate self-regulation and, in common with Koppen, they argue that these procedures, however informal they may look, must be considered as a new form of law, rather than simply being bargaining "in the shadow of the law".

The final two essays of this section debate, from very different theoretical perspectives, the possible effectiveness of the environmental officer introduced by German enterprise law. For Blecher, the role of the environmental officer can be effective if it can be integrated into new forms of management in the enterprise. In fact, he argues that total quality management, the form taken by the "knowledge creating company", must be the key to understanding *any* form of the ecological responsibility of enterprise. He argues that as the enterprise becomes more open to flexibility and innovation, environmental law can profit by misreading total quality management as ecological quality management, referring to the dynamic legal coupling of the enterprise with all its environments. In other words, this is to argue that the enterprise's existing attention to its environment can be diverted by the appropriate legal mechanisms into an attention to the natural environment. Breunung and Nocke, by contrast, show an extreme scepticism towards the entire concept of ecological responsibility, and hence towards the possible success of the institution. Using autopoietic theory, they develop the argument that as economic institutions, enterprises are bound to follow the profit motive. Thus, even while they appear to be open to ecological considerations, this is always dependent on this being profitable at the same time. In more general terms, they suggest that the problem is the concept of ecological responsibility itself, for in its application to the behaviour of firms, it

> blurs the boundary between politics/law and the economy to such an extent that ultimately, if the blend is looked at closely, it makes either the concept of the economy or that of politics superfluous (p. 291).

The answer is to be found in a more rigorous theoretical approach to economic, political and legal institutions.

So we come to the final section where some attempt is made to analyse the theoretical implications of a theory of self-organization. In a complex and wide-ranging essay Ladeur discusses the issue of how environmental law can cope with the situation of uncertainty caused by the impossibility of developing sufficient information about chemicals and their effect on the environment. He suggests that environmental law has traditionally developed by means of the concept of experience that allowed the law to accumulate knowledge about dangers. This, he argues, is a concept of law that is completely unable to cope with either the imbalances in the social distribution of knowledge or the form of the modern enterprise. The first of these problems requires a transformation in the way that the law conceives of property in order to take account of the importance of information in modern society. In his conception, property should be valued for "its dematerialized function of generating knowledge" (p. 316). In relation to the second, he argues that as the enterprise is proceduralized, attaining higher levels of flexibility and capacity for innovation, it follows that the law must also be proceduralized ("a second-order proceduralization") in order to hold itself open to reflection and revision.

Ost's essay challenges the paradigm of autopoiesis, suggesting that it tends to focus on the efficacy of the means, rather than the desirability of the result. In its place, he develops the idea of the law as a game, as a means of restoring some sense of "social interactions in which the firm is an important, though not exclusive, actor" (p. 339), and of the sense of play and improvisation that characterizes them. The paper takes the form of a warning: beware that the compromise between enterprise and government is not simply the show of compromise; that the impact study merely offers a front of objectivity; that the encouragement of green consumers is not simply another marketing strategy. His conclusion is that self-organization can only work within a broader normative framework, before the impartial judge, the *tertium* excluded by autopoietic theory.

These extended theoretical pieces are succeeded by a short meditation on the nature of responsibility. For Hijikata the challenge of the debate on ecological responsibility is whether it can have any significance outside the Western world. He shows how Japanese religious culture is based on a different conception of the individual and his or her relation to nature. The question that he poses is that of how ecological responsibility can function in a culture that lacks the risk-calculating subject of Western law.

The section is brought to a conclusion with an essay from Power, which once again demonstrates the rewards of an interdisciplinary approach. In his discussion of the ability of accounting techniques to foster the ecological responsibility of enterprise, he is able to develop some of the proposals made by lawyers earlier in the volume, whilst cautioning that accounting initiatives might be more placebo than panacea if not used carefully. In a discussion of autopoiesis, he claims that the "categories of law, science and economy are too crude to characterize the reality of institutional differentiation and interaction" (p. 372). This, is a function that is carried out by "hybrid bodies of expertise", such as accounting, which give rise to different communicative potentials and restrictions. He then addresses the role of accounting as a representational media, and examines its capacity to effect ecologically relevant transformations of corporate behaviour—a power that it may already possess because it already organizes technology and information flow within and outside the organization. He argues that the power of accounting is something that is not only informational but also expressive and symbolic, and that the success of green accounting initiatives may depend on their ability to use this to create new informational categories and symbolic linkages—in something akin to the risk-pool discussed in the first essays in the volume.

Finally, the volume concludes with critical and cautionary remarks by Selznick. He reviews the general issues of ecological self-organization discussed in the volume, and embeds the discussion in a broader framework of institutional theories. He points out that the institution must be seen as a carefully designed mode of social integration—a fact which forces us to confront the moral significance of institution building and the internal morality of institutions. For Selznick this takes the form of the requirement that we build moral competence, in the sense of greater accountability, into the institution. Following from the insight that the question of self-regulation is really one of external regulation, he concludes that the challenge posed by the theory of institutions is that of establishing regulated forms of openness, able to look both into the institution, and from the institution out.

SECTION II

ECOLOGICAL LIABILITY AND ITS EFFECTS ON SELF-ORGANIZATION

The Invisible Cupola: From Causal to Collective Attribution in Ecological Liability

Gunther Teubner[1]

London/Florence

LOOSENING CAUSAL LINKS

When I speak about the cupola, it is not Brunelleschi's architectural master-piece crowning the Florentine duomo that appears before my eyes. Rather, I see the ugly architecture of organized crime, the brutal super-secret "cupola" on top of the Mafia's hierarchy which overshadows social life in Italy. Does this cupola exist or not? Over the recent years, the Mafia's cupola has been the object of a bitter struggle between Italian judges. While the lower courts have been sending numerous mafiosi to prison because they were part of the invisible cupola controlling "*tutti i grandi delitti*", the court of the next instance has stubbornly denied the cupola's sheer existence and has set the Godfathers free. In February 1992, in a landmark decision, la Corte di Cassazione for the first time acknowledged the cupola's legal existence. Is the cupola a phantom—a fantasy product of paranoic judges? Or is it a social reality that lawyers have to face if they want to understand organized crime and macro-criminality? In my view it is neither one nor the other; neither lofty legal fiction nor hard social reality "out there". The cupola is an artificial construct of legal architecture that serves one overriding purpose: it allows the punishment of individuals whenever certain facts can be proven that indicate their membership in this "organization". Whoever is a member of the cupola, is as such responsible for the mafia's crimes. The legal construction of the cupola makes the causal attribution of individual crimes superfluous by replacing it with collective attribution. It transforms individual liability into collective liability.

Environmental Law and Ecological Responsibility: The Concept and Practice of Ecological Self-Organization
Edited by G. Teubner, L. Farmer and D. Murphy
© 1994 John Wiley & Sons Ltd

The contours of a similar cupola are emerging from some recent legal constructions of ecological liability. Ecology is a world of tremendously complex interacting causes. The complexities of causation in the three ecological media—air, water, soil—have frustrated lawyers in their attempts to construct causal links between individual actions and ecological damage. Accordingly, they tend to rely less and less on the underlying architecture of causation, resorting instead to rather audacious legal constructions that weaken the structures of individual causation. Causation-in-law, *prima facie*, enhanced *res ipsa loquitur*, reversing the burden of proof, probabilistic causation, joint and several liability in multiple causation, enterprise liability, market share liability, Superfund liability—all these new forms of "risk liability" (Robinson, 1985) tend to reduce or even eliminate individual causal linkages between acts and damages and to replace them with an overarching cupola of quasi-collective responsibility for ecological damages (Bush, 1986: 1480ff.; Abraham, 1987: 859ff.)

While lawyers are busily constructing the new cupola, they are at the same time anxiously trying to keep it invisible. They hide it behind the conceptual façades of "probabilistic causation", "risk liability", and "lost value" which stress the individual nature of risk contributions and do not speak about collective attribution (King, 1981; Robinson, 1985: 749; Rosenberg, 1984: 866ff.; Celli, 1990: 652ff). George Priest, one of the leading architects of the new constructions of risk liability emphatically denies that these changes represent "a decline in commitment to individual responsibility or, perhaps, a shift of expectation toward an impersonal or a collective responsibility." On the contrary,

> far from incorporating a diminished view of individual responsibility, the shift of the law's purpose toward risk control represents a vastly expanded commitment to standards of individual responsibility. (Priest, 1990: 214).

Against these anxious attempts to hide the ecological cupola, I want to argue in this chapter that it makes a lot of sense to make the cupola visible. It is crucial to the further development of ecological liability to understand more precisely under what circumstances and in what way courts and legislatures are shifting liability away from the individual responsibility of single actors towards a new collective responsibility of risk networks. As one commentator puts it we need to confront the "emerging shift in tort law from individual to group responsibility and its legal, political, and philosophical implications" (Bush, 1986: 1473). It is important to understand both the implications for legal doctrine and the real-world consequences of this new architecture. It is far from clear whether the law of ecological liability is able to control this process of risk collectivization in such a way that the obvious loss of individual responsibility is outweighed by the gains of collective responsibility. This depends mainly upon the dynamics of self-organizing processes in the real world and upon the way the law perceives and reacts to them.

I shall put forward the following four propositions:

(a) By loosening the causal links between acts and ecological damages, the courts have come to realize how inadequate individual attribution of responsibility is in a world of complex interacting causes. Ecological interdependencies press the law to substitute the dominant actor perspective by a systemic perspective. The law tends to create new forms of risk pooling, and even, in some cases, outright formal organizations of risk management which seem to be more adequate for the characteristics of certain ecological risks.

(b) The new risk pools created by liability law pose a challenge to the capacity of legal doctrine to design their external relations and internal structure. What are the limits of the organization? Who is a member? What organizational rights and duties are imposed? What are the rules that govern the internal redistribution of liability? These are some of the difficult new problems for the law of ecological risk associations.

(c) With the re-entry of these legally created risk pools into socio-economic reality, self-organizing processes of collective action are set in motion. Their effects on ecological responsibility are rather ambivalent. On the one hand, there are clearly negative side effects—moral hazard, free riding, loss of individual incentives. On the other hand, new forms of collective risk management are emerging which may compensate for those side effects and may even create incentives for collective ecological innovation.

(d) The question of whether law is in a position to partially direct these self-induced developments via institutional design will be decided by the strength and quality of self-organizing processes. The negative effects of collective action can be partially compensated for by a secondary re-individualization within the risk pools. More importantly, tendencies toward collective risk management can be strengthened if the law helps to organize those corporate actors that reallocate risks, monitor behaviour and develop new technologies. This may make it necessary to blur the line between "private" liability mechanisms and "public" regulatory institutions.

COLLECTIVE ELEMENTS IN ECOLOGICAL LIABILITY

Lawyers tend to underestimate systematically the dramatic changes that occur when, under the pressure of ecological damages, the causal links are broken. They like to see it as a merely technical evidential problem which can be solved by reducing the rigid legal requirements of proof (e.g. Nicklisch, 1991:

346ff.). Either one replaces full proof by statistical evidence or by the means of legal techniques such as *prima facie*, reversing the burden of proof, or rebuttable and non-rebuttable presumptions, or one shifts the evidential standards. The most lawyers are willing to admit is that causation as a legal concept, which has historically already changed from causation-in-fact to causation-in-law, is undergoing a new change towards probabilistic causation and risk responsibility (Robinson, 1985; Celli, 1990). In any case, the changes are regarded as being limited to the concept of causation itself with the general principle of individual responsibility remaining intact (Priest, 1990: 213ff).

It is as if these lawyers blind themselves to avoid facing the ecological cupola. As I have already said, whenever the law loosens legal requirements for causation it is necessarily creating constellations of collective liability. Individual actors are made responsible for deeds which other actors may have committed. Their responsibility is no longer exclusively connected to actions of their own that caused the damage, but is mediated by the overarching cupola that covers them and other actors. They become part of one and the same risk-creating community—the eco-mafia, as it were.

This is already true of the seemingly harmless cases in which legal certainty is reduced from full to probabilistic proof of causation. To the extent that full causation differs from statistical causation, individual actors are burdened with an additional responsibility for actions that other actors have committed and over whose actions they have no control. This realm of collective liability is growing the more we move toward reversal of the burden of proof or towards legal presumptions of responsibility. If a defendant is not in a position to prove that his actions could not have caused the environmental damage, he also bears the responsibility for the actions of others.

What we find in these constellations is an asymmetric collective responsibility, a kind of "horizontal" vicarious liability. With "vertical" vicarious liability in hierarchical organizations, employers are made responsible for the actions of their employees. Here the law, by imputing causation where no causal link can actually be proven, makes one individual vicariously responsible "horizontally", for wrongs that have been committed by some other members of a group of actors.

This asymmetric responsibility becomes symmetric in the cases of multiple causation with contribution, market share responsibility, Superfund liability, and, more generally in cases of risk liability. Here, it is the actual membership in a class of risk-bearers that makes individual actors mutually liable for the actions of each other, independently of the factual causal chain that led to the damage. People like to call this technique "risk contribution" for it can once again be attributed to individuals (Celli, 1990). However, our context is not penal law where sanctions can be imposed for the mere creation of a danger independently of actual harm. Our context is financial liability for ecological damages. Here the question is: Who pays for the damage? And the answer is: The collective is responsible. If you happen to be a member, you have to pay.

This is fully-fledged collective liability. Membership, not action, determines liability.

Thus, what Priest (1990: 214) and others euphemistically call the new risk law's "vastly expanded commitment to standards of individual responsibility", is, in reality, only the desperate attempt of sworn-in individualists to cope *post factum* with the perverse effects of their having manipulated the law of causation. The sacred principles of individual causation have been violated, even if they do not admit it. The original sin against the holy principles of individualism consisted in accepting probabilistic and statistical evidence (Bush, 1986: 1493). The inevitable effect is the creation of a "liability collective" which in principle has no boundaries. One should not try to downplay this effect, but instead face the radical consequence that under such a concept of risk contribution every action (and omission) in the whole society is risk generating (Priest, 1990: 215). And, as a consequence of this, it becomes necessary to construct artificially new secondary criteria that define the somewhat arbitrary boundaries of a risk community and to re-individualize responsibility via a similarly arbitrary risk apportionment within such a risk community. This is what the legal architects are actually doing. First they collectivize liability, and then they try to do away with the consequences of their own action by re-individualizing it as much as possible. After creating *de facto* a collective liability regime they have to define intra-organizational rules of loss allocation in order to determine to what degree a member has to bear an individual share of the collective's responsibility.

How are we to interpret these dramatic changes in ecological liability law? Obviously, they cannot be limited to problems of causation alone. They give rise to the more fundamental question of whether for ecological risks with a complex web of interacting causes, "it is no longer tenable to attribute them to individual decisions (rational, intuitive, or habitual)". Should we not, instead, experiment with a "strictly sociological approach" to these ecological risks (Luhmann, 1993)? Such an approach would identify networks of communication as the exclusive bearers of risk, rather than individual or corporate actors. It would no longer focus on individual preferences, individual risk perception, individual choices and individual responsibility but on communication about ecological risks as self-organizing social processes. Driven to its extreme, such an approach would expect that the attribution of risk to communication in modern society takes place even in cases where a decision cannot be identified at all (Luhmann, 1991: 130).

It is the very interwovenness of ecological risks in these constellations that drives the fundamental assumptions of liability law into a deep crisis. While the reasons for the crisis, and its symptoms have been identified, it is rather unclear what direction ecological liability law is going to move in. Specifically, three conspicuous features of modern technology have been identified that render causal isolation for the purposes of individual attribution almost senseless (Bechmann, 1990: 128ff.; Bechmann, 1991: 222ff.; Wagner, 1990: 27ff.).

There is, first, the so-called butterfly effect. Small-scale technological effects gradually accumulate and lead to sudden catastrophic changes. Secondly, there are problems of interference. Several technologies may have unanticipated effects in their combination with each other. Thirdly, situations of highly improbable coincidence arise when two or more causal chains intersect in a non-foreseeable way.

These problems of causal isolation make it almost impossible to identify the traditional elements of liability rules. These are the conceptual symptoms of the structural tensions between ecological risks and liability law:

(a) How can we identify the "individual actor", the "action", and the "injury" if we face a gradual long-term development with the interference of several risks?

(b) How can we disentangle a causal link between action and damage if causation is multiple, interwoven and circular?

(c) How can we define the boundaries of the potentially dangerous actions?

(d) How can we identify the victims as social units if they form an amorphous mass (ecological damages, future generations)? (Rabin, 1987: 27ff.; Brüggemeier, 1991: 297ff., and see Chapter 4; Luhmann, 1993; Schmidt, 1991: 378ff.).

Let us try to determine more precisely the collective element in ecological liability. Obviously, it is not only collectivization in the sense of the well-known shift to impersonal liability, from the traditional highly personalized liability to a standardized liability of "role bundles" of legally constructed "persons". Rather, it is individual attribution to "persons" as such which is increasingly being doubted. However, since a very specific collectivization seems to occur it is also too simple to speak—as Bush (1986) and Abraham (1987) do—only of a general shift to "collective liability" (Köndgen, 1991: 105).

Nor can the new ecological liability be subsumed within the standard cases of collective liability—vicarious liability and respondeat superior—in which either managers in the upper echelons of a hierarchy or the whole corporate actor are made liable for actions of organization members (Prosser & Keeton, 1985: § 69). These are attempts by the law to change the risk perception of whole organizations instead of individual actors (see Brüggemeier, Chapter 4). But in our case of ecological liability there is no pre-existing overarching organization, no hierarchy, no corporate actor which would co-ordinate action and serve as the point of attribution.

Equally, situations of "piercing the veil" in corporate law, or the extension of liability beyond the traditional boundaries of the corporate limits of liability or contractual privity in contract law, are not comparable (for ecological liability of groups, see Hofstetter, Chapter 5; for the limits of contractual privity, see Adams & Brownsword, 1990). In all these constellations the law still refers to socially pre-existing corporate actors or at least, some stable co-operative arrangements and makes the encompassing collective unit responsible for the members' actions. Admittedly, in these cases, the law does redefine and reshape the collective units according to its liability policies, frequently against the intent of the actors involved. But these extensions of individual liability into collective liability do not and cannot go beyond the boundaries of a pre-existing collective social arrangement.

However, when the new law of ecological risks attaches liability to markets, to "bubbles" of air pollution, and to contaminated sites, it does not focus on actors, whether individual or corporate, and their risk perceptions. It totally abandons the actor perspective and focuses on risk communication as such. Risk communication is generated within social configurations that cannot easily be identified as organized systems. And the law attaches liability to this risk communication, to a set of activities as such (a product market, a contaminated site, an air pollution bubble) which do not necessarily aggregate into the organized will-formation of a corporate actor. *The crucial difference seems to be that the law does not refer to pre-existing social units but creates anew networks of ecological risks.* It does not attempt to change the risk perceptions of existing collective actors created as a result of organizational decision making. Rather, it tends to influence ongoing, non-co-ordinated risk communication in a social area which is defined by attributing a specific level of ecological risk. And whenever individual or corporate actors enter such a communicative space they become compulsory members of such a risk pool by force of the liability law. They become collectively liable without regard to their psychological will-formation and their social aggregation, and without regard to the causal links between their individual action and the damage.

In some cases, the law even goes one step further. It seems as if modern ecological liability in the new American Superfund (Stewart, 1991), as well as in the most recent discussion of ecological risk associations (*Umweltgenossenschaften*) and ecological funds in Germany, is inspired by Old Bismarck with his compulsory associations of the *Berufsgenossenschaft* from 1884 (Kinkel, 1989: 297; Rehbinder, 1989: 161; Wagner, 1990: 52ff.). Here, law not only creates compulsory risk pools for certain categories of actors for liability purposes, but it also creates fully-fledged corporate actors—compulsory associations which have a corporate capacity for action. Risk pools are not only made liable for failing certain ecological standards but also for failing to take corporate action in collective risk management. Here, we face the merging of collective liability with the active regulation of collective innovation (Rose-Ackerman, 1990: 746).

Now we understand better why some lawyers emphatically warn against manipulations of causation-in-law (Epstein, 1985: 1377; Medicus, 1986: 781, 785; Abraham, 1987: 898). What began as a simple equitable demand—not to leave uncompensated victims in cases involving complex causal networks—becomes a fundamental transformation of individual liability into collective risk pooling. Those lawyers see this manipulation of causation as an "overshooting", as

> creating wrong incentives for prevention and falsifying resource allocation, since
> it burdens actors who have not caused the damage with costs (Rehbinder, 1989:
> 157; see also Assmann, 1988: 111).

However, this argument overlooks two crucial points. First, it would not only
be unfair but even more inefficient if victims remained uncompensated in those
cases of complex interwoven causes. There would be no incentive for prevention at all, and resource allocation would be even more misdirected
(Rehbinder, 1989: 157). The choice at hand is not collective liability versus
individual liability since the complexities of causation make individual liability
virtually impossible. The real choice in these cases is collective liability or no
liability at all.

Second, the critics indeed hit the mark with their claim that actors bear costs
for damages which they did not cause, but they ignore the cupola. It is as if
they do not want to see the potential of collective attribution. This has to do
with their ideology of methodological individualism which compels them to
systematically devalue collective action by dissolving the corporate actor in a
nexus of individual contracts. They cling to individual attribution in a situation where, as they well know, it has become totally inadequate, nay, where
it is impossible to establish individual causal links. If one admits, however,
that in these situations risk, action, causation and responsibility can be
attributed to risk pools created by *fiat* of the law then one sees clearly that the
law burdens precisely those collective actors who have actually caused the
damage. Incentive creation and resource allocation must then be rethought, so
that they do not only deal with individuals and pre-existing corporate actors
but also with the newly created risk pools.

THE CONTOURS OF THE CUPOLA

When lawyers are replacing the single causal links of individual responsibility
with the overarching cupola of collective liability, they are confronted with a
whole new set of construction problems; the questions of how to design the
inner architecture. How far can one expand the outer contours of the ecological cupola without the whole edifice breaking down? Who is covered by
the roof of the cupola as a member? What duties does a cupola dictate for its
members? How does the cupola distribute losses among its members?

The first question for legal doctrine is how to identify the risk pool. When
the causal links are broken, one of the hardest tasks for a collective liability
is to find reasonable boundaries for limiting the range of risk creating activities. Any social communication can potentially contribute to ecological risks,
and thus the ecological cupola seems to cover the whole society. "Almost every
human action will increase the probability of occurrence of loss in all contexts.

It follows, thus, that under modern conceptions of risk, no action is ever truly innocent" (Priest, 1990: 215). And the main question becomes: "How far do we expand enforcement of tort standards based on probabilistic measures of harm?" (Robinson, 1985: 796).

Desperately seeking a corporate actor is how the courts initially tried to cope with multiple causation in different countries, after they came to realize that the individual actors had been lost. Apart from attempts to narrow the problem down to the technical question of proof of causation, the more conscious doctrinal attempts to design collective liability were looking for workable criteria that would identify an acting collective, a purposive organization or at least planned co-operation, which could be made responsible as a whole.

In the United States, under the original tort doctrine of "concerted action" joint liability could be imposed upon:

> all those who in pursuance of a common plan or design to commit a tortious act, actively take part in it, or further it by co-operation and request, or who lend aid or encouragement to the wrongdoer, or ratify and adopt the wrongdoer's acts done for their benefit. (Prosser & Keeton, 1985: 323).

This doctrine tries to identify a collective through certain characteristics of individual actors (intention to co-operate) and structural characteristics of action (interwovenness). It thus sticks as closely as possible to the original actor model: "the act of one is the act of all, and liability for all that is done is visited on each" (Prosser & Keeton, 1985: 346). However, the structures of ecological risks do not follow the logic of co-operation and, therefore, strain the doctrine of concerted action. Express agreements to damage the environment are not common, tacit understanding is difficult to prove, and seems anyway to be alien to the typical constellation of multiple causation, and especially of ecological risk causation.

The courts then attempted to develop the idea of collective liability and invented so-called "enterprise liability" by further loosening the requirements for conscious, planned and purposive co-operation (Sheiner, 1978: 995–1006). In *Hall* v. *DuPont*, the court was satisfied that there was a "joint enterprise" between otherwise independent actors only where there was "joint control of risk" through common adherence to industry safety standards and the delegation of the functions of safety investigation and design to a jointly sponsored trade association (345F Supp. 353 (EDNY 1972) at 375–6). Although a formal joint venture is not required to prove "joint control", at the very least evidence should be provided of shared research, joint testing of products, and joint legislative lobbying (*Connor* v. *Grand Western Savings & Loan Association* (1968) 69 Cal. 2d 850). A minimal requirement for enterprise liability is an "insufficient, industry-wide standard of safety as to the manufacture of the product" (Sheiner, 1978: 995). But again, this doctrine fails to catch the typical risk structure and even turns out to be counter-productive. It cannot cover "parallel conduct" on the part of the manufacturers (Bush, 1986: 1483; Spitz, 1990: 626) and even has the strange effect of privileging parallel conduct in relation to collective attempts of joint risk control. Attempts at joint risk control are punished by imposing enterprise liability while competitive individual "parallel conduct" without any attempt at collective risk management does not give rise to liability. Enterprise liability thus has the perverse effect of creating disincentives to industry-wide co-operation and risk control.

Following this, it seemed only logical to give up the vain search for a corporate actor and to choose the market itself as the new collective unit on which to base a collective liability, despite its non-co-operative and competitive character. This is what happened in the famous "market-share liability" (Abraham, 1987: 861ff.). Any attempt to find an overarching organization, be it a firm, a corporate group, or a co-operative network which co-ordinates action in a "joint enterprise", is dismissed in favour of the search for the "relevant" market.

Whenever you enter a market, you become responsible for the ecological risks which the market is creating. This is a very strange idea which seems to contradict the well-founded principles of collective liability, presupposing co-operation, joint activities, and mutual control (Bush, 1986: 1477). The competitive market is the very opposite of a joint enterprise as a plausible basis for collective liability (French, 1982; 1984). How can I be made liable for the actions of others if there is not any co-operation among us and if I have no control whatsoever over their actions which could justify common attribution? This contradicts the commonsensical understanding that "companies are liable only for the harm their own product does, not for their rivals' damage" (*The Economist*, 29 February 1992, p. 16). And the fact that market share liability limits my liability to my market share does not really make things better. I will still be made responsible for the damaging actions of others and the limitation to the share only anticipates the existing apportionment of losses within the risk pool (Bush, 1986: 1485; Abraham, 1987: 862).

Looking to doctrinal developments in my own country there are similar tendencies which produce similar dilemmas (Medicus, 1986; Assmann, 1988; Brüggemeier, 1991; Köndgen, 1991). The German *Bürgerliche Gesetzbuch* recognizes joint liability for multiple tortfeasors whenever each tortfeasor could have caused the whole damage and the action was committed jointly (§ 830 I2 *BGB*). Initially, the courts demanded a co-operative link between the actors, at least in the sense of mutual knowledge of the dangerous activities in the mind of each of the tortfeasors. But this link has been progressively loosened so that today a certain "unitary spatial or temporal connection" is sufficient for joint liability. The Federal Court (*Bundesgerichtshof* (*BGH*)) even goes so far as to reverse the burden of proof whenever the interaction of two causal chains makes the identification of individual causes impossible (*BGHZ* 66, 70 (Steinbruch) 1976), a result which is criticized by academic commentators as unprincipled equity (Köndgen, 1991: 101ff.). In the discussion of a new Environmental Civil Liability Act (*Umwelthaftungsgesetz*) the giving up of the requirement of "alternative causation" in certain key ecological areas was seriously considered in favour of joint and several liability. However, the final version is much more limited (*Gesetz über die Umwelthaftung vom 10.12.1990, BGBl I S. 2634 (UmweltHG)*). It creates a presumption of causation for an individual operator under certain conditions (§ 6 *UmweltHG*) and extends this presumption to multiple operators (§ 7 *UmweltHG*).

The German doctrinal developments do not lead to exactly the same results as their American counterparts. However, they have one thing in common. Liability law artificially creates a collective in situations of non-co-operative behaviour. The courts no longer search for pre-existing purposive co-ordination of action as a premise for collective liability. They have given up linking collective legal responsibility to social collectives, or *de facto* co-operative behaviour, let alone linking it to the existence of a fully-fledged corporate actor. They no longer attempt to identify a collective in social reality, but bluntly impose a collective by the force of law.

This innovative move resolves one problem at the cost of producing a new one. If there is not a pre-existing collective in social reality, then what is the underlying principle that guides the artificial and authoritative imposition of a risk pool by law? The American experience would suggest that the market be chosen as the risk creating unit. But which is the "appropriate market": the local, the regional, the national, or the global market? What happens in market share liability if the ecological damage occurred in an area geographically distant from one industrial polluter in the product market? And why market and not industry liability? In the (in)famous *DES* case, the national US-market was a convincing solution for identifying the dangerous medical products, but is the market a generalizable criterion for cases of ecological liability?

The German solution suggests a more abstract "spatial or temporal unity of risky activities" or an equally abstract "non-traceable interaction of causal chains". But both criteria reveal the usual fault with German abstractions. They may be theoretically convincing but are too vague and too general to be useful in practice.

I suggest seeing the underlying principle as a legal policy that defines the relevant risk pool as *an ecological problem area according to its suitability to collective risk management*. The crucial criterion is neither ecological interwovenness nor pre-existing co-operative links but the capacity for risk management. This is, admittedly, "opportunistic" attribution (Luhmann, 1991: 129), though not causal but collective. For the purposes of collective liability, the law can identify concrete areas of ecological risk (a lake, a river, a site) with an eye to creating a social arrangement that can deal with these risks. This can be in two senses. The first is that of dealing with past damages. The law delineates the collective so as to make compensation possible in cases of multiple causation and to create a sufficient finance pool for covering the losses and spreading the risks ("deep pocket", "risk spreading"). In the second sense, and perhaps more importantly, risk management means the collective influencing of future behaviour (Bush, 1986: 1553ff.). The law designs the contours of the risk pool in such a way as to create a realistic basis for an active and joint risk prevention in an area where ecological problems are concentrated. In both aspects, the law isolates a social space for collective responsibility, combining ecological and social criteria so that a joint ecological risk technology has a chance to develop and operate. In cases of conflict, it could even cut through ecological interdependencies if reasonable limits for a social arrangement could be identified.

Admittedly, this formula does not provide clear means for setting the conceptual limits of the collective, as the more traditional search for pre-existing hierarchical or co-operative arrangements was able to do. A policy-mix would be responsible for a "reasonable" pooling of risks. The situation resembles the strategic policy-mix that insurance uses when creating different risk categories among the insured ("risk classification", Abraham, 1988: 949; Eubank, 1991: 194). The formula is wider than the American doctrine's concentration on markets with their individual risk contributions, but narrower than the spatial and temporal unit of German doctrine which gives no criteria by which the "unit" can be defined. The formula has the advantage of allowing the identification of a whole range of different risk pools to which collective liability can be attached: not only product markets, but also ecological chains, contaminated sites, poisoned lakes and rivers, pollution bubbles and other ecological problem areas. Suitability for ecological risk management should, I would argue, always be the overriding criterion.

Under this formula, market share liability would only be one among several options for choosing an ecological problem area. In the case of hazardous products that are distributed in a consumer market, the ecological risk area would seem to be best identified by the market itself:

> ... the industry rather than the individual manufacturer should be the focal point for liability because it can best allocate risks, distribute costs, and take preventive measures (Sheiner, 1978: 1002–4).

But the subsequent question of how to identify the "appropriate market" (Spitz, 1990: 619ff.), in terms of identity of the product and the geographical extension, should not only be answered by reference to its economic properties (substitutability of the product, marketing approaches, and density of transactions) but openly by criteria of ecological risk management. Equally, it is frequently unclear whether the "appropriate market" is local, national or global. Under our formula this choice would not be made independently of such considerations as: Is the market widely enough defined as to be able to financially absorb the damages? Will the definition of the market lead to an acceptable spreading of risks? Is it, on the other hand, narrow enough to allow the actors involved to co-operate and to create a decentralized joint risk control? Is there a realistic chance of collective risk management?

The *Hall* v. *DuPont* case clearly showed that suitability for risk control was a concern by emphasizing that its opinion only applied to:

> ... industries composed of a small number of units. What would be fair and feasible with regard to an industry of five to ten producers might be manifestly unreasonable if applied to a decentralized industry composed of thousands of small producers (378).

What market share liability does is to induce a partial horizontal integration of firms. Liability law creates "joint ventures", in a manner of speaking, for collective risk management in a given market. This makes sense, as in the *DES* case, when a product market for hazardous goods exists. In other situations, however, the ecological risks are quite different. In an ecological chain, for instance, the typical hazards can only be identified if one takes different stages into account: raw material delivery, production, distribution, consumption and waste disposal. In these situations, it is not horizontal but vertical integration that is required. Liability law should create a vertical liability chain which might induce new forms of vertical risk management in the different phases of product transformation. There are Japanese and German experiences of ecological arrangements between firms in a whole production chain (Weidner *et al.*, 1990). Liability law should not hesitate to use the threat of drastic financial sanctions in order to facilitate such arrangements.

In the American Superfund a different principle again is at work (Stewart, 1991). Here, collective liability is no longer attached to socio-economic configurations, whether formal organizations, markets, or vertical production chains. Instead, the law defines geographical units. Contaminated sites are the new ecological problem areas to which certain actors have a close relation. The law creates a compulsory association between rather heterogeneous actors—landowners, producers of hazardous materials, transporters and site-managers—who are made jointly and severally liable for damages and clean-up operations. It is the social arrangement centring around a geographical risk-area that becomes the "unit" of risk management policies. And the Superfund law gives wide discretion to the regulatory agency, not only to define the broad risk pool of the contaminated site itself, but also to choose a core group of key actors within this pool that have the resources and the expertise to be especially suited to collective risk management (Stewart, 1991: 112).

"Bubbles" are comparable collective risk units also defined along geographical lines. Groups of air polluters are collected into bubbles in which, after the definition of global pollution limits for the whole bubble, individual pollution rights can be traded (Dales, 1968; Raufer & Feldman, 1987). If these limits are overstepped then problems of collective liability occur (Keeler, 1991). One answer might be pollution share liability. But, more interesting is the case of pollution right trading within the bubble leading to perverse effects such as "hot spots", where there is the undesired local concentration of pollution within the global limits of the bubble. In this case new collective liability problems occur that can be shifted to government as the initiator of the bubble (Roberts, 1982: 1026ff.; Peeters, 1991: 162). But with the same right they can be shifted to the polluter collective "bubble" itself (Boucquey, Chapter 3).

In Germany, there is a lively discussion of whether and, if so, how to create regionally decentralized ecological associations (*Umweltgenossenschaften*) which would combine collective liability with joint risk management (Bohne, 1987; Kinkel, 1989: 295ff.; Rehbinder, 1989: 161; Wagner, 1990). They could make use of old collective institutions, the so-called water associations (*Wasserverbände*), which regulated damages from the usage of water and mining. The associational purposes would be redefined according to new ecological needs:

> Creating risk pools of all polluters of one river or of all air polluters in one area offers unique possibilities for an ecological damage prevention which would take account of regional and functional characteristics. Since ecological damages mainly depend on the concentration of toxic material and the time span of exposure and since the greater part of emissions have regionally limited effects, regional institutions seem to be desirable for ecological damage prevention. (Wagner, 1990: 112).

This makes clear that it is the suitability for joint risk management which leads to the preference for regional and decentralized risk pools.

If the contours of the ecological cupola are thus defined by the ecological problem area concerned, this still leaves the question of which activities the cupola will cover: Who is a member of the risk pool? Once again, suitability for joint risk management would seem to be the overriding concern. It is not by chance that the German courts who were asked to make government liable for the *Waldsterben* (death of the forests) did not for a moment consider attaching collective liability to car-driving as the most obvious cause of the disaster in a typical ecological problem area (*Bundesgerichtshof—BGHZ* 102, 350, 362ff.). The reason for this exclusion is a simple policy consideration: suitability for joint risk management. The small scale risk contributions of millions of individual car-drivers aggregate into a large scale risk, and this peculiar situation seems to be the main underlying motive for not attaching liability. This is not the typical situation faced by the new collective liability which can induce smaller groups of risk contributors to undertake an active joint risk management whether on the basis of a collective deep pocket, or by creating an institution of joint risk control. Active joint risk management thus refers to "homogeneously composed, relatively small, interactive collectives"

(Wagner, 1990: 109; Kinkel, 1989: 296). This suggests a definition of liability pools which leaves out ordinary people and concentrates, instead, on "corporate, professional, and governmental defendants" which leads in turn to better prevention and better risk spreading (Priest, 1990: 219).

A similar principle is at work in market share liability. There a frequent concern is to attach liability only to "substantial" risk bearers (*Sindell* v. *Abbott Laboratories* 607 P 2d 924 (Cal. 1980)). The same is true for multiple polluters' liability in the German Water Resources Management Act (§ 22 *Wasserhaushaltsgesetz*). The American Superfund has a *de minimis* clause which allows the regulatory agency to settle financially with small scale contributors and to concentrate attention on the large and powerful players (Stewart, 1991: 112). In all these instances legal responsibility is attached to a certain capacity for social action.

Finally, there is the issue of organizational duties that are attributed to the individual members of the risk pool. For individual firms, an intra-firm "duty of ecological loyalty" has emerged parallel to duties of associational loyalty, fiduciary duties and good faith duties, that we are accustomed to in corporate arrangements. The German § 5 *Bundesimmissionsschutzgesetz*, for example, sets out the "duty of ecological organization" (Feldhaus, 1991: 931). Similar developments are going on at the inter-firm level. "Hot spots" in "bubbles" may once again serve as an example (Roberts, 1982; Boucquey, Chapter 3). In a market for pollution rights members are free to buy and sell as many pollution rights as they wish. But the institutional context imposes duties of loyalty which derive from the ecological "purpose" of the bubble. It is not, after all, just an economic institution which allows the courts to develop rules of unfair competition. It is an ecological institution created with the overriding purpose of protecting the environment against pollution by establishing a market of pollution rights. This legal purpose justifies the judicial development of ecological loyalty duties beyond the usual duties of loyalty in purely economic institutions. In the situation of a "hot spot", even if the overall emission standards are not overstepped, individual members are "estopped" from buying and using so many pollution rights that they create an unbearable local concentration of pollution. The further evolution of risk pools may reveal many other instances of similar ecological good faith duties for pool members.

The most worrying problem, but at the same time the most promising issue, in this context, has to do with "solidarity" within the ecological risk pool (Bush, 1986: 1473ff.). As we have said, collective liability means that individual actors are made responsible for the activities of others. If we translate this into the language of legal duties, it means that pool members are burdened with the duty of monitoring each other's behaviour. This seems to ask the impossible! Even those actors that make every attempt to reduce their own individual risk as much as possible cannot escape from this duty. They have to face the inevitable consequence of the loosening of the causal links: their individual risk sphere is no longer the single firm but the whole pool. This,

of course, has given rise to criticism in the name of fairness and efficiency and
leads to the condemnation of the whole collective approach to ecological liability (see above). Especially under the Superfund liability it seems "questionable from an efficiency point of view whether current owners of pieces of land
who had no influence on pollution should be jointly and severally liable with
the former owners and polluters of the land" (Hofstetter, Chapter 5, p. 107).
At the same time, advocates of risk liability seem to regard this as the main
motive for making the cupola invisible.

There is only one solution, only one way to fulfil this duty—to co-operate.
In our constellations, improvements in ecological safety are,

> ... a 'local public good' to the industry. A concerted industry effort to improve
> safety is required since, by definition, the problem is inherent in the nature of the
> product and is not the result of carelessness by individual producers
> (Rose-Ackerman, 1990: 745).

Thus, in the last instance, collective liability law creates, *de facto*, a duty
to co-operate in joint risk control, a duty to organize in collective action, and
a duty to create institutions that take over the collective management of collective risks, primarily through prevention research.

Of course, in the context of strict liability it is not very meaningful to speak
of a legal "duty" of co-operation, since liability is attached to the damage
whether this duty was fulfilled or not. But even in this context, the "duty"
reappears as a factual burden of prevention. If you enter a market governed
by market share liability and wish to reduce your own liability risk it is not
enough to reduce risks in your own firm. It is also necessary to ensure that a
collective institution exists that monitors the behaviour of all members in the
market.

It also makes more sense to think further about such a duty of co-operation
in liability for negligence (see also Brüggemeier, Chapter 4). Defining the
concrete duty of care in a situation of joint liability cannot be limited to
the measures taken by the individual firm. It also requires the definition of the
kind of measures that have to be taken so that risks are reduced within the
whole risk pool. The law has created a "duty to organize" according to which
the management of the individual firm has to make sure that a collective effort
is created among individual firms to organize personnel, material and technical
operations in such a way that environmental risks are reduced. And, finally,
thought must be given to the question of whether an adequate individual contribution to collective risk control can serve as a means of escaping this collective liability.

REAL WORLD EFFECTS

At this point, however, we should take care. After all, when talking about negligence and prevention we are no longer dealing exclusively with the symbolic

world of legal doctrine but with its real world effects. And this relation is not as close and direct as the assumption of norm-sanction-obedience in traditional legal doctrine suggests. Nor should we be carried away by the law-and-economics rhetoric which wants us to believe that small changes in the law, for example changes in the judicially defined level of due care, are directly translated into economic incentives for prevention (Tietenberg, 1989: 308ff.). Rather, we should listen carefully to what empirical research on the economic effects of liability law and its theoretical interpretation have to say about the more complicated relation between legal norms and corporate behaviour (e.g. Weber, N., 1987; McGuire, 1988). This suggests that we replace the over-optimistic model of "incentives through legal norms" by the more modest model of "social order from legal noise" (Förster, 1981: 17; Teubner, 1993: Ch. 5).

Empirical research on the effects of liability law suggests that the corporate world perceives changes in liability law, even dramatic ones like the change from negligence to strict liability, only as outside noise, as extremely vague messages, and not as the clear signals for the fine tuning of corporate behaviour that the law and economics literature tends to assume:

> All the firms viewed product liability as essentially a random influence, generating no clear signal as to how to adjust design behavior. ... we were struck in the companies that we visited by how few changes in law were transmitted to those involved in design decisions (Eads & Reuter, 1983: 107 and ix).

Sociological theorizing on law, politics and economy from Max Weber (1978: 319ff.) to Niklas Luhmann (1988: 324ff.) tells us that this lack of transmission is not a simple loss of information that could be easily corrected by improving the quality of communication. Rather we face here systematic communicative distortions which result from the inner logic of different worlds of meaning: ecological politics, the tort law system, the relevant market, the inner politics of formal organizations. No doubt, politics and law send signals to the economy which has to orient its actions toward these differences,

> However, this effect cannot be called political regulation of the economy, and this effect is itself not under political control. It depends on the context of the other system how the difference is constructed and how it is subsumed under the existing economic self-regulation programs (Luhmann, 1988: 337).

The corporate world will not observe legal norms as precise normative commands requiring obedience. Rather, this world perceives legal norms highly selectively and reconstructs them in a wholly different meaning context. Legal signals are reinterpreted anew, according to the inner logic of the concrete market and the concrete organization. In principle, each of these worlds reconstructs legal signals, but the same legal signal can reappear in a multiplicity of reconstructions (see Teubner, 1991: 129). And the choice between different reconstructions depends on the concrete situation. The world of economic transactions will reconstruct liability rules in a variety of ways: as mere cost factors, as economic property rights, as bargaining chips, and only rarely as changes of preferences. And the world of intra-organizational decision making reconstructs them again in a multiplicity of meanings: as organizational constraints, as internal power positions, as new elements in the goal set, as exclusively legal questions concerning the lawyers, as cost factors concerning the finance department, and only in rare cases as incentives for managers to change the monitoring of production and for engineers to change the design.

This leads to a deeper explanation of empirical findings which contradict optimistic assumptions about incentive-creation in the market:

> In reality, however, the connection between the law and product design is sufficiently weak that even quite major changes in the law would have little effect on the behavior of firms ... except to the extent that such change led to significant changes in the overall cost of product claims (Eads & Reuter, 1983: ix).

Due to the typical division of labour between different departments within economic organizations, the legal message usually gets lost before it can create incentives for different decisions (Stone, 1975: 201ff.; Scharpf, 1987: 117ff.). And in some organizations empirical research has even identified conscious strategies that made "substantial efforts to keep their ... liability problems separate from their ongoing operating decisions" (Eads & Reuter, 1983: 94). After such a twofold communicative distortion of legal messages in market and organization, the signals of legal liability tell at best: "Be careful or you will be sued" (Eads & Reuter, 1983: viii). And of course, this vague signal does not necessarily translate into preventive measures but into all kinds of evasive behaviour according to the dominant organizational policies.

The norm-incentive model has no systematic account of these complicated reconstruction processes in different worlds of meaning, in ecological politics, in liability law, in market transactions, and in intra-organizational dynamics. It subsumes them all within the impoverished language of economic cost considerations, and deals with real world deviations either by *ceteris paribus* clauses or by *ad hoc* adaptations of the models to "reality" (Tietenberg, 1989: 315ff.). This makes the norm-incentive model of limited use to our problem of how liability law relates to the real world when it begins to create a new collective ecological liability. We can better replace it by a model of recurrent "legal pressures" and "corporate responses" which give rise to new pressures and new responses in an infinite self-organizing process. Thus, we should be skeptical about our technical ability to design sophisticated cost incentives which are supposed to change the behaviour of rational actors within the universe of economic rationality. Instead, we should work with a multitude of self-organizing processes—ecological politics, the law of liability, the product market and the formal organization—that are separate from each other but at the same time structurally coupled to each other. They do react to each other, but only in a highly selective and rather unpredictable way (Blecher, Chapter 11). "'Sustainable developments', not efficiency emerges thereby as the ultimate paradigm" (Hofstetter, Chapter 5, p. 108).

We should be equally sceptical about the predictive power of elegant norm-incentive models. Consequentialism in liability law is possible and meaningful, but not in the sense of *ex ante* predictions as if liability law could effectively predict the effects of its changes with the help of economic models and react to these effects by anticipatory adaptations. Consequentialism *ex post* seems to be more realistic, in the sense that liability law institutions should become more sensitive to their empirical effects in the corporate world than they are at present and reshape their concepts according to their factual experience with corporate responses, creating new legal pressures and new corporate responses in a long-term "discovery process" based on "order from noise".

What can we say about the corporate responses to the new ecological risk liability? The retreat of the insurance industry from ecological risks? The invention of new insurance techniques that are adapted to joint and several liability? The emergence of institutions of joint risk management? The opportunistic behaviour of firms in situations of collective action leading to an inefficient level of ecological risk prevention? The completely defensive behaviour in massive litigation strategies? There is some empirical research available, especially in the field of product liability (Eads & Reuter, 1983; McGuire, 1988). But this is not yet sufficiently rich to enable us to speak about stable patterns of corporate responses to which liability law could react by issuing new pressures and awaiting new responses. And, if this is the case, then we can only retreat to the discussion of alternative scenarios in which, on the basis of the scarce experiences available, we speculate about possible courses of action in the parallel legal and economic processes.

SCENARIO I: OPPORTUNISM—SELF-INTEREST SEEKING WITH GUILE

While such a collective liability regime will have benefits in terms of corrective justice, it will at the same time create collective action problems to which corporate actors may react opportunistically. Is there in turn an adequate legal reaction to an opportunistic corporate response to a collective liability regime?

To be sure, compared to the lack of any liability under a strictly individualist regime, our new collective liability has its merits. First, the goals of corrective justice are served since under collective liability victims will be compensated. This would not happen if one stuck to fully-fledged individual causation. Second, one can expect a redirection of resources on the aggregate level because full cost internalization to the risk pool is achieved. In theory, all costs are brought back to the ecological risk pool. This will have an overall effect on prices and will make ecologically risky activities more expensive throughout the pool. Third, one can also expect a certain deterrence effect, if only a rather weak one. Each individual member of the risk pool is threatened by a certain share of collective liability, whether it be the risk of being singled out for joint and several liability in first tier litigation, the risk of being sued by the first responsible firm via second tier contribution, or the risk of facing market share or another form of direct apportionment of liability according to individual risk contribution. This situation may change the way that individual actors calculate their risks and lead to a higher level of prevention. Ideally, this threat would be translated into a joint effort by the polluters to minimize ecological risks. But here the famous collective action problems arise.

If co-operative links within the pool are weak, problems of "moral hazard" will emerge which are comparable to those of collective insurance which spreads individual risks to a risk community (Adams, 1985: 225ff.; Abraham, 1987: 863; Rehbinder, 1989: 151; Wagner, 1990: 45ff.). If they are not observed, individual firms will not reduce their specific risk contribution because this will not reduce their liability to the same degree. And since their liability relates to acts of others over whom there is no control they would spend even less on the prevention of ecological damage. Thus, the level of individual prevention would be lower than in a hypothetical situation of strict individual attribution. Empirically, it has been shown that this is a real danger. Keeler (1991) revealed that within a "bubble", a market for pollution rights in which only a global instead of an individual limit to pollution is set, the risk of violation of the standard is higher than in a highly individualized system.

Similar problems arise for collective measures of prevention. Although it would be in the interest of each pool member to engage in collective risk prevention in order to reduce the liability risk and the concomitant costs of compensation for each of them under certain conditions it is unlikely that they do so. Mancur Olson (1965) would teach us that if the number of pool members is high, co-operative links are missing, a sense of competitiveness prevails and "selective incentives" or massive negative sanctions do not exist, the pool members will not join in collective action even if it would reduce the costs for each pool member.

This Olson problem for collective risk control, together with moral hazard for individual prevention, has given rise to strong criticism of several new collective liability regimes—market share liability, Superfund under the Comprehensive Environmental Response, Compensation and Liability Act of 1980 (CERCLA) and more general joint and several liability in multiple causation (Epstein, 1985: 1377; Huber, 1985: 277; Abraham, 1987: 883ff.; Marino, 1991: 672ff.). Economically educated lawyers display a deep mistrust of collective solutions, even where corrective justice is served and resource allocation is efficient. They are harsh in their criticism of the weakness of prevention, although they are well aware that this criticism only holds if they compare collective liability to a strictly individualist liability regime which cannot work in complex ecological cases.

At this point, efforts to re-individualize collective liability make sense. If the corporate community reacts to a collective ecological liability with moral hazard, free riding and other forms of opportunism as "self-interest seeking with guile" (Williamson, 1985: 47) then the appropriate response of liability law would be to make the individual firms "see" at least their individual share of pollution, if not more, and to "redesign the sharing rule" (Marino, 1991: 672). These efforts cannot, of course, eliminate the collective element stemming from the breakdown of individual causation and the concomitant creation of a risk pool. However, they can reapportion the collective loss according to individual characteristics of the pool members.

The situation is comparable to the creation of a corporate actor. As a first step ("pooling"), actions, rights and obligations are attributed collectively to the collective actor as such; in a second step ("redistribution"), losses and gains are apportioned individually among the members according to individual contributions. The advantage of such a two-step procedure is that it can combine the distributional advantages of collective liability with the incentives of an individualizing apportionment. Collective liability makes sure that victims will be compensated in situations where no individual causation can be established; individual reapportionment creates incentives for individual actors to reduce the ecological risk.

Several liability techniques are available, the combination of which determines the balance between corrective justice, deterrence and allocation. Does it make sense to apply joint and several liability in first tier litigation with *post hoc* cost recovery in second tier contribution litigation among the tortfeasors? Or is direct allocation of risk shares through the courts more reasonable, as in market share liability? Should one combine strict liability and negligence for the imposition of collective liability and the individual apportionment? What are the criteria for apportionment: *pro capita*, market share, risk contribution, negligence? In the literature, several combinations of these techniques are discussed (Kornhauser & Revesz, 1989: 837ff.; Marino, 1991; Bodewig, 1985: 531ff.). Especially interesting are proposals for "weighted market share liability" that combine the criteria of market share with individual accident probabilities (Marino, 1991: 674). Of equal interest is the combination of negligence and strict liability in market share liability: primary liability to negligent actors in the market, secondary strict liability according to shares (Kornhauser & Revesz, 1989: 837ff.).

I would suggest the following combination. In the relation between the victim and the risk pool, strict liability should govern because it will further the distributional goals of corrective justice. In the relation between the pool and its members, the apportionment should be individualized as much as possible so that individual incentives for risk reduction are created. Negligence should be the governing regime; if this is not possible then apportionment according to individual risk contribution; if this is not possible then pollution share or market share; and if this is not possible then *pro capita* liability.

It is difficult to make the choice between joint and several liability with contribution among the pool members on the one side and direct apportionment through the courts in the first trial on the other side since both have grave disadvantages (Rosenberg, 1987; 220ff.). Joint and several liability is very generous to the victims, allows them to choose arbitrarily among the tort-feasors, gives full compensation to the victim from just one pool member, frees the first trial from any calculation of the individual pool shares, and leaves it to the first defendant to litigate the reapportionment among the pool members according to their share of risk and negligence in a series of consecutive suits. Its disadvantages are fairness problems for the first defendant who is burdened with the full risks of the first tier damage suit and its consequent second tier apportionment, and with tremendously high re-litigation costs (Weber, A., 1989: 1488ff.). The second one—direct share allocation through the courts in the first tier proceeding—does solve the fairness problem and saves the transaction costs of re-litigation. It burdens, however, the ecological victims with the problem of identifying and suing every pool member, and the additional problem of litigating a just share allocation among the pool members which has turned out to be a costly process (Rose-Ackerman, 1990: 743ff.). In the interests of ecological victims I would prefer the first solution. We should be aware, though, that both solutions have a grave fairness problem and produce unproportionally high transaction costs.

However, there is a deeper problem with all those efforts to make the individual apportionment as precise as possible:

> Fine-tuning damage awards to better reflect marginal harms, however, may be just so much academic hairsplitting because of a deep paradox that lurks beneath those cases where damages are exactly proportional to market share. ... the efficiency of a market-share test is very limited ... it will not generate efficient caretaking unless firms can collude (Rose-Ackerman, 1990: 745).

This is hard to swallow! On the one side, the most sophisticated procedure of individual damage apportionment leads to inefficient prevention. On the other side, collusive solutions that might be efficient in prevention will violate sacred principles of competition law. This paradoxical situation is the reason why Rose-Ackerman in the last instance argues that torts should be done away with in favour of a reliance on governmental regulation, whatever its vices (746; see also Menell, 1991).

But why not take "collusion" more seriously as the third path between torts and regulation? Is there not a chance for self-organization within the risk pool that could solve the fairness problems and diminish drastically the transaction costs of reapportionment, which necessarily come up when an outside authority, a court or a regulatory agency, imposes the rates of re-individualization. And above all, could not such a "collusion" create "efficient caretaking behavior" (Rose-Ackerman, 1990: 745)?

> In effect, the threat of joint and several liability motivates a collaborative solution among tortfeasors. Just as the defendants will collaborate to minimize their joint expenses as if they were a single person or entity in a concert of action case, so a group of independent firms may apportion liability through contract to avoid inefficient and unfair effects of joint and several liability (Rosenberg, 1987: 229ff.).

Indeed, self-organization of reapportionment is an attractive solution. If only the ecological risk pool were organized as a collective actor, the disadvantages of both solutions discussed could be avoided. The victim would have to sue only one defendant (the pool or one of its members, according to the legal construction of the pool), could ask for full compensation and would avoid touching any issue of reapportionment within the pool. These would concern internal self-regulation. Rules for apportionment could be agreed upon in advance and according to criteria that are easy to manage. This would save considerably on information costs and litigation costs. The costs of such a private risk allocation "are likely to be less than those entailed by post-accident judicial allocations using either apportioned liability or the increasingly common contribution rule" (Rosenberg, 1987: 231). Ideally, the rules of private apportionment would re-individualize the pool liability to such a degree that individual incentives for prevention are created. But what about joint prevention of ecological risks?

SCENARIO II: JOINT ECOLOGICAL RISK PREVENTION

Is this just the fantasy of socio-legal dreamers who do not learn the lessons from hardcore legal economics? There are at least some bits of empirical evidence according to which certain corporate responses to the pressures of liability law seem to contradict the prevailing economic cynicism about co-operative action. In some instances, institutions of joint risk management have emerged—whether or not they are compatible with Mancur Olson's theory and its authoritarian consequences. They took over the task of internal loss distribution, and in some cases did more than just that. It has been reported that, in some cases, they monitored the activities of the pool members via surrogate regulation. And perhaps the most promising aspect of such collective risk regulation is the chance for joint ventures in ecological innovation to develop "integrated ecological technologies" and turn away from the "end-of-the-pipe technologies" (see also Brüggemeier, Chapter 4).

However, the chances for collective risk management induced by liability law vary from context to context. Numbers of actors, structures of the market, power relations in the market, size of the risk pool, density of co-operative links, the existence of corporate actors initiating industrial cooperation (insurances, trade associations), an industrial culture of co-operation, and the role of public institutions in persuading or even forcing private actors into co-operation—these seem to be the decisive factors which influence the spontaneous development of joint ecological risk prevention. It is an open question of whether the law is capable of taking these differences into account in order to facilitate and support co-operative arrangements.

1986 was a year of ecological disasters—Chernobyl and Sandoz. In 1987 the Ciba-Geigy group formulated the corporate response to a wave of public accusation of the pharmaceutical industry, to the deterioration of its public image, to the threats of governmental action and to the pressures of liability law. It initiated the so-called RAD-AR ("Risk Assessment of Drugs—Analysis and Response") among the main pharmaceutical firms. Joint ventures of risk management were founded in each of the main countries concerned: USA, Canada, Japan, UK and Germany. One working group was the "Pharmaco-epidemiology Group" whose task consisted in gathering information on damages and the interconnection of damages with prescriptions. The other working group was the "Perception/Communication Group". Its task was risk perception and public relations (Burley, 1991: 152ff.).

This is the typical situation of an oligopolist market with a few powerful actors where one can indeed expect collective risk management to emerge. Given the enormous resources of the powerful chemical industry, joint efforts may be possible that go beyond a mere insurance function and take on tasks of research into ecological risks and develop technologies of risk prevention which could not be done by single firms alone.

In decentralized markets with numerous actors the chances for joint risk management are much smaller than in a situation of oligopoly. Co-operation is likely to develop only if there are pre-existing links of co-operation (Rosenberg, 1987: 232). It may be the case that holdings and loosely integrated groups of independent companies with decentralized decision-making have a chance to respond to the pressures of liability law by creating inter-firm risk management that takes care of ecological issues. Legal economics argues in favour of such a group risk management, as opposed to outside control through the courts or through regulatory agencies, since the "... monitoring cost for preventing the pollution would basically be lower on the part of the parent than on the part of the political community" (Hofstetter, Chapter 5, p. 107). Empirical findings suggest that this is a fertile field for intense inter-firm co-operation. Under the pressures of liability law, individual firms that have created environmental offices tend to shift these environmental offices from the single firm to the inter-firm level. In a group situation with a high degree of internal division of labour this shift is of special importance because "... there is a temptation to believe that the product as a whole is safe if each subsystem is safe". Thus, a co-operative effort between the single firms "may help surface especially subtle hazards caused by the interaction of subsystems in a technologically complex product" (Eads & Reuter, 1983: 95).

The situation is different again if the risk pool is defined along vertical lines in a product chain or in an ecological chain. Here, the chances for co-operative risk prevention are greater since there are pre-existing contractual links that can be exploited for arrangements that define liability risks and monitor behaviour. Actors "will in effect comprise an economically interdependent enterprise spanning the entire chain of production and marketing" (Rosenberg, 1987: 230). Marketing experts predict more vertical integration as a result of the new risk liability:

> The growth of market share liability could lead to greater cooperation within the channel as well as attempts by the most vulnerable channel members to control channel operations. Thus, the tendency toward vertical marketing systems is likely to be stimulated.
> Fewer, larger manufacturer-distributor combinations will be better able to withstand the financial impact of intra-industry joint liability lawsuits. The economies of such large-scale operations may even allow the participating firms to self-insure should intra-industry risks become unratable. Smaller firms unable to withstand the financial impact of such a suit will be either forced out of business or compelled to become members of substantially larger distribution channels.
> Current problems experienced by a channel member in seeking indemnification from other members will also lead to increased channel integration. ...
> As a result, manufacturers may begin to monitor the actions of their suppliers more closely, perhaps demanding assurances about the quality of the supplied component or assuming some of the testing and inspecting functions (Boedecker & Morgan, 1986: 74ff.).

Some experiences of Japanese and German delivery networks and distribution chains corroborate these findings. They are related to situations in which a vertical chain was transformed into a wheel-shaped relationship due to the existence of a hub firm that was dominating the whole chain (Rosenberg, 1987: 230). Some Superfund experience indicates that in such situations the hub firm tends to take over the role of the key corporate actor co-ordinating efforts of joint distribution of losses, monitoring behaviour of the others and planning risk prevention. Again, liability law plays a crucial role. Under Superfund the financial threats are so high and the risk pool is defined in such a way that wealthy firms almost automatically take over this co-ordinating role (Stewart, 1991: 112ff.).

In other constellations, trade associations tend to take over the role of the central agent. In the USA there is not a great deal of experience with these arrangements (Rosenberg, 1987: 231). There are some reform proposals that advocate government-controlled industry-wide self-regulation of ecological risks, especially through the creation of industry-wide risk funds (Eubank, 1991: 216ff.). The European tradition of corporatist self-regulation, however, has produced successful instances of private associations and semi-private quangos that take on the task of risk regulation and risk prevention. There are encouraging examples of governmentally sponsored corporate self-regulation in the field of work accidents with an elaborate system of financial distribution and an impressive record of "surrogate regulation". The professional co-operatives (*Berufsgenossenschaften*) are a successful example of semi-private regulatory agencies which serve as a model in the ongoing discussion about the creation of ecological co-operatives (*Umweltgenossenschaften*) (Wagner, 1990: 106ff.). The problem seems to be the legal definition of adequate risk pools. While the *Berufsgenossenschaften* define their risk pools according to branches of industries and are rather centralized unitarian organizations on the federal level, potential *Umweltgenossenschaften* should be organized according to ecological problem areas on a decentralized, regional level (Wagner, 1990: 111ff.).

Private insurance companies seem to be especially well equipped for tasks of a collective risk management in an ecological problem area (Abraham, 1988: 954ff.; Eubank, 1991: 174). They have the professional experience of risk spreading and can collect risk related information in order to reapportion costs on a highly individualized basis. There are also empirical instances, especially in the health sector, where insurance companies played an active role in monitoring and developing techniques of risk prevention. However, the new ecological liability, especially in its collective forms of enterprise liability, market share liability and Superfund joint and several liability have driven the insurance industry into a deep crisis (Eubank, 1991: 197ff.). In the USA, the insurance industry simply retreated from insuring ecological liability, despite the fact that ecological insurance promised to be the super-profitable business of the 1980s (Brockett, Golden & Aird, 1990).

It is currently an open question whether this crisis is due to a profound incompatibility between collective liability, especially joint and several liability, and the fundamental principles of insurance as some authors have suggested (Eubank, 1991: 197ff., 209ff.). It is equally plausible that the insurance industry is going through an experimental period, after which new forms of risk calculation and even a new type of insurance organization will emerge. In any case, at the moment alternative forms of insurance seem to be necessary in order to cope with the specific structures of collective risks. The risk retention group of Superfund is one possible answer, another is mandatory insurance for a whole industry, and a third is the institutionalization of an industry-wide fund for ecological risks (Eubank, 1991: 216ff.). This last solution seems to be especially appropriate for atomistic, highly competitive markets where industry-wide co-operation has no structural chance and where joint and several liability and market share liability seem rather inadequate.

Finally, the above-mentioned problems of collective action and moral hazard may make "hybrid" forms of regulation necessary. In a mixture of private law liability and public law regulation, governmental agencies are combining their regulatory powers with the weapons of liability law in order to organize collective risk control. While such hybrid regimes for the public control of private self-regulation in ecological affairs are clearly making progress on the intra-firm level (Feldhaus, 1991: 928ff.), their chances on the inter-firm level remain unclear. Of course, the American Superfund for cleaning up sites contaminated with hazardous toxics is the most exciting experiment today. And, for us, one point is crucial. The costs of litigation are a staggeringly high amount compared to the sums that effectively went into clean-up activities (Comment, 1988: 289). This highlights the central importance of collective arrangements, "contract allocations", that are struck out of court among the firms involved. As one sympathetic commentator observes:

> In effect, contract allocation tailors legal regulation of toxic substance risk-taking to the individual needs of the parties and context. Its flexibility, in contrast to the more formal and rigid rules of judicial allocation, promises benefits in lower costs, swifter enforcement against cheating, and more protection for confidential information ... legal regulation of toxic substance risks may often be achieved effectively by creating incentives for, and by all means allowing, private contract and enforcement as an alternative or supplement to centralized command and control decision making by courts and other government agencies (Rosenberg, 1987: 237).

PALERMO OR FIRENZE?

These scattered experiences of "hybrid regulation", combining regulatory activities with private joint arrangements, suggest a perspective of ecological neo-corporatist arrangements which rely on public institutionalization and public control of collective inter-firm self-organization. The idea is not to replace *in toto* individual liability or governmental regulation with a new form of collective risk management. It is, rather, to define a limited and specific area of ecological risks where the joint risk management of private actors will complement individual liability and regulatory activities. It focuses on those situations of ecological risk in which individual attribution of causation is no longer feasible, but where the identification of a relatively small class of polluters is, nonetheless, still possible. A collective liability of this group should be combined with incentives to institutionalize collective risk control in a way that compensates damages, reallocates individual risk contributions, monitors the risky behaviour of the group members, takes joint preventive measures and engages in technological innovations of risk control. Of course, this collective liability cannot cover diffuse large-scale ecological risks generated by a large number of actors within a great time-horizon or over great geographical distances. In this case society-wide ecological funds and ecological taxes are clearly preferable (Wagner, 1990; Hohloch, 1992). Equally, it should not replace classical individual liability in cases of clearly identifiable causal links, but provide for a legal remedy when the causal links cannot be attributed to individual actors but only to a group of actors.

It seems as if the ecological cupola is changing its shape. It may yet develop from the sinister threatening hierarchy of the eco-mafia into an institution protecting the environment. Is the *cupola palermitana* gradually being transformed into the *cupola fiorentina*?

NOTES

1 I would like to thank Lindsay Farmer and Declan Murphy for critical comments.

REFERENCES

Abraham, Kenneth S. (1987) Individual Action and Collective Responsibility: The Dilemma of Mass Tort Reform. *Virginia Law Review*, 73, 845–907.
Abraham, Kenneth S. (1988) Environmental Liability and the Limits of Insurance. *Columbia Law Review*, 88, 942–988.
Adams, Michael (1985) *Ökonomische Analyse der Gefährdungs- und Verschuldenshaftung*. Heidelberg: Decker & Schenck.
Adams, John N. & Brownsword Roger (1990) Privity and the Concept of a Network Contract. *Legal Studies*, 10, 12–37.
Assmann, Heinz-Dieter (1988) Multikausale Schäden im deutschen Haftungsrecht. In Fenyves A. & Weyers H.-L. (Eds). *Multikausale Schäden in modernen Haftungsrechten*. Frankfurt: Metzner, pp. 99–151.
Bechmann, Gotthard (1990) Großtechnische Systeme, Risiko und gesellschaftliche Unsicherheit. In Halfmann J. & Japp K.P. (Eds). *Riskante Entscheidungen und Katastrophenpotentiale: Elemente einer soziologischen Risikoforschung*. Opladen: Westdeutscher Verlag, pp. 123–149.
Bechmann, Gotthard (1991) Risiko als Schlüsselkategorie der Gesellschaftstheorie. *Kritische Vierteljahresschrift für Gesetzgebung und Rechtswissenschaft*, 74, 212–240.
Bodewig, Theo (1985) Probleme alternativer Kausalität bei Massenschäden. *Archiv für die civilistische Praxis*, 185, 505–558.
Boedecker, Karl A. & Morgan Fred W. (1986) Intra-industry Joint Liability—Implications for Marketing. *Journal of Public Policy and Marketing*, 5, 72–82.
Bohne, Eberhard (1987) *Umweltschutzgenossenschaften*. Manuscript.
Brockett, Patrick L. *et al.* (1990) How Public Policy Can Define the Market Place: The Case of Pollution Liability Insurance in the 1980s. *Journal of Public Policy and Marketing*, 9, 211–226.
Brüggemeier, Gert (1991) Jenseits des Verursacherprinzips? Zur Diskussion um den Kausalitätsnachweis im Umwelthaftungsrecht. *Kritische Vierteljahresschrift für Gesetzgebung und Rechtswissenschaft*, 74, 297–310.
Burley, D.M. (1991) Risk Assessment and Responsibility for Injuries Associated with Medicines. In Howells G.G. (Ed). *Product Liability, Insurance and the Pharmaceutical Industry: An Anglo-American Comparison*. Manchester: Manchester University Press, pp. 146–155.
Bush, Robert A. Baruch (1986) Between Two Worlds: The Shift from Individual to Group Responsibility in the Law of Causation of Injury. *University of California at Los Angeles Law Review*, 33, 1473–1563.
Celli, A.G. (1990) Toward a Risk Contribution Approach to Tortfeasor Identification and Multiple Causation Cases. *New York University Law Review*, 65, 635–692.
Comment (1988) Liability Insurance Coverage for Superfund Claims: A Modest Proposal. *Modern Law Review*, 53, 289.

Dales, John H. (1968) *Pollution, Property and Prices*. Toronto: Toronto University Press.

Eads, George & Reuter, Peter (1983) *Designing Safer Products: Corporate Responses to Product Liability Law and Regulation*. Santa Monica: Rand.

Epstein, Richard (1985) Two Fallacies in Joint Tort Law. *Georgia Law Review*, 73, 1377–1388.

Eubank, Katherine T. (1991) Paying the Cost of Hazardous Waste Pollution: Why Is the Insurance Industry Raising Such a Stink? *University of Illinois Law Review*, 173–217.

Feldhaus, Gerhard (1991) Umweltschutzsichernde Betriebsorganisation. *Neue Zeitschrift für Verwaltungsrecht*, 927–935.

Förster, Heinz von (1981) *Observing Systems*. Seaside, Cal.: Intersystems Publications.

French, Peter A. (1982) Collective Responsibility and the Practice of Medicine. *Journal of Medicine and Philosophy*, 7, 65.

French, Peter A. (1984) *Collective and Corporate Responsibility*. New York: Columbia University Press.

Hohloch, Gerhard (1992) Ausgleich von Umweltschäden in Teilgebieten durch Entschädigungsfonds: Rechtsvergleichende Anmerkungen. *Umweltrecht*, 3, 73–78.

Huber, Peter (1985) Safety and the Second Best: The Hazards of Public Risk Management in the Courts. *Columbia Law Review*, 85, 277–337.

Keeler, Andrew J. (1991) Non-compliant Firms in Transferable Discharge Permit Markets: Some Extensions. *Journal of Environmental Economics and Management*, 21, 180–189.

King, Joseph H. (1981) Causation, Valuation and Chance in Personal Injury Torts Involving Preexisting Conditions and Future Consequences. *Yale Law Journal*, 90, 1353–1397.

Kinkel, Klaus (1989) Möglichkeiten und Grenzen der Bewältigung von umwelttypischen Distanz- und Summationsschäden. *Zeitschrift für Rechtspolitik*, 8, 293–298.

Köndgen, Johannes (1991) Multiple Causation and Joint Tortfeasors in Pollution Cases According to German Law. In Dunné J.M. van (Ed). *Transboundary Pollution and Liability: The Case of the River Rhine*. Rotterdam: Vermande, pp. 99–106.

Kornhauser, Lewis A. & Revesz, Richard L. (1989) Sharing Damages Among Multiple Tortfeasors. *The Yale Law Journal*, 98, 831–884.

Luhmann, Niklas (1988) Grenzen der Steuerung. In Luhmann N. *Die Wirtschaft der Gesellschaft*. Frankfurt: Suhrkamp.

Luhmann, Niklas (1993) *Risk: a Sociological Theory*. Berlin: de Gruyter.

Marino, Anthony M. (1991) Market Share Liability and Economic Efficiency. *Southern Economic Journal*, 57, 667–675.

McGuire, Patrick E. (1988) *The Impact of Product Liability*. New York: Conference Board.

Medicus, Dieter (1986) Zivilrecht und Umweltschutz. *Juristenzeitung*, 41, 778–786.

Menell, Peter S. (1991) The Limitations of Legal Institutions for Addressing Environmental Risks. *Journal of Economic Perspectives*, 5, 93–113.

Nicklisch, Fritz (1991) Die Haftung für Risiken des Ungewissen in der jüngsten Gesetzgebung zur Produkt-, Gentechnik- und Umwelthaftung. *Festschrift für Hubert Niederländer*. Heidelberg: Winter, 341–352.

Olson, Mancur (1965) *The Logic of Collective Action*. Cambridge: Harvard University Press.

Peeters, Marjan G.W.M. (1991) Legal Aspects of Marketable Pollution Permits. In Dietz, F. *et al.* (Eds). *Environmental Policy and the Economy*. New York: Elsevier, pp. 151–165.

Podgers, James (1980) DES Ruling Shakes Products Liability Fields. *American Bar Association Journal*, **66**, 827.

Priest, George L. (1990) The New Legal Structure of Risk Control. *Daedalus*, **119**, 207–227.

Prosser, William L. & Keeton Page (1984) *Prosser and Keeton on the Law of Torts*. St. Paul. Minn.: West Pub. Co.

Rabin, Robert L. (1987) Environmental Liability and the Tort System. *Houston Law Review*, **24**, 27–52.

Raufer, Roger K. & Feldman, Stephen L. (1987) *Acid Rain and Emissions Trading: Implementing a Market Approach to Pollution Control*. Totowa, N.Y.: Rowman & Littlefield.

Rehbinder, Eckard (1989) Fortentwicklung des Umwelthaftungsrechts in der Bundesrepublik Deutschland. *Natur und Recht*, **11**, 149–163.

Roberts, Manley W. (1982) A Remedy for the Victims of Pollution Permit Markets. *The Yale Law Journal*, **92**, 1022–1040.

Robinson, Glen O. (1985) Probabilistic Causation and Compensation for Tortious Risk. *Journal of Legal Studies*, **14**, 779–798.

Rose-Ackerman, Susan (1990) Market-Share Allocations in Tort Law: Strengths and Weaknesses. *The Journal of Legal Studies*, **19**, 739–746.

Rosenberg, David (1984) The Causal Connection in Mass Exposure Cases: A "Public Law" Vision of the Tort System. *Harvard Law Review*, **97**, 851–929.

Rosenberg, David (1987) Joint and Several Liability for Toxic Torts. *Journal of Hazardous Materials*, **15**, 219–239.

Scharpf, Fritz (1987) Grenzen der institutionellen Reform. *Jahrbuch zur Staats- und Verwaltungswissenschaft*, **1**, 111–151.

Schmidt, Eike (1991) Effizienzbedingungen für privatrechtlichen Sozialschutz. *Kritische Vierteljahresschrift für Gesetzgebung und Rechtswissenschaft*, **74**, 378–385.

Sheiner, Naomi (1978) DES and a Proposed Theory of Enterprise Liability. *Fordham Law Review*, **46**, 963–1007.

Spitz, Stephen A. (1990) From *Res Ipsa Loquitur* to Diethylstilbestrol: The Unidentifiable Tortfeasor in California. *Indiana Law Journal*, **65**, 591–636.

Stewart, Richard B. (1991) Recent Developments in the Field of Liability for Hazardous Waste Under CERCLA and Natural Resource Damage in the United States. In Dunné J.M. van (Ed). *Transboundary Pollution and Liability: The Case of the River Rhine*. Rotterdam: Vermande, pp. 107–128.

Stone, Christopher (1975) *Where the Law Ends: The Social Control of Corporate Behavior*. New York: Harper & Row.

Teubner, Gunther (1991) Autopoiesis and Steering: How Politics Profits from the Normative Surplus of Capital. In Veld, R. in t' *et al.* (Eds). *Autopoiesis and Configuration Theory: New Approaches to Societal Steering*. Dordrecht: Kluwer, pp. 127–141.

Teubner, Gunther (1993) *Law as an Autopoietic System*. London: Blackwell.

Tietenberg, Tom H. (1989) Indivisible Toxic Torts: The Economics of Joint and Several Liability. *Land Economics*, **65**, 305–319.

Wagner, Gerhard (1990) *Kollektives Umwelthaftungsrecht auf genossenschaftlicher Grundlage*. Berlin: Duncker & Humblot.

Weber, Anne D. (1989) Misery Loves Company—Spreading the Costs of the CERCLA Clean-up. *Vanderbilt Law Review*, **42**, 1469–1509.

Weber, Max (1978) *Economy and Society*. Berkeley: University of California Press.

Weber, Nathan (1987) *Product Liability: The Corporate Response*. New York: Conference Board (Rept. no. 893).

Weidner, Helmut, *et al.* (1990) Die Umweltpolitik in Japan: Ein Modell für die EG? *IFO-Schnelldienst Nr. 16/17* 33–43.

Williamson, Oliver (1985) *The Economic Institutions of Capitalism: Firms, Markets, Relational Contracting*. New York: Free Press.

CHAPTER 3

Hot Spots in The Bubble: Ecological Liability in Markets for Pollution Rights

Nathalie Boucquey[1]

Florence

INTRODUCTION

If one speculates about the purposes of the organization of "bubbles" or pollution markets in the USA, it would seem that these systems have been conceived in order to improve environmental conditions. Indeed, their establishment rests upon the definition of standards limiting the admissible pollution rates in the area where they take effect. Within this area, the individual distribution of the pollution permits would depend on the game of supply and demand between the polluting firms.

From the viewpoint of ecological responsibility however, these markets are raising interesting questions. A paradox appears: pollution markets would constitute the specific context of certain damages due to localized pollution increases that the literature has dubbed "hot spots" (see notably Tietenberg, 1990: 30; Roberts, 1982: 1022; Peeters, 1991: 162; Noll, 1991: 75). These localized increases stem from the market principle itself, since the individual distribution of pollution emissions is no longer defined by a norm. Indeed, the administration restricts itself to the formulation of an emissions standard that just has to be observed on the global level of the market region. In certain areas of this region, however, pollution could concentrate in a disproportionate way and create "hot spots". Even if the administrative standard is not necessarily overstepped, the damages implied seem perverse in the context of environmental regulation.

This paradox compels liability law to construct appropriate compensation and prevention mechanisms. In this chapter, I will try to sketch out a liability

Environmental Law and Ecological Responsibility: The Concept and Practice of Ecological Self-Organization
Edited by G. Teubner, L. Farmer and D. Murphy
© 1994 John Wiley & Sons Ltd

system likely to meet these requirements. On the occasion of their permit swaps within the market, the polluters could be subject to certain duties our liability mechanism would sanction. The legal formulation of these duties could be justified by the ecological purpose of the general regulations organizing pollution markets: even if the permit transfers are not individually controlled, the polluters do not have the advantage of total freedom when they buy and sell permits.

Since there is a quantitative standard limiting the admissible pollution rates in the market region underlying the existence of the mechanism, I consider that such a system can be called an ecological institution and that the rules organizing it belong to the category of social regulation. Following Teubner, this feature allows the judicial expression of "duties of ecological loyalty" whenever liability problems arise from the functioning of the market (Teubner, Chapter 2). This idea will prove to be useful for drawing up a compensation mechanism for the damages resulting from the hot spots. On the other hand, the duties of ecological loyalty are important from the viewpoint of possible corporate responses to the liability questions we will discuss. They could constitute typical "organizational duties", making it easier to define the contours of an "ecological cupola" appropriate to a joint management of the risks in the market. On the level of the market, the judicial existence of ecological loyalty duties would justify the constitution of corporate organs assuming the function of an active prevention of the risks included in the permit transfers.

In order to justify the judicial expression of duties applying to permit transfers and the corporate responses of the pollution market to this liability mechanism, the ecological purpose of the regulations organizing such a market should be firmly established. Nevertheless, it is also important to see that these rules look to the protection of the economic interests of the firms they apply to, and that they leave them with sufficient room for manoeuvre. A pollution market is conceived on the basis of a global standard limiting the admissible pollution in a given region, but the global character of this constraint aims at applying it to a limited number of firms: those with the lowest abatement costs. The other partners of the market can keep the same pollution rate or even increase it, if the standard limit is observed at the level of the region. The economic interest of the enterprise seems thus to carry a considerable weight in this system, that is consequently known as "regulation by incentive" in contrast to "regulation by directive" (Peeters, 1991: 115) or "administrative regulation" (Stewart, 1988)[2].

In this chapter, I will firstly adopt the very general perspective of the conflict that can set the ecological purposes of the environmental rules against the economic interests of the enterprises. The principles of the organization of a pollution market will be defined in this context (environmental protection and the enterprise). I shall then set out the basic concepts of such a mechanism, in the course of which we shall see how the bubble notion is concretely realizing the idea of flexibility in environmental regulation (the bubble concept and the pollution markets). Finally, I will present the risks of this approach and the new questions of ecological responsibility that it is posing. Specifically, the paradoxical effects that can appear in terms of ecological damages within pollution markets will be tackled. A reference to the purposes of the regulations establishing these markets could be useful in order to solve the liability problems related to these effects (the environment of the bubble).

ENVIRONMENTAL PROTECTION AND THE ENTERPRISE

Environmental protection seems to be one of the main purposes of contemporary legal regulation, whether national or international[3]. In this connection the law has to judge typical conflicts of interests. Everybody has an interest in a safe environment and should have the freedom to live in such an environment. Nevertheless, if the law recognizes this interest and protects this freedom (Peeters, 1991: 157)[4], it usually simultaneously restricts other interests[5] and liberties[6]. In fact, environmental regulation usually applies to economic enterprises which are identified today as the main sources of pollution (Hijikata, Chapter 15)[7]. These conflicts thus typically set environmental purposes against the profitability of the enterprises (Ost, Chapter 14). The respective weights of these purposes can vary according to the importance of the ecological risk and the political tendencies of the legislative and administrative bodies.

In the legal balance, the ecological and economic purposes do not exert the same weight for all kinds of danger. Environmental protection rules privilege the ecological aim when they apply, for example, to very dangerous pollutants (Peeters, 1991: 115). In these cases, the legal rule will take the form of a command or of a prohibition (Ladeur, 1987: 7; Stewart, 1988: 112), or it will impose individualized emission or specification standards (see Majone, 1989: 125) on the polluter. However, when the ecological problem is less crucial (Stewart, 1988: 112), regulation will be more flexible towards the industry, insofar as its compliance costs will be taken into account.

I am presupposing, nevertheless, that the economic concern about environmental rules remains of minor importance compared to their ecological purpose. And in any case these rules seem to remain in the category of the "social regulation" (or of the "new social regulation") that is usually distinguished from "economic regulation" (Stewart, 1988: 115–116; Noll, 1991: 69; Majone, 1988: 164)[8]. Aiming mainly at protecting the environment, these rules can, however, take into account the interests of the enterprises which are their addressees. The relative importance they attach to these interests seems to depend principally on the seriousness of the relevant ecological risk: will this risk justify unconditional rules, depriving the enterprise of any "freedom of choice" (Peeters, 1991: 115) or will it admit a certain flexibility, providing the enterprise with the possibility to choose the cheapest means of compliance?

The idea of flexibility is usually discarded in the presence of an important ecological risk and more authoritative approaches are then recommended. But should not the certainty of this "limit of danger" (Ladeur, 1987: 7) be relativized? More flexible regulatory methods, leaving the enterprise with the possibility of opting for cheaper compliance strategies, can be recommended. Should not different types of uncertainty, which affect the representation of the danger, be taken into account beforehand[9]?

This is probably what alarmed the ecologists faced by certain administrative attempts at deregulation in the USA. Stewart notes that the processes of reduction of the costs related to many forms of regulation stirred sharp controversy with opponents even successfully blocking some of the deregulatory initiatives in court, even in the absence of statutes precluding deregulation. He argues that the courts acceded to these requests because they "seem to be especially sensitive to the increased risks associated with deregulation" as far as the environment is concerned (Stewart, 1988: 124). If deregulation represents a risk, is it not because the notion of danger is never perfectly defined in environmental matters, by reason of the complexity of the natural causality networks? Indeed, nobody contests the necessity of strict regulations whenever the danger is manifest, as in the case of the nuclear industry.

As social regulation, environmental law thus promotes a non-economic purpose, even if it seeks prevention of social damage that can express itself in terms of increased costs to the market. Legal rules thus apply to the main sources of contemporary pollution problems: industrial enterprises. From the economic viewpoint of these enterprises, regulation represents a cost likely to take different forms. Industry has exerted pressure on the administration in order to obtain more flexible rules that would leave the enterprise with the freedom to choose less costly ways of compliance with the standards. In the context of the bubble policy, these pressures have on several occasions been directed towards the Environmental Protection Agency (EPA), and one could, in this sense, speak of "regulatory capture" (for some definitions of this notion, see Majone, 1989: 164; Robert, 1990: 111; Stigler, 1975: 114; Teubner, 1984: 330–331).

In 1975, the EPA interpreted s. 111 of the Clean Air Act of 1970 (CAA) as exempting enterprises from federal performance standards requirements for their new plants, so far as the global level of emissions remained unchanged (netting)[10]. The CAA requires the EPA to impose federal performance standards on the new sources of pollution (New Source Performance Standards (NSPS), s. 111(b)(1)(B)). Section 111(a)(2) also includes in the "new sources" category those sources whose modification has started after the publication of the regulations imposing those performance standards. In order to avoid the necessity of defining new standards for every modification of fittings inside a plant the EPA considered that there was no "modification" (involving the requirements of s. 111) when the general emissions level of the plant remained unchanged. Owing to this interpretation, enterprises that respected a certain global emissions level for the entirety of their fittings would be exempted from NSPS. But the courts, in the case of *Asarco Inc.* v. *EPA*[11] for example, rejected this administrative innovation, which can be seen as the first attempt of the EPA to apply the bubble concept. They considered that the approach was incompatible with the technology forcing purpose of the CAA (Stewart & Krier, 1982: 90–91; Cook, 1988: 64–66; see also the notion of intra-plant bubble, below)[12]. However, as a result of this initiative by the iron and steel industry, the "conceptual seeds" of the bubble system had been sown (Cook, 1988: 64). Subsequently, the EPA regulations of December 1979 met with remarkable success, for they allowed the states to organize pollution markets on the level of "multi-plant" or "regional" bubbles. Here again, it has been noted that pressure came from industry (Cook, 1988: 75).

These regulatory flexibilities can thus be understood as the results of the phenomena of regulatory capture. But they were justified by a concern regarding the efficacy of administrative action: by integrating economic reasoning into the regulation, would not one increase its chances of success? This was the idea of the Carter administration reforms with regard to the "off-set" technique. Before 1977, the CAA did not allow the states to authorize the construction of new pollution sources in the areas where the national ambient air quality standards (NAAQS) were not attained (Growth Ban of 1970). In 1976 however, the EPA authorized the setting up of new sources in these areas, insofar as they could be compensated for by even larger emission reductions by the existing sources and one could observe a gradual improvement of air quality. Concretely this meant that enterprises wishing to settle in non-attainment areas had to obtain emission reductions (more than proportional to their own emissions) from the enterprises located in these areas (Raufer & Feldman, 1987: 18). And in 1979 the EPA authorized the latter to "bank" their emissions reductions in order to sell them later to the firms which would settle in the area. Commentators have recognized in this easing "the interests of a new administration bent on reform by bringing economic reasoning to decision-making for regulatory programs" (Cook, 1988: 69).

The ecological objective prevails in the CAA (planning air quality standards) and is supposed to prevail in the EPA regulations. However, the reform can also be understood as an attempt to reconcile this purpose with economic growth rationality: banking allows the setting up of new sources whenever the overall economic situation best lends itself to this. Such a provision thus secured the economic acceptability of the environmental rule, because it did not run counter to growth. We should note that a banking mechanism underlies the concrete feasibility of a pollution market system, since it allows the reduction of transaction costs (Cook, 1988: 66–9; see generally Stewart & Krier, 1978: 494–7, 593–5; 1982: 92; Roberts, 1982: 1025; Maloney & Yandle, 1984: 247).

Economic interests thus justify increased flexibility in the environmental rules. Their influence is, however, limited by the notion of ecological danger.

Beyond a certain limit it is generally admitted that the regulations take industrial interests into account in an inadequate way. Up to that point, the respective weight of ecological and economic purposes in the legal balance could depend on an eventual industrial capture of the agencies and on the importance administrators attach to economic reasoning, considered as a means of carrying out ecological purposes. However, for the purposes of the argument in the next section, let us suppose that this "limit of danger" represents a firm criterion. We will then try to understand the issues at stake in the drawing up of pollution markets and the concepts used by these markets. Then, in the final section, we shall see that such a limit is fundamentally uncertain, when we examine the specific questions of liability that are posed for this system.

Environmental rules permitting the organization of pollution markets are considered to use an incentive technique (Peeters, 1991: 151; Noll, 1991: 74; Hahn & Noll, 1990: 352; Ladeur, 1987: 7). As far as the environment is concerned, the regulatory incentive methods are defined as "producing emissions control targets at least cost" (Noll, 1991: 73). From the viewpoint of the polluting firms this means that these methods aim at minimizing the compliance costs induced by the rule. Alongside the ecological purpose there is thus also a place for the field of economic considerations, which leaves the enterprise with a certain flexibility in the observance of the environmental standards by ascribing quantitative emissions thresholds. At the outset, one can see that these limitations represent variable costs, following the nature of the fittings concerned (defective, used or more modern fittings) and depending on the more or less polluting character of the firms concerned (Maloney & Yandle, 1980: 50).

This led to the questioning of the first EPA definition of the concept of "source" in the CAA. This referred to the objects of the different emission standards which the states had to impose in order to meet the federal air quality requirements[13]. The EPA[14] initially considered that it referred to every pollution unit. The same firm could thus comprise several "sources", and on the basis of this interpretation, the states imposed standards on the firms in their respective territories for each of their particular production units, thereby effecting uniform reductions. These standards would be accompanied by "Control Technology Guidelines", defining the abatement methods for each type of source (Maloney & Yandle, 1984: 246; 1980: 49–50). Thus, enterprises did not have any flexibility in the observance of the standards. This situation provoked the criticisms of economists and the industrial world. The economic analyses revealed the necessity for softer rules that would allow firms to vary their emissions controls in accordance with the possibilities of their plants, and to opt in that way for the cheapest apportionment of these controls.

On the basis of economic analyses, the idea of flexibility was developed further. The standards would leave the possibility of trading off the pollution control requirements between several enterprises. The less polluting firms in the region would take charge of the regional requirements. The more polluting firms would be exempted from these duties but would contribute financially by buying pollution permits. This would minimize the compliance costs of the enterprises without affecting the global purpose of ambient air quality. The EPA was inspired to use the bubble concept in the implementation of the CAA by these economic analyses. This has made it possible for industrial interests to progressively counterbalance the ecological purposes of the statute.

THE BUBBLE CONCEPT AND THE POLLUTION MARKETS

Just before the EPA declaration of the 11 December 1979, Maloney and Yandle (1980) had systematized the results of the economic analysis of pollution control based on data gathered in 1976 by engineers of the DuPont company. The study compared three types of standards from the viewpoint of the respective abatement costs. The results of the study established the greater efficiency of the bubble systems compared with the initial system (uniform standards for particular production units). These results also showed that the regional bubble implied reduced costs compared with the plant-bubble. In the EPA declaration, the bubble concept was introduced with the announcement of two new options in its policy of air pollution reduction: the recourse to standards bearing on a whole plant (plant-bubble system) and to regional standards (regional bubble or pollution market) (Stewart & Krier, 1982: 92).

The plant-bubble concept

This originates in the CAA provisions related to the economic growth in the areas that did not attain the national ambient air quality standards (Offset Policy, CAA, ss. 171–173). As we have seen, the EPA had initially interpreted the "source" concept as referring to an individual production unit. In 1977, the CAA was amended, primarily in order to suppress certain obstacles to economic growth. One of these obstacles was the Growth Ban of 1970 that had prohibited the construction of new "sources" in the non-attainment areas. The provisions of 1977 gave this up but subjected the installing of new sources in a non-attainment area to diverse conditions (Offset Policy). Under the Reagan administration, the EPA changed its interpretation of the source notion in these provisions, and introduced the bubble concept. In this context, the source was taken to be the unit constituting the object of the constraint defined by the standard (Stewart & Krier 1982: 90). This interpretation made it possible to formulate standards bearing on a whole plant (plant standards) and represented an easing of the regulatory constraints in order to favour growth. In 1984, this interpretation was confirmed by the Supreme Court, considering that the 1977 amendments had modified the EPA's role[15]. If this agency had been responsible, before 1977, only for the protection of the air quality ("new

social regulation"), this was henceforth to be counterbalanced by the requirements of economic growth (more economic regulation). With this interpretation of the source concept, the EPA would allow more flexibility to firms
wishing to settle in a non-attainment area or to extend their production units.
This would undoubtedly favour the interests pressing for more economic
growth. The court's opinion is interesting, because it reveals the increased
weight exerted by industrial interests.

The regional bubble concept

The theoretical economic model

The model was initially formulated by Dales (1968) for the particular case of
water pollution. He argued that, "at the outset, the government would define
a market region and determine the overall level of emissions for each pollutant
within that region". Thus, the administration would not impose a well-defined
emissions rate on the polluters, but it would set "an upper limit L (in equivalent tons) to the amounts of discharge allowed into the environment of the
region for a given period of time" (Roberts, 1982: 1025). After that it would
"distribute a limited number of permits for each pollutant, each permit
representing the right to emit a particular amount of pollution for a specific
period of time" (Majone, 1989: 123). The sum of the permits for each pollutant would equal the total amount of emissions of the pollutant allowable
within the market region. Polluters in the region would be allowed to buy or
sell permits. The market would allow polluters themselves to determine the
most economical mix of pollution controls within a geographical region
(Roberts, 1982: 1026). A regional standard thus underlies the existence of the
market. And for this reason I consider that the rules organizing such a market
belong to the category of social regulation and that the market is an ecological
institution. This idea will prove useful for solving the liability questions connected with the pollution problems that can specifically appear in a permit
market.

   This system rests upon the attribution of a cost to the pollution emissions,
even if a standard ascribes, at the outset, an actual physical limit to the use
of the environmental resource[16]. The enterprise is confronted with an alternative (Ladeur, 1987: 9; 1988: 321; Noll, 1991: 73): either it goes on with its
pollution emissions, but it is then in charge of emission costs (buying of
permits); or it reduces emissions by installing better abatement systems,
though it will then have to sustain the abatement costs implied. With this
alternative, the enterprise has to opt for emission costs or for abatement costs.
From the viewpoint of the individual enterprise, the incentive to reduce pollution thus seems similar to the mechanism of an emissions taxes system. There
is, however, a difference. In the emissions taxes system, the money produced
serves the public interest: "the cost of (an environmental protection) policy to

a company is abatement costs plus taxes, not just abatement costs" (Noll, 1991: 74). Indeed, the taxes come to finance the public environmental policy. On the other hand, in a pollution market system, the emission costs serve the private interests of the buying firm. They are part of the costs it admits in order to keep the same level of production. They are not paid to the state but to other companies that will have to reduce their production if they sell their rights (or invest in technological innovations). This serves the interests of the buying firm in competition terms. The higher political legitimacy accorded to the pollution market system, as compared with the taxes mechanism, is attributed to this difference (Noll, 1991: 74).

Organizational questions and conceptual foundations

A pollution market calls for transferable goods and for transfer mechanisms for those goods. In other words, one uses a monetary unit and circulation rules for this unit (Cook, 1988: 63)[17]. The monetary unit receives diverse denomi-nations: "pollution right" (Dales, 1968: 93), "pollution permit" (e.g. Stewart, 1988: 104), "Verschmutzungsrecht" (in Germany), "emissions reduction credit" (ERC, proper to the Emissions Trading system), "discharge permit" (proper to the Transferable Discharge Permits system), and so on. This unit can be conceived in different ways. Certain approaches rest upon an initial dis-tribution of the unit by the administration, which can take the form of an auc-tion or of a repartition according to established administrative criteria (Peeters, 1991: 154; Dales, 1968: 93). Other approaches attribute the initia-tive of the creation of the right to the polluting firm, this initiative being fol-lowed by an administrative certification constituting the monetary unit. The current Emissions Trading programme of the EPA is based on this last technique[18]. Whatever conception one is using, the four conditions used for the creation of the ERC in the Emissions Trading system seem to be general requirements. The emissions reduction corresponding to the right must be surplus (compared with the regional standard), enforceable, permanent and quantifiable (Cook, 1988: 64).

Some recent legislative and administrative reforms have established certain important conceptual foundations, or circulation rules, for the functioning of a pollution market. The EPA regulations of January 1979 are especially interesting. Notably, they do away with the prohibition of banking as far as offset operations are concerned. Despite some differences in the economic policy purposes, the offset technique shares with the market technique the idea of an emissions rights transfer between several firms. In the offset regulations of January 1979, the EPA authorized the firms that produced emissions reduc-tions to bank them, should the offset operation not follow immediately[19]. On the basis of this authorization, banks have been set up in some American cities (San Francisco, Seattle and Louisville) (Stewart & Krier, 1982: 128; Raufer & Feldman, 1987: 18), the existence of which seems to establish the concrete feasibility of a pollution permits market (Cook, 1988: 63). In the absence of such a system, the transaction costs (costs represented by the search for a partner) borne by the partners in the market would be too high compared with the savings stemming from the permits transfers. From the technical viewpoint, the concrete functioning of a bank in a market system should not pose any *a priori* problems[20]. The current Emissions Trading programme of the EPA explicitly allows the enterprises to store emissions credits in order to use them or to sell them later on (Hahn & Noll, 1990: 354).

Finally, if one poses the question of the market partners, it would be interesting to ascertain whether the environmental protection associations could buy pollution rights on the market. Independent of the question of finance, this would afford them an important power when facing the industry.

THE ENVIRONMENT OF THE BUBBLE

Since 1970 the EPA has been entrusted by Congress with the formulation of federal standards regulating the use of certain pollutants in order to protect public health and welfare. The states are responsible for the local implementation of these standards. In 1979, and under the influence of industrial pressures, the EPA authorized the states to limit themselves to the definition of regional standards. This means that the environmental constraint does not apply to any particular plant, but to the group of the firms in the region. This system is thus protecting the industrial interests in the face of the ecological purpose of the regulations: the enterprises may make agreements in order to realize the cheapest distribution of the abatement requirements. However, the functioning of a pollution market can raise certain questions of responsibility.

In such a system, the fixed emission thresholds are applicable to the group of firms constituting the regional market. This collectivization poses the first problem of responsibility. Indeed, the application of the sanctions of s. 113 of the CAA (provisions for cases of overstepping of the standard) becomes quite problematic in this context. The hypothesis of a collective civil penalty then seems to be the only possible solution. However, the collective dimension of this sanction seems to include a risk of the lack of responsibility of the actors concerned. Certain economic studies seem to confirm this fear revealing the increased risk of a violation of the standard in a market system (Keeler, 1991b: 1361).

The second problem of responsibility derives from the fact that the organization of a pollution market also creates some risks of monopolization of the permits by a limited number of firms. The firm excluded from the market would then undergo a damage in competition terms (Misiolek & Elder, 1989). However, the liability hypothesis that we are especially interested in consists in another type of concentration that is usually designated by the term "hot spot". With the regional standards, the emissions authorizations are not normatively localized: their geographical distribution will depend on the free play

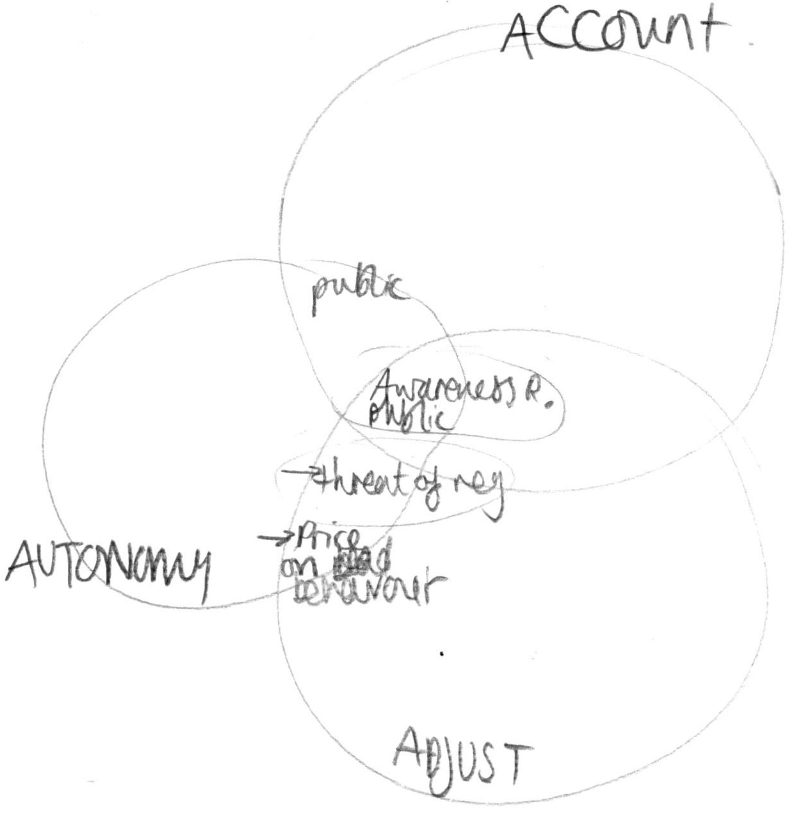

public

mca.
threat

ACCOUNT

public

Awareness R.
public

→threat of neg

AUTONOMY →Price
on
behaviour

ADJUST

BTM
S.45 — S.58 oldfield

→ enforced self reg.
→ ~~Explicit imp. options~~ ⟸
→ ~~Autonomy - edit.~~
→ ~~Examples in the laws.~~
→ self reg thresholds   → write conclusion.

→ ~~check all refs.~~
→ read through   1112.

Weds
~~Explicit options~~
~~Autonomy.~~

~~Examples in the law.~~

Thurs.
→ Include tech based + BAT.
→ ~~Examples in the law.~~ ////
→ ~~check self ref text/thresholds.~~

→ Include tech
    -based.
    autonomy + BA

Fri
→ write conclusion
→ check all refs.
→ cover letter.

of the market. This can provoke localized pollution hoardings, generating specific damages for the persons residing in those areas. The question is then that of who will bear the burden of compensation for the damages in such a case.

First liability problem: public damage

In a pollution market system, the duty stemming from the air quality standards has a collective dimension. In other words, the emissions limitations do not apply to any individual enterprise, but to the group of the sources in the market. If the standard is overstepped, the CAA orders the EPA to sue the contravener for an injunction in order to put a stop to the infraction and/or for civil penalties (s. 113(b)). If the violation is committed knowingly, its author is liable to a criminal action (s. 113(c): fine or imprisonment). Problems arise in the face of the regional collectivization of the duty defined by the standard: the moral element required by the criminal suits is quite problematical, since a bubble doesn't "think". Equally the collective aspect of the duty defined by the standard makes it difficult to apply the civil sanctions.

Injunction

The basis of the injunction is the overall standard, which does not apply to a particular firm, but to the regional collectivity. It would, therefore, be quite unjustifiable to address it to the most polluting firm in the group. The formulation of a collective injunction could be another possibility. However, since it would not be addressed to any particular source, its force risks being significantly watered down. In the meantime, the harmful emissions can increase and worsen the importance of the exceeding of the standard. The efficacy of the injunction thus looks quite problematic in the context of a pollution market.

How could the injunction continue to perform its function in this collective context? The discussion is now connected to the key problem of collective attribution in ecological liability. However, the hypothesis of a public damage contains a specificity: it would not necessarily be impossible to identify the individual source of the damage. One could determine scientifically the respective amounts of pollution emitted by each firm in the bubble, and establish in that way the extent to which each has contributed to the global breach of the standard.

This problem is addressed in the chapter by Teubner. In this he deals with problems that make it difficult to proceed to individual attributions by reason of causal identification questions. According to Teubner, these very problems compel liability law to give up the causal perspective of a socially pre-existing actor and to focus on risk communication as such (Teubner, Chapter 2). The law would then identify concrete ecological risk areas with the prospective idea of a social arrangement dealing with these risks. These new social spheres would be defined by the concept of joint ecological risk management. By contrast to this, the hypothesis of the collective regional standard being overstepped does not discard the possibility of individual causal attributions through scientific expert reports. Nevertheless, I will show that an optimal risk prevention could require the constitution of a corporate organ. Indeed, the judicial context may exclude the use of expert reports for individual causal attributions because an injunction must be formulated quickly. To wait for the results of expert reports would represent an important waste of time and a dysfunction of this legal mechanism. However, if the courts contented themselves with addressing a global injunction to the collectivity of the firms in the market, the process of enforcement of the decision would be significantly slowed down. The courts should then institute a summary individualization of the collective injunction. This would not discount subsequent contributions on the basis of the actual infractions commited by every firm in the bubble. Expert reports could then be taken into account in this way.

In order to respond to the requirement of a provisory individualization of the injunction, Teubner has proposed a mechanism of pollution share liability. The courts would order emissions limitations according to the respective number of permits possessed by each firm in the market. The surplus pollution amounts ought then to be reduced following the distribution of the permits among the firms: those possessing a lot of permits would be presumed to pollute more and thus also to have overstepped the standard to a greater extent. The solution rests upon the inadequacy of a causal perspective. This is not because a scientific causal attribution is impossible, but because it would introduce dysfunctions into the judicial mechanism of the injunction. The courts would then rely on presumptive attribution criteria.

From the viewpoint of the prevention of risk, however, an individual attribution of liability is to be wished for, on the basis of concrete proofs of causality. Indeed, as long as the attribution rests upon presumptive criteria (reduction according to pollution share), the problems of moral hazard can occur: the firms would not actually reduce their individual contributions, because this would not reduce their liability to the same extent.

The collective injunction of the CAA together with the individual presumptive attributions of pollution share liability allow the courts to order emissions reductions in a summary and individualized way. However, the problems of moral hazard show that it is of importance to identify the actual sources of the infraction in order to prevent it. Two possibilities would then be open, on the basis of scientific expert reports establishing the extent to which each firm had contributed to the global overstepping of the standard.

First, a judicial bundle of uncoordinated contributions could be observed. On the basis of pollution share liability, the judicial injunction would have ordered certain firms to reduce their emissions without consideration of the infractions actually committed The firm whose lump contribution exceeded the actual contribution could then invoke these expert reports in an action against the firm whose responsibility was actually greater. This first solution could introduce a preventive deterrence effect that could incite the firms not to overstep their authorized emissions rates. It has two disadvantages, however. On the one hand, the lack of co-ordination of these contribution procedures would imply high transaction costs (proceedings and expert reports). Moreover, this situation is excessively reliant on the fact that the parties involved have to undertake legal proceedings, the time-consuming nature of which may be a disincentive to action.

The second possibility, which might avoid these disadvantages, is the concept of self-organized joint risk prevention proposed by Teubner. As we have seen, the judicial operation of the CAA collective injunction is not satisfactory in deterrence terms, even if it is improved by the provisory individualizations of a pollution share liability. Nevertheless, it is suggested that this legal mechanism could trigger self-organizing processes of risk management within the bubble. An "ecological cupola" would then take shape. The definition of its contours would not pose any problem: the pollution market corresponds to a pre-constituted economic infrastructure. The relevant "unit" of a risk management policy is already defined along the geographical lines of the market region. The collective enforcement of the regional standard thus solves the question of the identification of an ecological problem area (Teubner, Chapter 2). A pollution market is a sphere of economic communications and an ecological problem area at the same time. The judicial liability mechanisms could give this social space a new dimension if they were reinforced by systems of active environmental risk prevention. Through the medium of the courts, one could now observe a re-entry of the environmental purpose at the level of these social spheres. The present liability question reveals the necessity for a joint prevention of the risks within the bubble: this could result in the crystallizing of an ecological organization, grafted onto the economic institution of the market.

Let us now re-situate this conceptual approach within the present liability case. In consequence of an overstepping of the regional standard, a court has ordered an injunction, individualized according to the fictitious criteria of the distribution of the pollution permits in the market. As a result of the urgency, the actual responsibilities of the various firms could not be taken into account. Nevertheless, the firms whose lump responsibility exceeds their contribution will probably try to get compensation, by invoking expert reports establishing their actual emissions rates. This could trigger the deterrence effects that are lacking in the first phase of the procedure (collective injunction and pollution share liability). However, we are still facing the costs induced by the lack of co-ordination of these contribution procedures. These problems could be solved in a more creative way by the constitution of an ecological association at the level of the bubble. The association would undertake to organize expert reports establishing the actual extent to which every firm has overstepped its emissions rates. The results of these reports could be invoked in legal proceedings. Specialized in expert reports, this association would at the same time be able to identify the specific technical problems posed by the emissions limitations of the standard. This competence would also enable the association to define technological guidelines adapted to the particular problems of each firm. In that way, the concern for a creative prevention of the ecological risks implied by the pollution market would be met by a corporate response.

Civil penalties

If the air quality standard is overstepped, the EPA can also sue the contravenor for civil penalties (CAA, s. 113(b)). In the context of a pollution market, the addressee of the standard is the market itself. Considered as sanctioning, the civil penalties should be collectivized, but this could pose several problems. Could one solve these new questions by using the corporate solution defined above?

If the civil penalty regime seems applicable when the regional standard is overstepped in the market, one can nevertheless question the deterrence effect such a mechanism could produce. Once again, the collective dimension of the duty defined by the standard could provoke a lack of responsibility of the actors in the market, especially when they are more or less numerous. A study by Keeler confirms this fear (1991b: 1361). According to his findings, pollution risks are greater in a market system than in a system of uniform standards, and the legal limits would be exceeded to a greater extent. The study attributes the increase of the pollution level to the insufficiency of the enforcement provisions related to the use of the permits, that is, that the enterprise oversteps the pollution level admitted by its permits. Thus, one cannot rely on the deterrence effect that would result from the penalty enforcing the standard[21]. The collective dimension of the penalty seems to be the very reason for the general exceeding of the standard. Here again, we are facing a problem of moral hazard: the share of each firm in collective responsibility does not correspond to its actual role in the infraction.

The problem is that the collective civil penalty seems to be the only possible sanction under the current legal system. Indeed, the EPA could not use the statutory provisions to institute individual proceedings against every polluter: this competence rests upon the standard that applies to the market collectively. The legal context of the statute thus does not justify any individualized recourse based on the actual role of every firm in the overstepping of the standard. Moreover, this would contradict the very principle of a pollution market, discarding the idea of a normative distribution of pollution emissions. The EPA could renounce its statutory competence, and bring a common law action against the polluter that had contributed most to the overstepping of the global standard. This possibility, however, is blocked by the principle of pre-emption of the federal common law nuisance actions by the statutory provisions (civil penalties of the CAA) (Roberts, 1982: 1029). Thus, if the action aims at sanctioning an overstepping of the regional standard and at getting a financial compensation, the only possibility for liability is a collective civil penalty.

Collective responsibility could take different forms: it could be distributed along the mechanisms of a joint liability, of a several liability or of a pollution share liability. With these three mechanisms however, we are still faced with a fictitious distribution of liability, deprived of any deterrent effect. Here again, concrete proofs of causation are required in order to make individual attributions. These attributions would take place as a result of judicial allocation of contributions between the members of the bubble. The lack of co-ordination of these decentralized procedures and of the expert reports could imply high transaction costs. This problem could be solved if a corporate organ was taking the responsibility for organizing expert reports in a decentralized way. This would also enable the association to develop technological abatement guidelines that would fit the technical specificities of every firm in the bubble. Prevention would then take on more creative features.

The liability hypothesis we were considering is triggered by the overstepping of a regional standard of air quality, on which public health and welfare are dependent. Besides this social purpose, a place is left for the economic interests of the enterprises, hence its regional scope. But this is what causes certain problems for the enforcement of the public interest purpose: the application of the statutory liability provisions is problematic, and their deterrent effect becomes uncertain. However, the studies showing the increased possibilities for overstepping the standard in a market system reveal the importance of appropriate means of prevention. The priorities of the bubble could otherwise be inverted. If the standard is principally conceived of as an instrument for environmental protection, its regional feature (serving the interests of the firms in the market) would represent a specific risk-factor for the environment. The ecological purpose could be re-introduced through the judicial mechanisms triggered when a damage occurs since collective liability provisions are then applicable. Nevertheless, the courts do not take up the function of active risk prevention. They are not entitled to develop technological guidelines. But the judicial processes could motivate the crystallizing of a corporate joint risk management at the level of the bubble. By partially assuming the functions of liability law, this self-organizing mechanism would re-introduce the ecological purpose of the standards at the corporate level of the market[22].

Second liability problem: private damage

Independently of any overstepping of the regional standard, a pollution market can also give rise to specific problems of private damages[23]. These risks arise because the market mechanism no longer imposes uniform duties on individual plants. Pollution concentrations and emissions authorizations are no longer localized in a normative way. Their geographical distribution will depend on the free play of supply and demand. The polluting firms with high abatement costs will buy pollution permits from the firms for which these costs are lower. All the buying firms could be located in a particular area of the region. Hence the possibility of localized pollution hoardings, exceeding the average pollution level prior to the establishing of the market. These localized concentrations are known as "hot spots" (Peeters, 1991: 162; Tietenberg, 1990: 30; Noll, 1991: 75; Roberts, 1982: 1026–1028)[24]. The persons residing in the vicinity of these hot spots will undergo increased pollution rates compared with the rates admitted by the uniform standards. These persons will be the victims of the damages which the localized increase of pollution rates can provoke.

Here we are faced with an interesting question of conflicts of interests. As we have seen, the organization of a pollution market derives from a concern to improve the economic interests of the enterprises affected by the regulation. Public purposes, profitable to society in general, are thus pursued. The improvement of the air quality is estimated in global and regional terms and the reduction of the abatement costs of the enterprises will translate into reduced costs for the consumers (Roberts, 1982: 1027)[25]. However, these general benefits are matched by disadvantages for a minority. The pollution concentrations are likely to cause diverse damages to the inhabitants of the area concerned. The reduction of the quality of their environment can have repercussions for their health or the value of their property.

The concentration phenomena proper to the free play of the market do not only express themselves in terms of pollution. They can also affect the conditions of competition. Since the number of available permits is limited by the regional standard, some firms can monopolize them and, in this way, exclude their competitors from the market (Misiolek & Elder, 1989). Peeters describes the damage that the excluded firms then undergo:

> A concentration of pollution rights on one polluter might make it impossible for other potential polluters to get pollution rights. This will impede their efforts to carry out their production plans. Instead of creating more flexibility for polluters, the opposite can be a consequence of a permit market (Peeters, 1991: 162).

This observation is interesting for the problematic of the conflicts of interests we have outlined. This can be summarized as follows. We have already seen that while the regional aspect of the standard was protecting the economic interests of the polluters, this very aspect could provoke ecological damages (overstepping of the standard). The ecological purpose of the regulation is thus affected. In this case, the question of the damages leads to a paradox, revealing a perverse effect of the mechanism: the market should protect the economic interest of the polluter, but at the same time it could actually lead to the harming of this very interest.

Attribution of the burden of compensation of the damage

We can now take a limited hypothesis of a private damage suffered by a particular person because of a localized concentration (hot spot), presupposing moreover that a causal connection can be established between this damage and a particular permit swap between two firms[26]. We are thus restricting ourselves to the hypothesis of a permit swap between a firm A and a firm B, both belonging to the same regional market. This exchange causes an increased pollution concentration in a particular area of the region, that is, in the area where the buying firm is located. An inhabitant of the area catches a disease and manages to prove that it results from this increased pollution.

In this case could the victim successfully institute proceedings against the two firms that were parties to the transfer, in order to obtain compensation for the damage he or she has suffered? Several uncertainty factors, related to the proof of the moral element, have to be taken into account. The thesis of an intentional nuisance would probably be turned down by the courts for the following reason:

> A court may deny recovery for an intentional nuisance if the conduct is "reasonable". State and federal regulations authorizing the defendant's pollution level would have a strong bearing in determining reasonableness. Furthermore, in determining reasonableness, the court may weigh the burden to plaintiff against the utility of defendant's conduct. In this case, benefits to society would include not only defendant's production but also the lower cost of pollution abatement and the research incentives created by the market (Roberts, 1982: 1030).

Even the proof of negligence can be problematic, insofar as pollution concentrations cannot always be foreseen. The defendants could establish that they had taken all the precautions required by prudence, and that the concentration resulted from a climatic hazard[27]. In certain cases the particular transfer of a pollution permit could correspond to the creation of a situation of special danger on one's own land. The compensation would then correspond to a regime of strict liability where the victim does not have to prove the existence of negligence (Wade, 1977: 631). This could make it possible to obtain compensation from the buyer of the permit. Nevertheless, the complementary perspective of the public authorization of the transfer and of the organization of the market by the state should be considered here.

From the perspective of state liability, it seems important to distinguish between the transfers that need an individual governmental authorization and those that just have to meet general criteria. The EPA regulations of December 1979 authorizing the states to introduce pollution markets foresaw a procedure of public approval of each transfer[28]. However, the weight of this procedure was such as to discourage the transfers and for this reason it was criticized by the industry. The State of New Jersey tried to answer this criticism by adopting general authorization procedures in a state implementation plan (SIP). In April 1981, the EPA approved this initiative, and it announced a generic policy in the meantime: as far as the particular transfers complied with general administrative criteria, they would not require any individual approval (Cook, 1988: 76; Stewart & Krier, 1982: 92).

This declaration, however, does not exclude a procedure of individual authorization in certain cases. States are still allowed to organize two regimes: the general administrative criteria in question can attribute the regulation of the transfers to either one regime or the other. There would then be two possible alternatives. Either the authorization would be the principle, matched by general criteria of admissibility of the transfers (the lawfulness of a particular transfer would then depend on its conformity to general criteria) or; prohibition can be the principle, with possible exceptions, in the form of individual governmental authorization. In this case, the competent authority would not have to observe any general legal criteria, and would have the freedom to evaluate the particular circumstances of the transfer (Peeters, 1991: 157). The choice between these two possibilities would depend on the importance of the ecological risk.

Let us provisionally adopt the perspective of a situation of transfer where a particular public authorization is required. Does this authorization procedure not lend itself to certain preventive provisions for protection of the third parties that the transfer could affect (Peeters, 1991: 158)? Before authorizing the transfer, should the administration not make it possible for the third parties to assert their interests, and if necessary to oppose it?

The administration could carry out this duty by instituting publicity provisions and by admitting the participation of the third parties in the decision-making. This very concern to protect the third parties could justify other duties once the decision is taken. For example, the administration could register the transfers (publicity in order to inform the third parties of the localization of the permits). Alternatively, it should comply with the judicial decision enjoining it to withdraw its authorization, or to make it subject to certain conditions (Peeters, 1991: 158, 160). If it did not institute these protection provisions accurately, the administration could be considered responsible for certain damages caused by pollution concentrations. The victim could then prove that he or she could not challenge the transfer, as a result of the absence or the insufficiency of the publicity provisions. The administration could also be considered responsible in the case that these provisions were accurately applied, but where the authorization decision had not taken into account the information deriving from the consultation of the third parties.

However, the EPA seems to be moving towards an easing of the transfer procedures and the individual administrative authorization is no longer always required. How can responsibility be conceived if damage occurs after a transfer that does not require any individual authorization? We have already tried to outline the hypothesis of an individualized liability, that is a liability of the two firms party to the transfer, or of the buying firm. However, should we not take notice of the public organization of the pollution markets? Does this public organization not represent a specific risk factor that the attribution of liabilities should take into account? The localized pollution concentration provokes damage that seems to be partly due to the programme of a government that is using the market instrument in its environmental policy. This is the premise that Roberts adopts when he argues for governmental compensation based on the "takings doctrine". The constitutional basis of this is the Fifth Amendment of the federal Constitution, which forbids the public appropriation of private property without just compensation.

In the field of environmental policy, the application of this doctrine was discussed on the occasion of the decision in *South Terminal Corp. v. EPA*[29]. In this particular case, the EPA had drawn up transportation control plans (TCP) in order to reduce the air pollution due to automobile circulation[30]. For that reason, the plaintiff, a car park operator, had been obliged to limit the occupation rates of his car parks, and this had reduced the amount of his income. On the ground that these regulations constituted a taking for public use, this operator claimed a compensation of the lost opportunity from the EPA on the basis of the Fifth Amendment. The circumstances of the case required a particular formulation of the takings doctrine, since the plaintiff was not deprived of his property, although its use was affected by the TCP. Despite its rejection of the legal definition proposed for the case in point, the court formulated an hypothesis that could be of interest to environmental law. According to the court, "a right to use or burden property in a particular and permitted way may be transferred from the original owner to another person, or to a governmental body". And the court went on to consider that such a hypothesis can be termed a taking and require a compensation for the original owner.

Roberts has drawn inspiration from this formulation in order to justify the use of the takings doctrine for indemnifying the victims of the hot spots:

> The government program would create an air easement over the hot spot victim's land ... . In the case of a hot spot, the landowner would lose control of that easement through his superadjacent air. The government would transfer control of that easement *to a limited market* that would allocate pollution without regard to the landowner's interests (Roberts, 1982: 1034, my emphasis).

His thesis then consists in saying that the hot spot represents a charge for the property that should be compensated by the government. Indeed, the government would have conceded an easement (on the air adjacent to the victim's property) to this "other person", that is to the market.

If "the market" is gaining from such an environmental policy, should we not formulate a new hypothesis of liability? The group of the firms in the bubble is gaining from the generic authorization of the transfers inside the regional market. An insurance argument, consisting of "placing liability on those best placed to bear the costs" (Smith, Chapter 6[31]) could therefore justify the contribution by each firm to a collective compensation fund[32].

Finally, the ideas of governmental responsibility and corporate responsibility of the firms in the bubble could be conceived in a cumulative way. This possibility is also envisaged by Roberts. In the context of the takings doctrine, governmental responsibility would not discard the idea of a contribution by the polluters in the region, collected in the form of a tax (Roberts, 1982: 1039–1040).

The collectivization of the risks inherent in a pollution market can take the form of governmental liability (takings doctrine) or of the contribution to a common compensation fund. But would this risk a lack of responsibility of the firms in the market (Teubner, Chapter 2)? The individualization of responsibility is, of course, a problem, even if the victim of a localized pollution concentration can prove that the damage is the direct result of a particular transfer between two firms. The circumstance of the organization of the market by the state would justify a mitigation of their responsibility. In other respects, this organization is profitable to the group of firms in the market and this could justify their contribution, even when they are third parties with respect to the particular transfer that has provoked the damage.

These collectivizations look appropriate for the indemnification function of the law, but can they meet the requirement of a prevention of the pollution risks? Do we not need another type of legal mechanism, in order to strengthen the consideration of the environment in the decision processes of the enterprises? In our hypothesis, the relevant decision the law should try to influence is when two firms transfer a pollution right in the context of generic authorization, presupposing of course that a causal connection between the damage and a particular transfer can be established.

This will nevertheless preserve the collective moment. As we have seen, a pure individualized attribution is problematic in our case, since one has to take account of the organization of the market by the state, of the generic authorization of the transfers, and of the background of the other firms in the market. Collectivization could concretely correspond to the organization of a common compensation fund by the government or by an organ representing the regional group of the firms of the market which would simplify the indemnification of the victim[33]. Nonetheless, these collective tasks of organization and participation should provide a place for a new individualization of liability. Otherwise, the firms exchanging pollution rights would not consider the concentration risks included in their transfers. Such a responsibilization could take the concrete form of a contribution to the fund (by analogy with s. 107 of the Comprehensive Environmental Response, Compensation and Liability Act of 1980 (CERCLA)). The fund would first compensate the victim and second it would ask for a contribution from the two firms which exchanged the permit.

In a sense, these aims of re-individualization and of responsibilization are political and cannot justify the attribution of liability, but is it not also possible here to formulate certain legal duties? Thus, we are trying to justify legally the contribution of the two parties in the transfer into the collective fund.

For Peeters, the development of a public authorization procedure constitutes an appropriate framework for certain provisions for the protection of third parties. Preventive publicity provisions allowing them to set themselves against the transfer can be made in this context, a participation to the procedure can be organized, and the decision of authorization itself can be controlled by the judicial power (Peeters, 1991: 158). These provisions would limit the autonomy of the partners in the transfer and they would make it possible for third parties to set themselves against the transfer by invoking the pollution risks it involves. However, we have seen that the EPA's policy is directed towards an easing of the procedures. If the transfer does not need any particular authorization, does it mean that the parties are exempted from any duty to prevent the risks? Of course, the absence of public procedure discards the possibilities of participation and of judicial recourse. Publicity, however, remains possible: the polluters interested in the transfer will be able to declare their intentions beforehand. Registration of the transfers constitutes another mechanism for protection of the third parties by informing them of the localization of the permits and of the risks of pollution in a certain area. By registering their transfer, the parties could thus contribute to the protection of these persons. Even if the transfer does not take place in the context of a public authorization procedure, the damages it risks provoking to the environment can still be taken into consideration. Concretely, preventive publicity and registration should not pose any feasibility problems. But can these arrangements be justified by the existence of a legal duty? If the parties have not instituted these protection provisions (and in the absence of explicit rules in this sense), by what right could the law attribute to them the responsibility for the damage provoked by the transfer?

The reasoning of the *South Terminal* decision seems to provide a possible answer to this question. In this decision, the court rejected the argument of the plaintiff, which compared the transport plans of the EPA to a taking without just compensation. By imposing the TCP, the EPA had tried to reduce air pollution. By the following argument, the court refused to use the taking concept. First, the conditions of the taking are specified in the abstract. One can speak of a taking when "a right to use or burden property *in a particular and permitted way* is transferred from the original owner to another person, or to a governmental body" (Stewart & Krier, 1978: 453, my emphasis). The circumstances of the case are then confronted with this condition, and it is established that the condition was not realized—one could not be expropriated of a right that one has never possessed and one could not conceive that this right be transferred to another person. The regulation would then just ratify a pre-existing duty. It would just express an implicit duty. By limiting explicitly the admissible density of cars in the car parks, the TCP provisions would not change anything in the existing legal order. Commenting on this decision,

Stewart and Krier specify the context of this existing legal order, arguing that one could find the idea that "one's property rights do not include the right to use one's land in ways seriously harmful to others". This idea allows the judge to discard the hypothesis of a taking. If the regulations are explicitly formulating certain duties, it does not change anything in the anterior legal situation. Stewart and Krier note that the American courts usually use this kind of interpretation "where the regulation can rationally be related to the protection of public health or some other substantial welfare objective" (1978: 455).

Can one compare this argument with the particular question of the duties of the firms exchanging a pollution permit? The regulations in the *South Terminal* decision would then be seen as express rules of protection of the third parties in the context of the permit transfers. If this correlation was justified, one could conclude that the absence of rules of transfer (organizing a public authorization procedure and integrating express provisions of protection of the third parties into this procedure) would not mean that the parties are exempted from any preventive duties of protection. As we have seen, the parties could comply with this requirement by watching over publicity and registration of their transfer.

This was the issue at stake in the correlation between the rules for the permit transfers and the regulations in *South Terminal*. We still have to question whether this correlation is justified. Stewart and Krier consider that the TCP regulations of *South Terminal* are aimed at protecting the public health and welfare. In a pollution market, the express rules aim at protecting the citizens facing the localized pollution concentrations that can result from the transfer, since the regional standard in a pollution market aims at implementing a federal ambient air quality standard. The purposes of such a standard are the public health and the public welfare (CAA, s. 109(b)). These purposes are thus the same as those of the TCP regulations. Of course, the regional standard is not necessarily overstepped in the hypothesis of a localized pollution concentration, but one could consider that the public health and welfare purposes of the standard are effectively endangered by localized pollution increases, to which individual damages can be added.

The judicial liability processes this solution implies could set off the crystallization of ecological self-organization within the bubble. The context of the pollution market could then give rise to corporate organs, taking the function of an active prevention of the risks upon themselves. Competent for the market as a whole, these organs would organize the concrete methods of publicity and registration of the transfers. This would make it easier for the partners to comply with their legal duties. At the same time, the ecological purpose of the regulations establishing pollution markets would be given a corporate dimension.

NOTES

1 I would like to thank Gunther Teubner for his valuable help. I am also grateful for the useful comments of Christine and Eric Heilmann.

2 One also distinguishes "market-oriented regulatory instruments" from a "traditional administrative approach" (see Majone, 1989: 117, 125). Or one speaks of an "approche régulatrice", using strategies designed to encourage a global normative conformity rather than of a "régime de coercition", attributing penal, administrative or civil sanctions to particular transgressions (Robert, 1990).

3 In the opinion of the European Court of Justice, for example, environmental protection is one of the main purposes of the European Community. See the decision *Procureur de la République* v. *Association de défense des brûleurs d'huiles usagées*, Case 240/83, 7/2/1985.

4 In the USA, the courts have often affirmed that there is no constitutional right in this matter (Piette, 1988: 439; Stewart and Krier, 1978: 300–312).

5 The standards of environmental protection can define global pollution thresholds, limit the emissions of a particular source, or specify the techniques of emissions reductions. These purposes of pollution reduction can also be attained with fiscal charges imposed on each pollution unit. In both cases, the regulation will imply increased production costs for the enterprise. These costs can take the form of investments in abatement technologies or of financial contributions (Majone, 1989: 121–2, 125; Noll, 1991: 73–74).

6 In the European Court of Justice decision cited above, see n. 3, the plaintiff was invoking the general principles of the freedom of trade and industry, free competition and the free circulation of goods.

7 It is possible, of course, to identify other sources of pollution, such as car traffic, and a car is not a firm! Here again, however, the law turns towards the enterprise. The concept of "indirect source" in s. 110 of the CAA is an example. Related to the state implementation plans (SIP) of the national ambient air quality standards (NAAQS), this section authorizes the states to include programmes of control of the indirect sources in their implementation plans (s. 110(a)(5)(A)(i)). The notion of indirect source is defined in the same section: "the term 'indirect source'" means a facility, building, structure, installation, real property, road or highway which attracts, or may attract, mobile sources of pollution. Such term includes parking lots, parking garages and other facilities subject to any measure for management of parking supply ..." (s. 110(a)(5)(C)). On the debates that preceded the adoption of this provision, see Stewart and Krier, 1978: 456–458; 1982: 79–80.

8 Matheny and Williams (1991: 2) distinguish these notions by referring to their purposes. Economic regulation tends to "restore efficiency in economic markets", while social regulation promotes "equity, or non-market, political goals in society". These authors add the category of the "new social regulation", that aims at capturing the "externalities as they occur through subtle failures of the market" and they include the environmental regulation in this category.

9 We are concerned here with the damages that manifest themselves in the long term. These dangers could threaten the market, insofar as it is characterized by a fundamental uncertainty: the choices of the consumers are unpredictable, and an increased demand for a particular resource can appear unexpectedly. On the other hand, these dangers could threaten even more precious values, such as public health: this is the case when certain physical causality networks attributing the disappearance of a natural resource to the use of a particular industrial process subsequently appear.

10 The netting technique makes it possible to avoid "new source review requirements

by ensuring that the 'net' increase in emissions is insignificant" (see Raufer and Feldman, 1987: 18).

11 578 F. 2d 319 (DC Cir. 1978).

12 The idea of technology forcing means that the administration watches over the development of less polluting techniques, by the imposition of standards specifying the production methods to use or by the concrete definition of the abatement processes (for a critique of this idea, see Ladeur, 1987: 13; 1988: 319–321).

13 Section 109 of the CAA entrusts the EPA with defining national ambient air quality standards (NAAQS). These standards are called "primary" when they aim at protecting public health, and "secondary" when they take care of public welfare. They consist in defining maximal concentrations of diverse pollutants that cannot be overstepped anywhere in the USA. The sources of air pollution cannot thus apply them as such. The states' role begins here: s. 110 entrusts them with conceiving and developing plans for the implementations of the federal standards (for more details, see Piette, 1988: 442ff.).

14 Under s. 110(a)(2) the EPA has the power to approve or disapprove the state implementation plans.

15 *Chevron USA Inc.* v. *Natural Resources Defense Council Inc.*, 467 US 837 (1984). The decision is commented on by Matheny & Williams, 1991: 12ff.

16 This represents a difference compared with a charges model including emissions taxes. But we still cannot distinguish our system from the charges-and-standards model (see Majone, 1989: 121, 125; Noll, 1991: 73).

17 Note that different market programmes exist. For example, the Emissions Trading programme is distinct from the Transferable Discharge Permits system (TDP).

18 "A pollution source creates an emission reduction credit when it controls an emission point to a greater degree than required to meet legal obligations, and applies to its state control authority for certification of the excess control" (Cook, 1988: 64).

19 The previous offset regulations (1976) had prohibited the banking (see Stewart & Krier, 1978: 495; Cook, 1988: 68).

20 Raufer & Feldman define the emissions banking concept as, "The storage of qualified emission reductions credits for later use in offset, *bubble* or netting transactions" (1987: 18 my emphasis).

21 We can note that this analysis could lead to the same conclusion for a collective injunction as well.

22 Would the firms in the bubble have a "duty to cooperate" in the association? The concept of "factual burden of prevention" seems more appropriate in this context, since s. 113 of the CAA organizes a regime of strict liability (see Teubner, Chapter 2).

23 If the standard is not violated, the citizen's suits under s. 304 of the CAA cannot be applied (see Roberts, 1982: 1029). The victim of the damage thus could not appeal to this provision in order to get compensation (see Roberts, 1990: 109).

24 Roberts notes that the probability of a hot spot is in proportion to the size of the market region: "Larger regions will contain more potential emission traders and greater disparity between marginal costs. In addition, a larger market region increases the likelihood that the seller and buyer of pollution rights are not in the same immediate vicinity and that emission reductions by the seller do not cancel out emission increases by the purchaser" (1982: 1027).

25 Some authors also consider that this system promotes technological innovation (see notably Noll, 1991: 79; Stewart, 1988: 112).

26 The hypothesis thus excludes the problems related to public damages corresponding to a violation of the regional standard, be there private damage or not. It also excludes the cases where it is impossible to establish any causal connection between the damage and a particular selling, and where the pollution sources that have contributed to the localized concentration are multiple. These three problems are, however, quite frequent in environmental law and in the context of pollution markets. First, the pollution costs are often widely distributed, and insignificant on the individual level. Second, it is usually difficult to prove that damage to the health or to the property value results from air pollution. And finally, the localized concentrations we are speaking about can result from a combination of circumstances, bringing multiple pollution sources into play (see Stewart & Krier, 1978: 603; Roberts, 1982: 1030; Teubner, Chapter 2).

27 The uncertainty and the unpredictability due to the complexity of the natural causality networks are at the base of countless problems for environmental law (see, for example, Ladeur, 1987: 7ff.; Majone, 1989: 129–130).

28 Any individual transfer had to be approved by the state and by the EPA, according to the revision procedure of a state implementation plan of the national standards.

29 504 F. 2d 646,679 (1st Cir. 1974).

30 The TCP in question had been formulated by the EPA in consequence of a default by the state (that should have done this in its implementation plan of the federal standards) (see s. 110(a)(2) of the CAA; Stewart & Krier, 1978: 441–443).

31 Where he also notes that such insurance arguments have "now become the principal justifications for imposing liability in tort cases".

32 Since our hypothetical liability consists of a very limited private damage, the individual victim should be entitled to call on the collective fund. The existing collective fund mechanisms organizing such a possibility are very rare. For example, the funds of the Federal Water Pollution Control Act of 1972 (FWPCA) and of the Comprehensive Environmental Response, Compensation and Liability Act of 1980 (CERCLA) are only open to the public authorities. In Maine, private victims can obtain compensation from the collective fund of the Oil Discharge Prevention and Pollution Control Act of 1970. This mechanism, however, is limited to water pollution. From the viewpoint of our problem, the 1973 Japanese law for the compensation of pollution-related health damages is interesting (Stewart & Krier, 1978: 605ff.).

33 For example, if the buying firm is the first cause in the occurrence of the damage, but cannot proceed to an immediate payment, it will then be possible to get an advance from the fund.

REFERENCES

Cook, Brian J. (1988) *Bureaucratic Politics and Regulatory Reform. The EPA and Emissions Trading.* New York: Greenwood Press.

Dales, John H. (1968) *Pollution, Property and Prices.* Toronto: Toronto University Press.

Hahn, Robert W. & Noll, Roger G. (1990) Environmental markets in the Year 2000. *Journal of Risk and Uncertainty*, 3, 351–367.

Keeler, Andrew J. (1991a) Noncompliant Firms in Transferable Discharge Permit Markets: Some Extensions. *Journal of Environmental Economics and Management*, 21, 180–189.

Keeler, Andrew J. (1991b) Marketable Pollution Permits with Incomplete Enforcement. *American Journal of Agricultural Economics*, 72, 1361.

Ladeur, Karl-Heinz (1987) Jenseits von Regulierung und Ökonomisierung der Umwelt: Bearbeitung von Ungewissheit durch (selbst-)organisierte Lernfahigkeit—eine Skizze. *Zeitschrift für Umweltpolitik und Umweltrecht*, 1–22.

Ladeur, Karl-Heinz (1988) Umweltrecht und Technologische Innovation. *Jahrbuch des Umwelt- und Technikrechts*, 305–333.

Majone, Giandomenico (1989) *Evidence, Argument and Persuasion in the Policy Process*. New Haven: Yale University Press.

Maloney, Michael T. & Yandle, Bruce (1980) Bubbles and Efficiency. *Regulation*, 49–52.

Maloney, Michael T. & Yandle, Bruce (1984) Estimation of the Cost of Air Pollution Control Regulation. *Journal of Environmental Economics and Management*, 244–263.

Matheny, Albert & Williams, Bruce A. (1991) Realism and the New Social Regulation: Redefining Judicial Review of Administrative Action. *Paper presented at the first international meeting of the Law and Society Association*. University of Amsterdam: 26–29 June.

Misiolek, Walter S. & Elder, Harold W. (1989) Exclusionary Manipulation of Markets for Pollution Rights. *Journal of Environmental Economics and Management*, 16, 156–166.

Noll, Roger G. (1991) *The Economics and Politics of Deregulation*. Florence: The European Policy Unit at the EUI.

Peeters, Marjan G.W.M. (1991) Legal Aspects of Marketable Pollution Permits. In Dietz, F. *et al.* (Eds). *Environmental Policy and the Economy*. New York: Elsevier Science Publishers.

Piette, J. (1988) La protection de l'environnement au Canada et aux Etats-Unis. *Les cahiers de droit*, 29, 425–445.

Raufer, Roger K. & Feldman, Stephen L. (1987) *Acid Rain and Emissions Trading Implementing a Market Approach to Pollution Control*. Totowa N.J.: Rowman & Littlefield.

Roberts, Manley W. (1982) A Remedy for the Victims of Pollution Permit Markets. *Yale Law Journal*, 92, 1022–1040.

Roberts, Pierre (1990) Les sanctions et les politiques d'application des lois de protection de l'environnement au Canada et aux Etats-Unis. *Déviance et Société*, 14, 103–113.

Stewart, Richard B. (1988) Regulation and the Crisis of Legalisation in the United States. In Daintith, T. (Ed). *Law as an Instrument of Economic Policy: Comparative and Critical Approaches*. Berlin: de Gruyter.

Stewart, Richard B. & Krier, James E. (1978, suppl. 1982) *Environmental Law and Policy*. Indianapolis: Bobbs-Merrill.

Stigler, George J. (1975) *The Citizen and the State*. Chicago: Chicago University Press.

Teubner, Gunther (1984) Verrechtlichung—Begriffe, Merkmale, Grenzen, Auswege. In Kübler, F. (Ed). *Verrechtlichung von Wirtschaft, Arbeit und sozialer Solidarität. Vergleichende Analysen*. Baden-Baden: Nomos.

Teubner, Gunther (1991) Beyond Contract and Organization? The External Liability of Franchising Systems in German Law. In Joerges, Ch. (Ed). *Franchising and the Law. Theoretical and Comparative Approaches in Europe and the United States*. Baden-Baden: Nomos.

Tietenberg, Thomas H. (1990) Economic Instruments for Environmental Regulation. *Oxford Review of Economic Policy*, 6(1), 17–33.

Wade, H.W.R. (1977) *Administrative Law*. Oxford: Clarendon Press.

CHAPTER 4

Enterprise Liability for Environmental Damage: German and European Law

Gert Brüggemeier
Bremen

INTRODUCTION

This subject has three aspects—liability, environmental liability and enterprise liability—and this chapter will seek to analyse their interrelation rather than add to the numerous, detailed investigations of each part in isolation. The problem of enterprise liability (*Unternehmenshaftung*) as the imposition of liability for damage caused by industrial activities exclusively (or primarily) on the enterprise or the corporate entity as the supposedly "correct" bearer of risk, is the central issue and point of reference of this chapter. In American law this term is closely associated with the concept of so-called "strict liability" (James, 1955: 127–128) which has dominated the discussion on product liability in the USA since the beginning of the 1960s (Priest, 1985). In German law, by contrast, enterprise liability developed along two lines: organizational liability in tort (§ 823 (1) *BGB*) and no-fault liability (*Gefährdungshaftung*) in special statutes[1]. Rudiments of this dual approach may also be found in the EC law: tort liability of enterprises in the Proposal for a Service Liability Directive[2], and no-fault liability in both the Product Liability[3] and the Draft Waste Liability Directive[4]. These different approaches to enterprise liability will be reconstructed with reference to the example of environmental responsibility. In doing this we will use the term "production liability" to refer to a broad concept of liability for damage caused to person, property or environment by processes of industrial production.

The main aim of this chapter, however, is to develop a general and roughly effective concept of enterprise liability in tort for damage resulting from

Environmental Law and Ecological Responsibility: The Concept and Practice of Ecological Self-Organization
Edited by G. Teubner, L. Farmer and D. Murphy
© 1994 John Wiley & Sons Ltd

industrial processes independently both of public law regulation of activities/plants and of the special statutes creating strict liability. This concept comprises three elements:

(1) the duty of an enterprise to organize protection of the environment, with the burden of proof placed on the enterprise;
(2) an application of the quality management approach to the field of environmental protection;
(3) the enterprise's exemption from liability if it demonstrates its use of the "best available" environmental protection system.

LIABILITY LAW

Principles of liability law

As early as 1888 the Austrian economist Victor Mataja formulated principles of civil liability law which even nowadays, more than 100 years later, can be regarded as commonplace:

> No legislative act in the world can erase a damage which has occurred. The law is helpless against a completed act. The legislature can only pursue two goals vis-á-vis possible damage: It can (1) initiate preventive measures, as far as possible; and (2) assign the burden of damage which has occurred, to that party who is best fitted to assume it, according to the precepts of justice and economic efficiency. (Mataja, 1888: 19, author's translation).

Liability law is thus directed towards the twin objectives of damage prevention and the distribution of loss in accordance with considerations of social justice. Since the beginning of the 1960s, Mataja's successors have sought an effective conception of liability law from the perspective of optimum damage prevention (Calabresi, 1961/1970; Landes & Posner, 1987; Shavell, 1987). Liability for damage inflicted has to be allocated within the injuring party/injured party relationship: either fully internalized or externalized on one side or divided between both parties according to the quota in joint causation/joint fault. In European civil law systems, as well as in English and Scottish law, the rules for the allocation of liability outside of contract are provided by the concept of fault in tort law, where damage compensation is the central issue. And, with the exception of recovery for pain and suffering in personal injury cases, damages are only compensatory and not punitive.

A new type of liability without fault, the so-called "Gefährdungshaftung", gradually developed alongside this[5], and in the twentieth century has become the representative rule for the new technical-industrial risks. Towards the close of the nineteenth century, however, fault liability was, without exception, the predominant norm in all industrialized societies. With its individualistic and moralizing character it was the opposite of the comprehensive forms of modern enterprise liability. It relieved enterprises of the social and ecological costs of industrialization and served as a disguised form of national industrial policy in international competition. The principle *casum sentit dominus* (let the loss lie where it falls) took precedence over damage internalization.

It was not long though before alternative, broader, principles of inter-personal loss spreading appeared alongside liability law. The main device adopted was private insurance (Abraham, 1986). Once damage has occurred, the resulting loss will be more easily bearable if it is divided between several parties. The criterion of the "superior insurer" (who can better spread the risks of damage) became an important, if not the determining factor, in the distribution of damage through liability law. The damage was spread through first party (*Schadens- und Unfallversicherung*) and third party insurance (*Haftpflichtversicherung*), the latter becoming more generally acceptable in Germany following the introduction of strict liability for railway companies in the entire territory of the German Reich in 1871.

Private insurance was complemented by social insurance. State accident insurance financed by employers was used as a means of resolving the urgent social problem of accidents to industrial workers in Germany at the end of the nineteenth century. Social insurance thus eased the hardships of early capitalism and corrected the deficiencies of liability law[6]. A final institution of inter-personal risk distribution was the stock markets. The limited liability for the incorporated company allowed the spreading of economic risks amongst a plurality of stockholders.

The instruments that regulate societal damage today were thus, without exception, already available at the beginning of this century. Nevertheless, the contemporary liability law cannot be compared with the law at the beginning of the century. The principle of *casum sentit dominus* has been superseded by an almost manic move to damage internalization. Even the sacrosanct principle of the limited liability of corporations is coming under threat (Stone, 1980; Easterbrook & Fischel, 1985; Hansmann & Kraakman, 1991; Adams, 1991).

There are various reasons for this, but I would like to single out one dominant aspect, namely the fundamentally changed attitude towards technical risks and the pre-eminent importance of the handling of technical-industrial risks in the developed Western societies. The industrial society has turned into a "risk society" (Beck, 1986)[7], and the liberal state into a welfare state (*Sozialstaat*) and a "protective state" (*Schutzstaat*) (see Ewald, 1986; Ritter, 1991; Isensee, 1983; Klein, 1989: 1633). Developed (post-industrial) societies are characterized by a balance between the market, the welfare state and environmental responsiveness, which may even constitute the structural preconditions for the modern "wealth of nations" (see Fikentscher, 1991).

In Europe today the debate on the protective state is primarily stimulated by the process of European integration. While the European Community was still described as Eurosclerosis in a rather pejorative manner in the 1970s, and particularly so in Germany, this attitude has now fundamentally changed. Thirty years after the foundation of the European Economic Community the process of European unification has developed an unexpected dynamism which also has a gravitational effect on non-Member States through the Single European Market[8] and the amendment of the EEC Treaty by the Single European Act of 1986[9]. What was originally conceived of as deregulation to boost free trade of goods and services has resulted in a highly differentiated re-regulation, enforcing a high level of protection for the citizens and the environment of Europe (EEC Treaty, Art. 100a (3)). The remarkable EC Directive on General Product Safety of 1992[10] marks only one step in this process (Joerges, 1990).

Liability law and risk regulation

The handling of technical-industrial risks in modern society is characterized by the three dimensions of state regulation by law[11]: preventive control of economic activities; reactive administrative control; and indirect forms of safety regulation[12], especially civil (and criminal) liability.

The first of these can be seen in the tightening of public admission controls over economic activities. This may take the form of either individualized and concrete controls, or abstract and general controls. Individual plants and products (motor vehicles, drugs, etc.) are subjected to licensing procedures (see Hart & Reich, 1990). The greater the potential risks, the greater the scale of the licensing procedures, the wider the spectrum of interest represented in hearings (of safety commissions, ethic committees, etc.), and the more extensive the expert opinions sought (in the cases, for example, of nuclear power plants, genetic engineering laboratories and so on). The prerequisites for the marketing of products can also be standardized abstractly. Thus Art. 3 of the EC Directive on Product Safety contains the general duty to market only safe consumer goods in the European Community. Basic safety requirements are formulated[13], which are then concretized through self-regulative procedures (dynamic reference; private technical standardization; certification and accrediting). In addition, the manufacturers may be subjected to information and classification requirements (especially in the case of foodstuffs).

Administrative control can be seen in the process whereby following the marketing of the products or the opening of a plant, both products and plants are subject to constant control and monitoring (see Micklitz, 1990). Public authorities may impose obligations, give public warnings, order the recall of products, prohibit marketing, and order plants to be closed down[14]. Threshold standards (Grenzwerte) and safety requirements may be corrected. State control is based on the extensive and systematic collection of information. Dangers are no longer warded off by policing (social and environmental protection mit der Pickelhaube), but through a differentiated network of measures employing state, quasi-state and private actors. Together these constitute the "risk communication configuration" (Teubner, Chapter 2).

These two forms of direct regulation by public law are complemented by the third element, that is the civil liability of enterprises as a distinct type of indirect state control through the courts by means of private law enforcement.

Complementarity of liability law: a European perspective?

These types of state risk regulation are not alternative but complementary approaches[15]. It is only through this complementarity that the societal goal of damage prevention can, to some degree, be reached.

The efficiency of direct (preventive as well as reactive) state control of risk-laden activities has inherent limitations. These can be the interference of powerful individual economic interests in the legislative and administrative decision-making process, unavoidable enforcement deficits, budget and financing restrictions, and capture syndrome, where the supervisors tend to identify themselves with the supervised[16]. This situation is frequently denoted by catch phrases such as "regulatory paradoxes" (Sunstein, 1990a; 1990b) or the "regulatory trilemma" (Teubner, 1987).

The indirect control which is effected through private enforcement has the advantage of higher context sensitivity and flexibility. Liability law can react more directly to the phenomena in a given individual case than state law-making agencies at whatever level. Liability law is more responsive. Legal action is not taken until the "suitable solution" for the conflict of interests has been found. In consequence, the liability law which has been developed through case law is regarded as an efficient form of law production in both the USA (Landes & Posner, 1987) and Europe (Schäfer & Ott, 1986). Moreover, decisions in test cases have a quasi-legislative function. The legislative process is increasingly influenced by the results of competent and professional judge-made case law. To a large extent, then, in liability law the judge's dependence on the legislation has thus been replaced by the legislator's dependence on the judge.

However, indirect control through liability law enforcement also has its limitations. Civil procedures are expensive and have to be instituted by private plaintiffs. The facts are frequently not clear, especially in cases of environmental liability. Assessing the required information takes a lot of time and money and the outcome of the procedure is, at best, open (hence the saying that "Before court and at sea one is alone in God's hands"). Furthermore, liability law sanctions will often be without effect if the defendant company has gone bankrupt, its liability is restricted by incorporation or by its available assets, or if a financially weak subcontractor or subsidiary has been involved[17].

These disadvantages apart, focusing on the essential complementarity of public and private law, opens up three important insights. First, that however efficiently it may be conceived, as a matter of principle liability law has no more than an additive function. It would be asking too much to expect liability law to bear the whole burden or even the largest part of the societal task of damage prevention. Second, risk regulation under public law is becoming transformed into private liability law. Public law regulation grants private causes of action. Public law standards function as "protective statutes", providing statutory duties in civil liability, thus complementing the general duties of care in tort and enabling enforcement of public law by civil litigation. Third, liability law is critical of regulation under public law. It enables recursivity of the legal system and increases its ability to learn. This touches upon the issue of the "unity of the legal system", which has long been considered as a problem by European legal scholars (Selmer, 1991). Even if a risk-laden economic activity has been approved by public law or a product has been declared merchantable (*verkehrsfähig*) this does not necessarily preclude liability, either in tort (presupposing breach of duty) or under strict liability (*Gefährdungshaftung*). Licensed freedom of activity is no longer synonymous with licensed freedom to injure. Where the authorities' decisions are made under conditions of uncertainty, official permits and compliance with abstract marketing requirements under public law represent only the first, and inevitably insufficient, step of control. These controls do not aim at the exclusion of liability but at the avoidance of risks. The actual damage may, but does not have to, lead to a revision of these abstract assessments. Whether the individual damage suffered by the injured party is to be compensated or not is a matter dealt with under the autonomous regime of private liability law[18]. It is this independent status which makes it possible for liability law to complement the regulation of risk under public law.

Changes in liability law

As a result of the increasing assumption of the task of protective policy by the judiciary, liability law has undergone profound changes in the course of this century (Brüggemeier, 1983; 1986). Tort law has changed from being "criminal law by means of private law" into a very complex and differentiated field. While the tort law revolution in US law is often said to date from around 1960 (Priest, 1990), Germany experienced a more continuous process of legal development through the case law of the *Reichsgericht* (*RG*) and the *Bundesgerichtshof* (*BGH*). Liability for negligence has thus established itself as an independent branch of fault liability. It is characterized by the concept of duties of care (*Verkehrspflichten*), by the objectification of the standard of care, and the derogation of the classic secondary liability of "masters"/ employers (§ 831 *BGB*) through enterprise-related duties. In addition it has organized an ingenious system of evidence facilitation *res ipsa loquitur* (shifting the burden of proof; reduction in the standard of proof; balance of probabilities). This developed form of negligence liability comes very close to strict liability (without fault). The most prominent example of this close connection between negligence and strict liability is product liability. Product liability law has to refer to a concept of "defect" in order to avoid absolute liability. Since this cannot be formulated entirely objectively (that is, independently of the manufacturer's conduct), a consistent product liability without fault (*Produktgefährdungshaftung*) is not feasible. In other words, the types of fault liability and strict liability which are adopted and separated in both legal doctrine and economic analysis have hardly any equivalents in legal reality. Negligence liability is quasi-strict liability. Strict liability is quasi-negligence, especially when referring to the notion of "defect" or when permitting the exclusion of liability for unavoidable events (*force majeure*).

ENVIRONMENTAL LIABILITY

Particularities of environmental liability law

Environmental responsibility is one of the most sensitive fields of liability law, where its limitations are clearly demonstrated and where the necessity of integration with other policy approaches is particularly evident. The particular difficulties arising in this area of liability law, as graphically described by Rabin (1987; see also Kanner, 1991: 1265), fall into four categories. The first are the "problems of identification". Unlike accidents or other suddenly occurring damages, a deterioration over a period of time or a disease or symptom which appears after a long period of incubation may be the result of chemicals transmitted through water, air, industrial products or food chains. What is more, the findings are frequently connected with more general social risks (nutrition, alcohol, nicotine and so on). It is extremely difficult to identify the symptom as an injury (*Rechtsgutsverletzung*).

The second group are "problems of source". If a violation has been identified as a possible cause of liability, as an injury, the equally complex problem of supplying proof of causation/pollution still remains. The causes and polluters are potentially numerous. It is particularly difficult to determine the requirements with which the abstract (epidemiological) and concrete (toxicological, medical) proofs of sole and partial causation ought to comply. Also, it is not clear who should be charged with supplying the information necessary to distribute the liability between those parties who are considered to be at fault (jointly and severally or *pro rata*).

Third, there are "problems of boundaries". Mass torts like Contergan/ Thalidomide, DES, Agent Orange/dioxin, asbestos, PCP and so on, which represent the borderline cases in tort law, are becoming the normal case in environmental liability law. Long-distance and cumulative cases (acid rain/*Waldschäden*) represent potential dangers, where the affected sections of the population as well as those responsible can no longer be defined socially, regionally or nationally. And, in these cases the extent of the damage is hardly assessable. The civil procedure with its clearly defined injuring party/injured party constellation cannot cope with these forms of chronic pollution.

Finally there is the important question of "problems of common interests" ("global common"/"common heritage"). To a very large extent, it is tomorrow's natural environment that is being destroyed today. However, there is no *locus standi* in traditional civil procedure for impairments to nature and the environment, ranging from endangered species to the ozone layer, that go beyond private property. At the same time, no institutions are yet available which would make it possible to organize a form of protection in trust of mankind's common heritage of natural environment in order to protect the interests of later generations.

All these emphatically underline the necessity for a fundamental ecological reform. This must take the form of an internationalization of liability law in the environmental field and, perhaps even more urgently, for liability law to be complemented, or perhaps even replaced, by collective compensatory systems and alternative incentives for prevention. As we shall see, the EC initiatives on a general environmental liability are primarily aimed at these ambitious targets.

German environmental law

It does not, perhaps, come as a surprise that attempts to create an efficient codified environmental liability law have mostly failed in the face of these challenges (Rehbinder, 1989; Brüggemeier, 1989). In recent times, the German legislature has tried to solve the problem through environmental strict liability.

A first example of this is § 22 of the Water Resources Management Act (*Wasserhaushaltsgesetz* (*WHG*)) of 1957 in which the German legislature codified a pioneering liability norm for a single sector of environmental protection—water pollution (Brüggemeier, 1991b). Here, it is not the pollution of water itself which establishes the claim to compensation; it is only the resulting individual damage which imposes liability. However, recoverable damages are not restricted to personal injury and damage to property. One of the most important examples of this is the compensation for rescue costs incurred by local authorities in preventing imminent water and ground water contamination. Liability is imposed for the discharge of substances (§ 22(1) *WHG*), and on plants from which dangerous substances are, directly or indirectly, transported into the water (§ 22(2) *WHG*). In cases of an unidentified quota of multiple causation the polluters are jointly and severally liable, provided that the individual pollution is proved to be substantial.

In contrast to this, the Environmental Liability Act (*Umwelthaftungsgesetz* (*UmweltHG*)), which was passed at the end of 1990 after much debate and publicity, and which entered into force on 1 January 1991, does not do justice to its name. Unlike under § 22 *WHG*, damage to the environment is not a prerequisite of liability. Instead, the strict liability (*Gefährdungshaftung*) of a certain class of industrial plants and agricultural undertakings, specified according to criteria of quality and size, is determined in relation to individual injuries (personal injury and damage to property). The environmental element is established solely by the fact that the emissions of an enterprise or plant should have been transported by an environmental medium (air, water, soil), irrespective of whether or not they have caused environmental pollution, as for example in the case of shock waves from an explosion. The terms of § 16 *UmweltHG*, however, are ecologically significant. According to this provision, restoration to the original condition or the cost required to accomplish this, can also be claimed in cases where the natural amenity is of low value (small natural ponds/biotopes), even if the cost is clearly beyond its value. Pure ecological damage (impairment of the environment beyond private property) cannot be claimed.

Thereafter the *UmweltHG* is concerned with classical plant liability for individual injuries. In the everyday cases of damage caused by the normal operation of one or more plants, the plaintiff bears the burden of proof of causation. It remains to be seen whether the courts will introduce alleviations of the burden of proof (balance of probabilities: abstract/epidemiological proof by the plaintiff; proof of lack of concrete causation by the defendant plant's operator). The law grants the injured party the right to demand information from the operators of the plant and from the supervising authorities. However, the "well-founded assumption" that the damage was caused by the plant in question is a prerequisite for this claim to information. This access to information is not comparable to the pre-trial discovery procedure in civil actions in the USA.

The question of liability in cases of an unidentified quota of multiple causation, which is a central problem in practice, has not been regulated explicitly. It has been left to the judiciary to find the answer to this problem through case law, where the alternatives are considered to be joint and several liability and *pro rata* liability. Considerations of efficiency certainly speak in favour of the latter, where there is liability in proportion to the probability of causation (Shavell, 1987: 227). Joint and several liability can be a crushing burden in cases of only partial causation of the damage. But *pro rata* liability raises problems of practicability in cases of unidentifiable quota. Nonetheless, considerations of justice plead for compensation of the victim in some manner.

Long-distance and cumulative damage can hardly be covered by liability law. Problems of identification, causation, and boundaries accumulate in this field. Awards for damage incurred can probably only be granted by way of collective compensation systems. Incentives for prevention have to be created by other means.

More advanced concepts of environmental liability can be found in the General Part of a Draft Environmental Code (*Umweltgesetzbuch* (UGB)) presented in 1990 by a group of scholars[19]. This Draft combines the elements of a classic strict plant liability with a polluters' liability for the handling of substances dangerous to the environment. It is more clearly committed to the regulatory philosophy of § 22 *WHG* than the *UmweltHG* and seems to take up ideas contained in EC initiatives as well as those of the Council of Europe[20].

The main pre-condition of liability is the causing of an impairment of the environment, that is to say, a negative change in the physical, chemical or biological condition of water, soil or air. This environmental pollution must result in individual damage. In other words, there are two key causal relations: that of emission (impairment to the environment); and that of environmental impairment (personal injury/property damage). Liability would thus fall on the (natural or corporate) operators of plants which posed a danger to the environment, and this is complemented by a polluters' liability for activities endangering water or soil, as well as for the transport of substances which are dangerous to the environment. It appears that the causation of an impairment of the environment by the plant or the polluting enterprise would have to be proved by the injured party according to the general principles of proof. With regard to the second element of causality, the causing of the injury to persons or property, the standard of proof is that of a preponderance of evidence (on the balance of probabilities).

If there is a reasonable indication that the operation of a plant has caused the damage, the injured party may claim for disclosure of information from the operator as well as from the licensing authorities. Recovery is provided for personal injury and damage to property only. Economic loss may be claimed only as consequential damages, though pure economic loss resulting from water contamination is an exception to this rule. In this case, the rules of § 22 *WHG* would be applied.

European product liability law is applied to the production of substances imminently endangering the environment, with rather problematic results. The "defectiveness" of the substance endangering the environment, as defined by Art. 3 of the EC Directive on Product Liability, is a prerequisite of liability (but can, for example, hazardous goods be "defective"?). Moreover, the manufacturer is also liable for developmental risks. The liability of the manufacturer of products manipulated by genetic engineering is one of the rare examples of this liability rule in German law[21].

If damage to property also constitutes an impairment to the environment committed by breach of duties under public law, costs for restoration to the original condition can be claimed. This rule would correspond to § 16 *UmweltHG*. Pure ecological damage (impairment of the environment where there is no damage to property (Rehbinder, 1988)) could be claimed by the public authorities (that is, nature conservation authorities) provided that there had been a serious dereliction of the duties of environmental protection under public law and that a significant impairment to nature has occurred. The claim is either for restoration to the original condition or compensation.

Compensation for immaterial ecological damage in cases of environmental impairment would not be awarded where restoration is impossible or is only possible in the long term.

Unlike the *UmweltHG*, the liability law section of the Draft Environmental Code would also regulate the extremely important cases of joint causation. Any party who had contributed in a relevant manner towards the damage would be liable in proportion to the damage they had caused. In dealing with this controversial subject, the Draft gives preference to *pro rata* liability over joint and several liability.

A particularly innovative feature is the concept of equity liability for chronic pollution, that is, cumulative and long-distance damage, in cases where there are no actual grounds for liability. This proposal is a reaction to the complicated legal and economic problems which would result from establishing a comprehensive environmental damage fund (Hohloch, 1992).

EC initiatives on environmental liability

For several years now, the EC Commission has been working on a general Directive on Environmental Liability[22]. So far, not even the first drafts of this directive have been presented to the public. A Green Paper has been announced, but has not yet been finished. The Commission's Communication of September 1991 summarizes the substantive problems and principles of environmental liability rather than presenting the concrete outlines of a directive. The remarkable 1989 proposal for a Directive on Civil Liability for Damage caused by Waste[23], amended in 1991[24], is, however, on the agenda of the EC Council.

The concept of waste used in this draft directive is highly subjective. The term refers to substances or objects the owner disposes of or has to dispose of[25]. The directive is aimed at the producing enterprise, and applies to all forms of waste produced by industrial activity (with the exception of radioactive waste and waste which is subject to special regulations). Waste avoidance is to prevail as a general principle of organization in the field of industrial production and the provision of services. According to Art. 3 every industrial producer of waste will be strictly liable for injuries to individuals (personal injury and damage to property) caused by his waste. Certain quasi-producers are put in the same category as waste producers, namely the EC importer of waste, the owner who is actually in charge of the waste when the damage occurs (if he cannot identify the producer) and the party disposing of the waste (that is the site operator to whom the waste has been lawfully handed over). The transporter is only liable under the EC Convention of October 1989 on Civil Liability for Damage caused during the Carriage of Dangerous Goods by Road, Rail and Inland Navigation Vessels (CRTD)[26].

Instead of there being the joint and several liability of all involved parties, these quasi-producers are responsible only in the place of the waste producer. Waste disposal which is organized on the basis of division of labour is thus not sufficiently regulated by liability law. By the lawful transfer of the waste to an authorized eliminator, the producer is relieved of liability under the Waste Directive. This would leave only the tort liability of the waste producer for negligent choice and, under certain circumstances, for insufficient control of the eliminator (control test).

The plaintiff would have to prove his injury and that it had been caused by the waste of the defendant manufacturer. There is no liability for damage caused by an "act of God" as defined by EC law. The comparative negligence of the injured party is taken into account. It should also be stressed that according to Art. 6(2), the fact that the waste producer holds a permit issued by the public authorities does not preclude liability. In view of the varying interpretations of the law within the Member States, the Commission deemed it imperative to state explicitly that the directive was not intended to restrict the extent of civil liability. Article 11(1) provides for the mandatory insurance of the waste producer and eliminator.

Liability would not only arise for individual damage, but also for ecological damage (impairment of the environment). Article 2(1)(d) describes impairment of the environment as, "any significant physical, chemical or biological deterioration of the environment insofar as this is not considered to be damage to property". As no private parties are involved, government agencies of the Member States (instead of the EC Commission) have to sue for injunctions or for recovery of these environmental impairments, which are conceived as "damage to the Community". They can claim for restoration to the original condition or, in some cases, alternative or compensatory measures or alterations of the costs required for these. The *locus standi* of non-governmental organizations/interest groups is regulated by national law. "Punitive" compensation for immaterial ecological damage in cases where it is impossible to restore nature to its original condition is expressly rejected.

The central question of how to proceed when the polluter cannot be identified, when the producer of the waste is insolvent, or when he has effectively protected himself against liability or employed a financially weak subcontractor or subsidiary, has also been explicitly left open. This problem is part of the comprehensive question of how to apportion the damage if it appears that no private law claim for compensation can be established, such as in the case of long-distance and cumulative damages. The EC Commission intends to tackle this problem in the more extended context of a general environmental liability law for all Member States of the Community and up to now, no concrete proposals relating to this issue have been published[27]. Liability, as defined by the Waste Directive, is concerned only with waste that is produced after the directive has been transformed into law by the Member States. Liability for damage from the past (*Altlasten*) is beyond its scope.

Finally, we can look at the 1991 Communication on Environmental

Liability. As has already been stated, this does not so much outline basic elements of a future general Directive on Environmental Liability as present a comprehensive assessment of the difficult problems in the realm of environmental responsibility. However, the elements of their broad approach to environmental liability can be detected. Damage prevention, control of the source of damage and the polluter-pays principle are cited once again[28] as the main guiding elements of future law. In contrast to the plant-oriented German *UmweltHG*, this approach favours a more comprehensive orientation to "dangerous" industrial activities. Second, only the internalization of impairment to the environment is taken into consideration (so-called pure ecological damage), although a number of questions, relating to damage valuation, *locus standi* and so on, remain to be clarified. The Commission assumes that individual personal injury and damage to property are largely covered by national legal systems. Third, the Community undoubtedly favours a strict liability (*Gefährdungshaftung*) channelling liability to the enterprise or the corporate entity responsible for risk management. And finally, it is emphasized that this European environmental liability law has to be complemented by an EC-wide system of collective compensation for damage. Environmental damage resulting from normal operations can no longer be attributed to individual polluters in their international and inter-regional network. Particulars of financing and claims settlement under such a European compensation fund are naturally still entirely open.

ENTERPRISE LIABILITY FOR ENVIRONMENTAL DAMAGE

Strict "production liability"

In general, the attribution of all damage to persons, property and environment (polluter-pays principle) and the unproblematic allocation of liability to the enterprise/corporate entity as the "correct" party to bear such damage (enterprise liability) are regarded as advantages of strict production liability (*Gefährdungshaftung*). This however is only true subject to certain reservations.

The application of the "polluter-pays principle" is frequently frustrated by the above-mentioned particularities (identification, source, boundaries, public interest) of environmental liability law. But even if these features are disregarded, production liability without fault is not conceivable as an "absolute liability". This would result in societal immobility. However, problems are caused by attempting a normative limitation of liability. With respect to a liability which is focused on "defects" no further explanations are required. In most cases, an attempt is made to define "specific risks" or types of hazards establishing liability (ultra-hazardous operations, dangerous animals, and so on). If this is not possible, enumerative (or open) catalogues of certain economic activities, plants or products are employed. Strict production liability

(*Gefährdungshaftung*) is therefore not practicable as a general principle of liability for damage caused by industrial production.

It is also still uncertain whether liability for so-called developmental risks would be desirable. On the one hand, it is impossible to react to unknown risks with preventive measures and an adjustment of the level of activities. On the other hand, the manufacturer or operator of the plant is in a better position to carry out research on the impacts of products and emissions (Panther, 1991). It is doubtful, however, whether *ex post* duties in tort to monitor products and emissions would create a sufficient incentive for this, or whether satisfactory results could only be achieved through a liability for developmental risks which comes close to an "absolute liability" and made the operating enterprises insurers of the injured persons. The results of empirical research on the economic effects of the common law strict product liability in the USA (Eads & Reuter, 1983; McGuire, 1988) seem to favour the former. In spite of this, thus far it is only in asbestos cases in the USA that some jurisdictions have taken steps in the direction of strict liability for developmental risks and thus precluded the defence of unfeasability[29]. The question that is increasingly being discussed is that of whether the potential injured party would not be the better insurer with regard to developmental risks—and risks which cannot be avoided at reasonable economic expense (Priest, 1990). Many injuries to health caused by pollution of the environment are already, partly due to lack of identification, absorbed by private and state health insurance.

The allocation of liability to the corporate entity also has its drawbacks. It is not the corporation but the corporate managers/officers and other enterprise participants who are responsible for the decisions and actions which cause damage. A strict production liability can only be successful in directing the firm's behaviour if liability incentives are passed on to the persons actually taking the decisions. This can be achieved indirectly through some form of internal tribunal or directly through the personal liability of corporate managers/officers[30]. If this does not happen, for example because of indemnification or insurance agreements, the result is a problem of moral hazard. The allocation of liability to the enterprise which is also protected by third party insurance and can restrict its liability through incorporation and/or segment its liability by building enterprise groups, renders corporate managers immune against the preventive effects of strict production liability (*Gefährdungshaftung*). It is then only the criminal law that can reach the responsible officers of the enterprise.

Organizational liability in tort: German law and EC law

The supposed advantages of strict production liability are traditionally opposed by two disadvantages of negligence liability. These are, first, that a need for some (however objective) fault is added to the central prerequisite of causality as an additional liability-establishing factor, and second that liability in tort is aimed at the individual "tortfeasor". The identification of malfeasance of participants in anonymous large-scale industrial enterprises meets with difficulties. In German tort law at least, which lacks an equivalent to the concept of vicarious liability/"*respondeat superior*" (Sykes, 1984), additional arguments are required to provide a complementary extension of liability to the employer/enterprise.

However, these disadvantages have not succeeded in obstructing the development of a modern enterprise liability in tort (Brüggemeier, 1991a). Combined with the concept of total quality and environment protection management, this may result in a roughly efficient concept of a "general production liability".

The *BGB*'s archaic regulation of the primary liability of employees (§ 823 *BGB*) and secondary liability of employers (§ 831 *BGB*) was soon modified by the judiciary in Germany. In the field of product and environmental liability, the development of duties of care has largely led to the socially and politically desirable exemption of employees from liability. Duties of care (*Verkehrspflichten*) are regarded as being enterprise-related. Corresponding to this, the so-called "duty to organize" was introduced by the judiciary. This lies primarily with the entrepreneur/corporate actor which, through its executive officers, has to organize personnel, materials and technical operations in such a way that "defects" are prevented. A radical tightening of the liability situation for enterprises was achieved through the shifting of the burden of proof for "organizational defects" which was introduced by the *BGH* for product flaws in 1968[31], and has since been extended to almost the whole range of production: failure to manufacture or to design properly and to instruct/warn at the point of sale[32]. It applies to both large and small enterprises and industry, craft trade and restaurants[33]. Only alleged breaches of the post sale duty of product monitoring are still exempt from the shifting of the burden of proof for negligence[34]. In 1984, the *BGH* extended these burden of proof principles to environmental liability[35]. *De facto*, the party injured by products or emissions of industrial plants has now only to prove causation. Everything else is in principle dealt with by the exculpation of the defendant enterprise.

It is noteworthy that this elaborate concept has since been adopted by European liability law. In November 1990, the EC Commission submitted a proposal for a Directive on Liability for Services[36]. At first, this directive attempted to transfer the concept of EC strict product liability to the service sector. However, due to massive resistance from the economic circles concerned, this attempt was abandoned. Instead, a negligence liability shifting the burden of proof for fault has been proposed. The principle of exclusive allocation of liability to the service enterprise which is characteristic of strict liability has, however, explicitly been retained. Exoneration from fault is equivalent to proof of a defect free business organization by the service enterprise[37].

From quality management to environment protection management

The liability of enterprises is also affected by the "revolution" in industrial organization structure through the quality systems (QS) movement. The QS concept no longer attempts to incorporate quality and safety "into" a product. Instead, it takes into consideration the fact that quality safety measures are more effective the earlier they are applied and, in consequence, the more they are aimed at the production process rather than at the finished product. This goes beyond the classical subject of QS, which was manufacturing process. Modern QS strategies influence every industrial operation, from research and development, construction and manufacturing up to purchasing, sales and management. In particular in its most advanced form, the so-called total quality management (TQM), the entire business organization is structured by given quality and safety requirements (Feigenbaum, 1983; Ishikawa, 1985; Juran, 1988; Masing, 1988). Its characteristic elements are: direct competence of executive officers; clearly defined responsibility, authority and delegation; interface control and co-ordination measures; existence of QS, which is usually based on international standards such as ISO 9000, CEN standards and so on; and the documentation of working processes and results. This comprehensive QS approach attaches particular importance to internal audits and meticulous synchronized quality control which make it possible to correct safety deficiencies and to improve quality standards during the production process.

This industrial quality and safety management meanwhile has been extended to the environmental impact of industrial operations. Product and process-related standards of quality assurance (DIN-ISO 9000–9004) are complemented by environmental protection-related standards (BSI Doc. 91/53 252, 53 256, 53 257). These technical provisions take up standards of safety regulation under public law. QS management is extended into environmental protection management (Friedman, 1990; Steger, 1988).

Following analyses of the American experience in this field, in 1992 the EC Commission presented a proposal for a regulation "allowing voluntary participation by companies in the industrial sector in a Community eco-audit scheme"[38]. The objective is to promote improvements in the environmental performance of industrial activities by:

(a) the establishment and implementation of environmental protection systems by companies;
(b) the systematic, objective and periodic evaluation of the environmental performance of such systems; and
(c) the provision of information on environmental performance to the public.

The environmental protection audit ("eco-audit") is the key element of this management concept[39]. Following the definition of the International Chamber of Commerce of 1988, the eco-audit is understood as a management tool "comprising a systematic, documented, periodic and objective evaluation of the performance of the organization, management system and equipment designed to protect the environment with the aim of: (i) facilitating management control of environmental practices; (ii) assessing compliance with company policies, including observance of the existing regulatory requirements" (Art. 2, lit. h).

Compliance with the eco-audit system standards can be certified by a logo. In addition, small and medium-sized enterprises are to receive special support. Such an environmental protection system[40] makes it possible to replace the outdated environmental protection policy of the "end of the pipe" technologies by the long-demanded concept of integrated industrial environmental protection.

In Germany, since 1990, § 52a of the Federal Emission Protection Act[41] has combined this self-regulatory approach with public law in a central field of environmental protection. § 52a *BImSchG* is directed at enterprises in the legal form of incorporated companies or partnerships which operate plants subject to permission. No specific form of industrial environmental protection is prescribed. The enterprises are, however, obliged to furnish information on the particular organization of environmental protection. They are also under a duty to advise the competent authorities which member of the board of officers/directors has been nominated to observe the "duties of the operator under the Federal Emission Protection Act".

Perspectives on a "general production liability" in tort

This post-Fordist organization of industrial production processes requires a reconsideration of the interrelation of strict liability and negligence liability with regard to enterprise liability. An organizational duty to ensure environmental protection is now imposed on the enterprise. The combination of state safety regulation "from above" with self-regulatory quality and safety management "from below", in the form of integrated quality and environmental protection systems, forms the background to a new dimension of the tort concept of the duty to organize. The tort liability of enterprises can finally develop into a liability of management defects. In combination with the shifting of the burden of proof for proper organization, this enterprise liability creates incentives to introduce "best available"[42] quality and environmental protection systems which are co-ordinated with the individual conditions prevailing in the respective enterprises. These incentives are complemented by a reward for quality and safety strategies in the form of favourable premium tariffs for third party insurance. In the case of damage nonetheless being caused, the enterprise can prove the "faultlessness" of its organization through its documentation. Quality and environment protection systems can exempt from liability, and where deficient they can result in liability and, in the mid-term, lead to an exclusion from the market. This tort approach is distinctly different from the negligence lottery of earlier days. Damage which occurs in spite of these protective measures has to be borne in most cases by the injured person or his/her first party or state health insurance. Only in problem sectors of society where the effects of damage are regarded as socially unacceptable, such as nuclear power plants, radioactive substances, genetic engineering labs and products, transport of dangerous goods, and hazardous waste dumping, is an additional form of damage compensation through strict liability to be established. The tort liability of organizations thus appears to be the regular form of enterprise liability with strict liability remaining the exception.

A further advantage of enterprise liability in tort is evident in the fact that it allows dual liability, that is that the responsibility of management for quality assurance and environmental protection can be taken into account. Defective planning by enterprises results in the ancillary personal liability of the executive officers. In the case that managers and owners are (partially) identical, as is often the case in small and medium-sized limited companies/ corporations, corporate law and tort law are seen to be in conflict. The corporate law principles of separation of assets and limitation of liability are here to a certain extent corrected by tort law. Nevertheless, it must be stressed that this is quite different from the problem of piercing the corporate veil (*Durchgriff*). Piercing refers particularly to owners/stockholders liability: an officer's liability extends only as far as responsibility for malpractice; whereas a person who acts as both owner and officer for his enterprise can be personally liable for his defective managerial conduct. The results of dual liability can only be welcomed from the point of view of damage prevention and allocation.

In German tort law the question of to which levels of enterprise hierarchy the personal liability of participants can or ought to be extended (to middle management, to the heads of the individual quality units, to the single employee) remains unsettled. As a result of the function-orientation of negligence liability, anybody who makes a mistake within his or her sphere of responsibility which causes damages to third parties can, in principle, be made liable (Brüggemeier, 1991a). Nonetheless, unlimited external liability of employees is socially unacceptable—even more so when the entrepreneurs are entitled to limit their liability through incorporation. The *BGH* recently stated that an employee can be personally liable under product tort law (§ 823(1) *BGH*), but that the burden of proof for fault is not to be shifted[43].

CONCLUSION

The general production liability of enterprises in tort comprising the three elements—duty to organize (with the burden of proof for the "best available" organization on the enterprise), quality and environmental protection systems (including eco-audits and documentation), and the exemption from liability when the enterprise provides proper environmental protection systems—appears to be a concept of liability which creates efficient incentives for a high level of social and environmental protection. This enterprise liability is complemented by the ancillary liability of executive corporate officers. The application of strict liability is to be limited to exceptional cases.

The system of civil responsibility in general, however, remains ill-suited to meet the particular challenges of environmental liability. The burden of proof on the injured party has to be erased, in cases both of negligence and strict liability. Regulation and standardization under public law, incentives through taxes and recovery through collective systems (social insurance, compensation funds) will continue to be of central importance in this field.

NOTES

1 The formal doctrine of German liability law confines *Gefährdungshaftung* strictly to the enterprise, whereas responsibility under *BGB* tort law in principle leads to dual liability (of the enterprise and of the corporate officers).

2 Proposal for a Directive on Liability for Services, OJ No. C12/8, 18.1.1991.

3 EC Directive Concerning Liability for Defective Products, OJ No. L210/29, 25.7.1985.

4 Proposal for a Directive on Civil Liability for Damage Caused by Waste, OJ No. C251/3, 4.10.1989; amended in 1991: OJ No. C192/6, 23.7.1991.

5 The first instance was the liability of railway companies in § 25 Prussian Railway Act of 1838. See Ogorek, 1975.

6 For the development in France; see Ewald, 1986.

7 See also Luhmann, 1991; Bechmann, 1991: 212.

8 See the White Book of the Commission to the European Council on the Completion of the Internal Market of June 14, 1985, COM(85) 310 endg.

9 OJ No. L169/1, 29.6.1987.

10 OJ No. L228/24, 11.8.1992; see Joerges & Falke, 1991: 159.10.

11 For the special field of environmental protection policy, see Rehbinder & Stewart, 1985.

12 Purely economic instruments of indirect control like taxes, subsidies, etc. are not discussed in this chapter.

13 See the Decision of the Council on a New Concept in the field of Technical Harmonization and Standardization, OJ No. C136/1, 4.6.1985. See Joerges, 1988: 157.

14 See the agencies' parameters provided in Art. 6 of the EC Directive on General Product Safety.

15 The former perspective still dominates the law and economics debate. See Shavell, 1984 and the contributions in Schuck, 1991. Thus, Rose-Ackerman pleads for an improved state safety regulation plus enterprise's exemption from liability (1991: 80ff.).

16 See J.K Galbraith's famous "life cycle theory" of agencies: "Regulatory bodies, ... have a marked life cycle. In youth they are vigorous, aggressive, evangelistic and even intolerant. Later they mellow and in old age—after a matter of ten or fifteen years—they become, with some exceptions, either an arm of the industry they are regulating or senile" (1980: 166).

17 This leads to difficult questions of "dual liability" (see Stone, 1980; Kraakman,

1985) and of liability of enterprise groups ("parent liability", "risk pools"). See the essays by Hofstetter and Teubner in this volume.

18 This corresponds with the situation in American law. See § 288C of the *Restatement (Second) of Torts*: "Compliance with a legislative enactment or an administrative regulation does not prevent a finding of negligence where a reasonable man would take additional precautions." (1965) St. Paul, Minn., vol. 2, pp. 39–40.

19 Kloepfer, Rehbinder & Schmidt-Aßmann. Professor E. Rehbinder drafted the section on liability law.

20 Council of Europe, Draft Convention on Civil Liability for Damage Resulting from Activities Dangerous to the Environment, DIR/IUR (91)3, July 1991.

21 See *Gesetz zur Regelung der Gentechnik (GenTG)* of June 20, 1990, *Bundesgesetzblatt* I, p. 1080.

22 Starting points have been the amendment to the Treaty of Rome introduced by the Single European Act 1986, and the Fourth Action Program on the environment published in 1987 (OJ No. C328/15, 7.12.1987). See Krämer, 1987.

23 OJ No. C251/3, 10.10.1989. See Wilmowsky & Roller, 1992.

24 OJ No. C192/6, 23.7.1991.

25 See the EC Directive on Waste, OJ No. L194/47, 25.7.1975; amended in 1991: OJ No. L78/32, 26.3.1991. See Dieckmann, 1992.

26 13 Transportrecht 83 (1990).

27 See Art. 11(2): "The Commission shall study the feasibility of the establishment of a European Fund for Compensation for Damage and Impairment of the Environment Caused by Waste".

28 See Art. 130r(2) of the EEC Treaty.

29 *Beshada* v. *Johns-Manville Prods. Corp.*, 90 NJ 191, 447 A 2d 539 (1982). Contrary to Beshada: *Anderson* v. *Owens-Corning Fiberglas Corp.*, 53 Cal. 3d 987, 810 P 2d 549 (1991).

30 This raises difficult questions of "dual liability". See Stone, 1980; Kraakman, 1985.

31 51 *BGHZ* 91 (1968).

32 116 *BGHZ* 60 (1992).

33 0116 *BGHZ* 104 (1992).

34 80 *BGHZ* 186; 80 *BGHZ* 199 (1981).

35 92 *BGHZ* 143 (1984). See Gmehling, 1989.

36 OJ No. C12/8, 18.1.1991.

37 What was regarded as a supposedly objective defect in the EC Product Liability Directive appears here as an element of negligence (Art. 1(3)).

38 OJ No. C76/2, 27.3.1992; see also the Commission's document COM(91) 459 endg. of 5 March 1992.

39 See also the contributions in the environmental audit volume *Cardozo Law Review* 12, 1215–1370 (1991).

40 Defined in the EC Draft Regulation as "the organizational structure, responsibilities, practices, procedures, processes and resources for implementing environmental management". Art. 2, lit. e.

41 Introduced by the 3rd Amendment of the Federal Emission Protection Act (*Bundesimmissionschutzgesetz (BImSchG)*) of 11 June 1990; *Bundesgesetzblatt* I 870. See Feldhaus, 1991.

42 To the "regulatory paradox of best available technology" see Sunstein, 1990a: 420–21. The burden of proof on the enterprise may help to reduce these alleged "self-defeating effects".

43 116 *BGHZ* 104 (1992). At the moment the Common Senate of the Highest Federal Courts in Germany has to decide upon the question of whether an employee who

causes damage to a third person while performing his or her job in principle has a right to indemnification from his or her employer. Political initiatives in the field of labour law tend to restrict employee's liability to gross negligence.

REFERENCES

Abraham, Kenneth S. (1986) *Distribution Risk. Insurance, Legal Theory, and Public Policy.* New Haven/London: Yale University Press.

Adams, Michael (1991) *Eigentum, Kontrolle und Beschränkte Haftung.* Baden-Baden: Nomos Verlagsgesellschaft.

Bechmann, Gotthard (1991) Risiko als Schlüsselkategorie der Gesellschaftstheorie. *Kritische Vierteljahresschrift für Gesetzgebung und Rechtswissenschaft*, 74, 212.

Beck, Ulrich (1986) *Risikogesellschaft. Der Weg in die andere Moderne.* Frankfurt/M.: Suhrkamp Verlag.

Brüggemeier, Gert (1983) Perspectives on the Law of Contorts. A Discussion of the Dominant Trends in West German Tort Law. *Hastings International and Comparative Law Review*, 6, 355.

Brüggemeier, Gert (1986) *Deliktsrecht. Ein Lehr- und Handbuch.* Baden-Baden: Nomos Verlagsgesellschaft.

Brüggemeier, Gert (1989) Umwelthaftung. Ein Beitrag zum Recht der Risiko-gesellschaft? *Kritische Justiz*, 22, 209.

Brüggemeier, Gert (1991a) Organisationshaftung. Deliktsrechtliche Aspekte innerorganisatorischer Funktionsdifferenzierung. *Archiv f.d. Civilistische Praxis*, 191, 33.

Brüggemeier, Gert (1991b) Liability for Water Pollution under German Law: Fault or Strict Liability? In Dunne, J.M. von (Ed). *Transboundary Pollution and Liability: The Case of the River Rhine.* Rotterdam: Erasmus University, 83.

Calabresi, Guido (1961) Some Thoughts about Risk Distribution and the Law of Torts. *Yale Law Journal*, 70, 499.

Calabresi, Guido (1970) *The Costs of Accidents.* New Haven/London: Yale University Press.

Cardozo Law School (1991) Symposium on Corporate Governance and the Environment: Beyond the Transactional Audit. *Cardozo Law Review*, 12, 1215–1370.

Dieckmann, Martin (1992) Der Abfallbegriff der EG-Rechts und seine Konsequenzen für das nationale Recht. *Natur & Recht*, 407.

Eads, George & Reuter, Peter (1983) *Designing Safer Products. Corporate Responses to Product Liability Law and Regulation.* Santa Monica: Rand Corporation.

Easterbrook, Frank H. & Fischel, Daniel R. (1985) Limited Liability and the Corporation. *University of Chicago Law Review*, 52, 89.

Ewald, François (1986) *L'état providence.* Paris: Grasset.

Falke, Josef & Joerges, Christian (1991) Die Normung von Konsumgütern in der Europäischen Gemeinschaft und der Richtlinien-Entwurf über die allgemeine Produktsicherheit. In Müller-Graff P.C. (Ed). *Technische Regeln im Binnenmarkt.* Baden-Baden: Nomos Verlagsgesellschaft, p. 159.

Feigenbaum, Armand V. (1983) *Total Quality Control.* New York: MacGraw-Hill.

Feldhaus, Gerhard (1991) Umweltschutzsichernde Betriebsorganisation. *Neue Zeitschrift für Verwaltungsrecht*, 10, 927.

Fikentscher, Wolfgang (1991) *Umweltsoziale Marktwirtschaft—als Rechtsproblem.* Heidelberg: C.F. Müller Verlag.

Friedman, Frank B. (1990) *Practical Guide to Environment Management*, 2nd edn. New York: Stewart/Lynn.

Galbraith, Kenneth (1980) *The Great Crash 1929.* London: A. Deutsch.

Gmehling, B. (1989) *Die Beweislastverteilung bei Schäden aus Industrieimmissionen*. Köln: Heymanns Verlag.

Hansmann, Henry & Kraakman, Reinier (1991) Toward Unlimited Shareholder Liability for Corporate Torts. *Yale Law Journal*, **100**, 1879.

Hart, Dieter & Reich, Norbert (1990) *Integration und Recht des Arzneimittelmarktes in der EG*. Baden-Baden: Nomos Verlagsgesellschaft.

Hohloch, G. (1992) *Entschädigungsfonds auf dem Gebiete des Umwelthaftungsrechts—Rechtsvergleichende Untersuchung vorhandener Fondslösungen*. Berlin: de Gruyter.

Isensee, Josef (1983) *Das Grundrecht auf Sicherheit*. Berlin/New York: de Gruyter.

Ishikawa, Kaoru (1985) *What is Total Quality Control?* New York: Prentice Hall Inc.

James, Fleming (1955) Products Liability. *Texas Law Review*, **34**, 44, 192.

Joerges, Christian (1988) The New Approach to Technical Harmonization and the Interests of Consumers: Reflections on the Necessities and Difficulties of a Europeanization of Product Safety Policy. In Bieber, R. *et al.* (Eds). *One European Market? A Critical Analysis of the Commission's Internal Market Strategy*. Baden-Baden: Nomos Verlagsgesellschaft, p. 157.

Joerges, Christian *et al.* (1988) *Die Sicherheit von Konsumgütern und die Entwicklung der Europäischen Gemeinschaft*. Baden-Baden: Nomos Verlagsgesellschaft. (English: Joerges, Christian (Ed) (1991) European Product Safety, Internal Market Policy and the New Approach to Technical Harmonization and Standards. *EUI Working Papers*, Law No. 91/10-14, Florence.)

Joerges, Christian (1990) Paradoxes of Deregulatory Strategies at Community Level: The Example of Product Safety Policy. In Majone, G. (Ed). *Deregulation or Reregulation? Regulatory Reform in Europe and in the United States*. London: Pinter. New York: St. Martin's Press, p. 176.

Juran, Joseph M. (1988) *Juran's Quality Control Handbook*, 4th edn. New York: McGraw-Hill Books.

Kanner, Allan (1991) The Evolving Jurisprudence of Toxic Torts: The Prognosis for Corporations. *Cardozo Law Review*, **12**, 1265.

Klein, Eckar (1989) Grundrechtliche Schutzpflichten des Staates. *Neue Juristische Wochenschrift*, **42**, 1633.

Kloepfer, Michael *et al.* (1990) *Umweltgesetzbuch—Allgemeiner Teil, Berichte 7/90 des Umweltbundesamtes*. Berlin: Erich Schmidt Verlag.

Kraakman, Reinier (1985) The Economic Functions of Corporate Liability. In Hopt, K.J. & Teubner, G. (Eds). *Corporate Governance and Directors' Liability*. Berlin/New York: de Gruyter, p. 178.

Krämer, Ludwig (1987) The Single European Act and Environmental Protection. *Common Market Law Review*, **24**, 659.

Landes, William M. & Posner, Richard A. (1987) *The Economic Structure of Tort Law*. Cambridge, Mass.: Harvard University Press.

Luhmann, Niklas (1991) *Soziologie des Risikos*. Berlin/New York: de Gruyter.

Masing, Walter (Hrsg.) (1988) *Handbuch der Qualitätssicherung*, 2. Aufl. München: Hauser Verlag.

Mataja, Victor (1888) *Das Recht des Schadensersatzes vom Standpunkte der Nationalökonomie*. Leipzig: Duncker & Humblot.

McGuire, Patrick E. (1988) *The Impact of Product Liability*. New York: Conference Board.

Micklitz, H.W. (Ed) (1990) *Post Market Control of Consumer Goods*. Baden-Baden: Nomos Verlagsgesellschaft.

Ogorek, Regina (1975) *Untersuchungen zur Entwicklung der Gefährdungshaftung im 19. Jahrhundert*. Köln: Böhlau.

Panther, Stephan (1991) Zivilrecht und Umweltschutz. In Ott, C. & Schäfer, H.B. (Eds). *Ökonomische Probleme des Zivilrechts*. Berlin: Springer Verlag, p. 267.

Priest, George L. (1985) The Invention of Enterprise Liability: A Critical History of the Intellectual Foundations of Modern Tort Law. *Journal of Legal Studies*, **14**, 461.

Priest, George L. (1990) The New Legal Structure of Risk Control. *Daedalus*, **119**, 207.

Priest, George L. (1991) The Modern Expansion of Tort Liability: Its Sources, Its Effects, and Its Reform. *Journal of Economic Perspectives*, **5**, 31.

Rabon, Robert L. (1987) Environmental Liability and the Tort System. *Houston Law Review*, **24**, 27.

Rehbinder, Eckard (1988) Ersatz ökologischer Schäden—Begriff, Anspruchsberechtigung und Umfang des Ersatzes. *Natur & Recht*, **10**, 105.

Rehbinder, Eckard (1989) Fortentwicklung des Umwelthaftungsrechts in der Bundesrepublik Deutschland. *Natur & Recht*, **11**, 149.

Ritter, Gerhard A. (1991) *Der Sozialstaat. Entstehung und Entwicklung im internationalen Vergleich*, 2 Aufl. München: Oldenbourg.

Schäfer, H.B. & Ott, Claus (1986) *Lehrbuch der ökonomischen Analyse des Zivilrechts*. Berlin: Springer Verlag.

Schuck, Peter H. (Ed.) (1991) *Tort Law and the Public Interest*. New York: W.W. Norton.

Selmer, Peter (1991) *Privates Umwelthaftungsrecht und öffentliches Gefahrenabwehrrecht*. Heidelberg: C.F. Müller.

Shavell, Steven (1984) Liability for Harm versus Regulation of Safety. *Journal of Legal Studies*, **13**, 357.

Shavell, Steven (1987) *Economic Analysis of Accident Law*. Cambridge, Mass.: Harvard University Press.

Stewart, Richard B. (1985) Environmental Protection Policy. In Cappelletti, M. *et al.* (Eds). *Integration Through Law*, vol. 2. Berlin/New York: de Gruyter.

Stone, Christopher D. (1980) The Place of Enterprise Liability in the Control of Corporate Conduct. *Yale Law Journal*, **90**, 1.

Sunstein, Cass R. (1990a) Paradoxes of the Regulatory State. *University of Chicago Law Review*, **57**, 407.

Sunstein, Cass R. (1990b) *After the Rights Revolution: Reconceiving the Regulatory State*. Cambridge, Mass.: Harvard University Press.

Sykes, Alan O. (1984) The Economics of Vicarious Liability. *Yale Law Journal*, **93**, 1231.

Teubner, Gunther (1987) Juridification—Concepts, Aspects, Limits, Solutions. In Teubner, G. (Ed). *Juridification of Social Spheres. A Comparative Analysis in the Area of Labor, Corporate Antitrust and Social Welfare Law*. Berlin/New York: de Gruyter, p. 3.

Teubner, Gunther (1991a) Beyond Contract and Organisation? The External Liability of Franchising Systems in German Law. In Joerges, C. (Ed). *Franchising and the Law. Theoretical and Comparative Approaches in Europe and the US*. Baden-Baden: Nomos Verlagsgesellschaft, p. 105.

Teubner, Gunther (1991) Unitas Multiplex. Das Konzernrecht in der neuen Dezentralität der Unternehmensgruppen. *Zeitschrift f.d. Unternehmens- und Gesellschaftsrecht*, **20**, 189.

Wilmowsky, Peter & Roller, Gerhard (1992) *Civil Liability for Waste*. Frankfurt/M: Peter Lang.

CHAPTER 5

The Ecological Liability of Corporate Groups: Comparing US and European Trends

Karl Hofstetter
Lucerne

INTRODUCTION

"Bhopal", "Amoco Cadiz" and "Seveso" have become buzzwords in the general debate about the ecological hazards of modern industry. In addition, they have a very specific common legal denominator: all three cases involved the question of parent liability *vis-à-vis* creditors of a subsidiary. Of course, the issue of corporate group or parent liability is not limited to ecological catastrophes. It can arise in connection with all tort and contract claims against a subsidiary. Yet, the problem is particularly acute in the field of ecological liability, as damages resulting from ecological disasters can run very high, thereby easily exceeding the assets of the primarily responsible subsidiary.

In this chapter I try to outline and evaluate some evolving trends in the USA and Europe with regard to parent liability for ecological damages. I will first describe recent developments in US law, mainly in connection with the Comprehensive Environmental Response, Compensation and Liability Act (CERCLA). Second, I will focus on concepts of parent liability in Europe. Third, I will deal with the need for an adequate paradigm to assess parent liability regimes. I will make reference to the economic analysis of law and discuss whether systems theory can provide additional insights. This epistemological excursion will serve as a basis for exploring the idea of self-organization in connection with parent liability law. Finally, I will conclude with some general comments.

Environmental Law and Ecological Responsibility: The Concept and Practice of Ecological Self-Organization
Edited by G. Teubner, I., Farmer and D. Murphy
© 1994 John Wiley & Sons Ltd

TRENDS IN US LAW

Piercing the veil in corporate law

The overarching legal mechanism in US parent liability law is that of "piercing the corporate veil" (Clark, 1986: 71ff.; Blumberg, 1987: 681–683; Hofstetter, 1990b: 591–593). Its uncertain and metaphoric name is partly reflected in its application: US courts refer to it in very different situations, and it is not easy to detect a coherent theory behind the uses of the concept. US piercing law requires basically that a parent company either controlled a subsidiary for illegitimate purposes (e.g. fraud) or controlled it in such a complete manner that the control seems illegitimate[1]. The courts use words like "sham", "shell", "agent" or "instrument" to describe the liability-triggering dependency of subsidiaries (Blumberg, 1987: 107). They look at a myriad of factors which may together indicate the critical level of control. Such factors are the percentage of stock ownership, the existence of interlocking directorships, the commingling of assets between the parent and the subsidiary, the total financial dependency of the subsidiary, its inadequate capitalization or the non-observance of its formal legal requirements (Blumberg, 1987: 138–140). Yet, the mere majority ownership and even the parent's involvement in the day-to-day business of a subsidiary does not, in itself, constitute a sufficient basis for piercing the corporate veil (see Blumberg, 1987: 185ff.).

In the famous case of *Lowendahl* v. *Baltimore & ORR*[2] a New York court described the required level of control in the following words:

> Control, not merely majority or complete stock control, but complete domination, not only of finances, but of policy and business practices in respect to the transaction attacked so that the corporate entity had at the time no separate mind, will or existence of its own.

Blumberg notes though, that there is a greater willingness on the part of US courts to pierce the corporate veil in tort cases than in others (Blumberg, 1987: 110–111). A recent judgment handed down in the case involving the "Amoco Cadiz" oil spill off the coast of France in 1978 further corroborates this view[3]. Even though the court for the Northern District of Illinois did not make specific reference to the piercing doctrine, it applied a similar instrumentality test and, moreover, seemed to gear it to the tort considerations in question. The court noted:

> 43. As an integrated multinational corporation which is engaged through a system of subsidiaries in the exploration, production, refining, transportation and sale of petroleum products throughout the world, Standard is responsible for the tortious acts of its wholly owned subsidiaries and instrumentalities, AIOC and Transport.
> 44. Standard exercised such control over its subsidiaries AIOC and Transport that those entities would be considered to be mere instrumentalities of Standard.

Furthermore, Standard itself was initially involved in and controlled the design, construction, operation and management of the Amoco Cadiz and treated that vessel as if it were its own.

45. Standard therefore is liable for its own negligence and the negligence of AIOC and Transport with respect to the design, operation, maintenance, repair and crew training of the Amoco Cadiz.

46. Standard therefore is liable to the French claimants for damages resulting from the grounding of the Amoco Cadiz.

A comparable cause of action, referred to as "multinational enterprise liability", was introduced by the government of India in its Bhopal complaint in the USA (see Westbrook, 1985: 321ff.). The action was dismissed on the basis of a ruling of *forum non conveniens*[4].

Parent liability concepts under environmental law statutes

CERCLA (42 USC § 9601–9657 (1982)) provides a regulatory programme controlling hazardous waste disposal on US soil. It also established a "Superfund" to finance environmental clean-up at hazardous waste sites (Superfund Amendments and Re-authorization Act (SARA) of 1986). Section 107(a) of the Act imposes liability on:

(1) the owner or operator of ... a facility,
(2) any person who at the time of disposal of any hazardous substance owned or operated any facility at which such hazardous substances were disposed of, ...
(3) any person who ... arranged for disposal or treatment, or ... transport ... of hazardous substances owned or possessed by such person .... .

From the beginning, the courts have emphasized that the statutory purpose of CERCLA requires a liberal construction of its liability provisions[5]. This tenet also influenced their approach to the question of parent liability under the Act. The courts have chosen two main routes to parent liability: either the statutory construction of the parent as an "owner or operator" or the piercing of the corporate veil[6].

In the first alternative US courts used the language and the purpose of CERCLA to reach beyond the constraints of piercing the corporate veil law. In *State of Idaho* v. *Bunker Hill Co.*[7] the court concluded that the parent and its subsidiary were "so intertwined" and the parent "so controlled the management and operations" of the subsidiary that the parent constituted an "owner or operator" for purposes of CERCLA. The court also found the parent to be an "owner or operator" because it was "intimately familiar" with hazardous waste disposal at the subsidiary's facility and had the "capacity, if not total reserved authority, to make decisions and implement actions and mechanisms to prevent the damage". Similar rationales were enunciated in subsequent decisions[8]. The District Court of Rhode Island in *US* v. *Kayser Roth Corp.*[9] first stated that a parent corporation cannot ordinarily be deemed an operator solely on the basis of its status as a shareholder. But the court here considered as critical the fact that the parent had exercised "pervasive control" ("practical total control") over its subsidiary through, among other things:

(a) total financial control, including collection of accounts payable;
(b) restrictions on the subsidiary's budget;
(c) its directive that subsidiary-government contacts, including in environmental matters, be funnelled directly through the parent;
(d) the requirement that the subsidiary's leasing, buying or selling of property first be approved by the parent;
(e) its policy that the parent approve any capital transfer or expenditures greater than US $5000.00;
(f) placement of parent personnel in almost all of the subsidiary's directorships or officerships, as a means of ensuring that the parent's corporate policy was exactly implemented and precisely carried out.

The court also considered "illustrative" the fact that the parent controlled all actions with regard to environmental matters, such as the release of hazardous substances and related investments[10]. In addition, the court emphasized that actual knowledge of the release is not required for establishing liability under CERCLA[11].

Even though the control test under CERCLA resembles that in traditional piercing law, it, by contrast, does not seem to require *indicia* of illegitimate conduct. A very firm grip on the subsidiary, particularly if extending to environmental matters, seems to be sufficient in order to render the parent liable as an "owner or operator".

The ruling in *US* v. *Fleet Factors*[12] further supports the view that *de facto* control over the environmental operations of a company can be determinative for extending CERCLA liability. The Court of Appeals of the 11th Circuit held that even:

> a secured creditor will be liable if its involvement with the management of the facility is sufficiently broad to support the inference that it could effect hazardous waste disposal decisions if it so chose.

Hence, the court observed that liability could even attach where secured creditors had the power and discretion to affect hazardous waste decisions of a company. This means that the mere existence or potential of control, not its exercise, may be dispositive.

Still, there seem to be limits to the scope of owner operator liability under CERCLA. In *Edwards Hines Lumber Co.* v. *Vulcan Materials Co.*[13], for example, Judge Easterbrook held that an independent contractor or joint venturer who designed the manufacturing facility and trained the facility's workers did not exercise control over the facility's operations and, therefore, was not an operator under CERCLA[14].

Before applying the second alternative, that is the piercing concept under CERCLA, the courts first have had to address the question of whether to apply state law or whether, instead, the development of federal common law was justified (choice of law). Citing the decision in *Re Accushnet River & New Bedford Harbor Proceedings*[15], the court in *US v. Kayser-Roth Corp.*[16] held that a uniform federal rule was warranted in view of the "overriding federal interest" in promoting environmental protection. It went on to state that piercing the corporate veil was an alternative route to rendering the parent company at hand liable as an "owner".

The court's ruling was two-pronged. It first relied on the basic principle that under federal law, "a corporate entity may be disregarded in the interest of public convenience, fairness and equity". In addition, it considered determinative the same *indicia* of control already cited in connection with the qualification of the parent as an "owner or operator".

CERCLA is, of course, just one of many environmental liability statutes in US federal or state law. Similar concepts of parent liability seem to emerge in the context of other statutes. A leading case involving the issue of group liability was *State Department of Environmental Protection v. Ventron Corp.*[17], which arose under the New Jersey Spill Compensation and Control Act. The New Jersey Supreme Court held that:

> Given the extended liability of the Spill Act, we conclude that the legislator intended that the privilege of incorporation should not, under the circumstances that obtain here, become a device for avoiding statutory responsibility. A contrary result would permit corporations, merely by creating wholly owned subsidiaries, to pollute for profit under circumstances when the legislator intended liability to be imposed.

The newly enacted Federal Oil Pollution Act of 1990 establishes, *inter alia*, strict joint and several liability of ship owners and operators for oil spills in US waters (Anderson & Wethmar, 1991: 1001ff.). Yet it does not contain a parent liability provision. Given the trends under CERCLA, and assuming that the *Amoco Cadiz* judgment might have some effect as a precedent, it can be expected that the Oil Pollution Act, too, will be subjected to a liberal standard of parent liability (Anderson & Wethmar, 1991: 1005–1007).

In summary, then, we can say that in the field of general corporate law the US courts still stick to a traditional piercing the corporate veil test which requires *indicia* of control and some form of illegitimate conduct. In tort law and under environmental law statutes, however, courts seem to have gone out of their way to be more liberal in the construction of parent liability. In this, they are able to adjust their parent liability concepts more specifically and flexibly to the policies underlying tort and environmental law.

TRENDS IN EUROPEAN LAW

Corporate law concepts

Unlike in the USA, where parent liability has had a relatively dynamic evolution in the context of specific liability laws and statutes, European trends are concentrated around corporate law, and as a result piercing the corporate veil is also well known in European jurisdictions. But corporate group law has moved beyond this concept (Hofstetter, 1990b: 576ff.), and in this move a leading role has been played by Germany.

The pioneering role of Germany in the area of corporate group law is undisputed. Even though the codification of the German *Konzernrecht* in 1965 turned out to be ineffective in various respects, most notably as regards the provisions for *de facto* concerns[18], this legislation seems to have inspired the German legal culture. In the area of parent liability, the German Federal Court has enunciated special rules for so-called "qualified *de facto* concerns". In *Autokran*[19] it held that where a parent company was "permanently and extensively" involved in the management of a subsequently bankrupt subsidiary, a rebuttable presumption exists that the parent did not show sufficient consideration for the subsidiary. Hence, unless the parent is able to defend itself successfully, it is directly liable to creditors for the subsidiary's obligations. In this decision the German Federal Court designed a way around the notorious enforcement problems in connection with parent liability provisions for highly integrated *de facto* concerns. It did this by resorting to two legal techniques: the shifting of the burden of proof to the parent company and the granting of a direct claim against the parent. In its subsequent *Tiefbau* decision[20], the court restricted the parent's defence to situations in which the subsidiary's "losses were caused by circumstances outside the parent's managerial control"[21].

French bankruptcy law, Swiss corporate law and the draft for a Ninth Directive in the European Community (Hofstetter, 1990b: 585–590) work with "*de facto* director" notions that put parent companies under the same fiduciary duties as corporate directors and officers. Parent liability requires that the parent acted like a director, that is to say, involved itself in the day-to-day operations of the subsidiary. Furthermore, creditors have to prove that in so doing the parent violated its fiduciary duties and thereby caused damage to the subsidiary. This approach to parent liability, even though promising, faces significant enforcement problems due to the heavy burden of proof placed on creditors. In highly complex and integrated modern corporate groups it can become almost impossible for outsiders to trace and prove past violations of the subsidiary's interests by the parent (Hofstetter, 1990b: 593ff.). Shifting the burden of proof to the parent company in analogy to the German qualified concern doctrine might, thus, be the only way to rebalance the situation[22].

Developments in the area of environmental liability law

The general emphasis on corporate law notwithstanding, European courts and scholars have approached parent liability from other angles as well. French labour law[23], or the Austrian *Eumig* decision[24], are examples outside ecological liability. With regard to environmental statutes there have, as far as I can see, been no specific developments (see Bender & Sparwasser, 1990: N 52, 86; Schmid, 1990: 1ff.; Tercier, 1990: 73ff., 161–162; Krämer, 1991: 85ff., 335ff.). Scholarship, however, has emphasized the potential of using general civil law constructions in order to render parents liable for their subsidiaries' obligations. This could also have implications for ecological liability. Eckard Rehbinder (1969: 85ff., 133ff.), for example, has elaborated a myriad of civil law devices that could lead to the liability of parent companies for specific claims against their subsidiaries. Particular plausibility goes to the concept of vicarious liability (Rehbinder, 1969: 529ff.) which could also play an important role in the field of ecological liability. Under this notion, which is also advocated by Swiss scholars (Hofstetter, 1990b: 590–591), parent liability could attach for tort claims against its subsidiary. It requires proof that the parent had management control over at least those activities of the subsidiary which caused the tort (Hofstetter, 1990b: 595–596).

An instructive case touching upon the parent liability issue developed in the aftermath of the Seveso catastrophe in 1976. The community of Seveso brought an action against Givaudan, the immediate parent of the primarily responsible Italian subsidiary Icmesa, but also against the latter's ultimate parent, Hoffmann-La Roche, in Switzerland[25]. The complaint against Hoffman-La Roche contained several causes of action. One was that Hoffman-La Roche had assumed the obligations of Icmesa by publicly declaring after the accident that Hoffman-La Roche always backed its subsidiaries in need. Another cause of action was vicarious liability. The plaintiff maintained that Icmesa had been an agent or subordinate of Hoffmann-La Roche and that the latter was therefore vicariously liable for Icmesa's negligence (Art. 55 of the Swiss Code of Obligations). An alternative legal construction brought forward by the plaintiff was based on Art. 55 of the Swiss Civil Code. It provides that companies are responsible for the torts of their directors and officers. The plaintiff maintained that the directors and officers of Givaudan, who were in charge of production safety at Icmesa, were also constructive officers of Hoffmann-La Roche, because they acted directly in the interest of the ultimate parent. The case was settled, however, before the Swiss courts could render an opinion on it.

In general, though, it can be said that, in contrast to the USA, European jurisdictions focus primarily on corporate law based parent liability concepts. They are thereby moving beyond traditional piercing notions. They show sensitivity to the specific risks for creditors in connection with the parents' management influence over their subsidiaries. Devices such as *de facto* director liability or the German "qualified concern" doctrine seem particularly well suited to deal with the general organizational risks faced by creditors in connection with corporate group integrations. As to ecological liability of parent corporations, the vicarious liability concept, as asserted in the Seveso action for example, exemplifies a second though relatively weaker trend in European law attempting to tackle parent liability by way of general civil law constructions.

IN SEARCH OF A PARADIGM FOR ECOLOGICAL PARENT LIABILITY

Economic analysis of parent liability law

In order to evaluate parent liability in environmental law and elsewhere, one needs a reference model. Traditional legal thinking does not suffice. It tends to be formalistic and is of little value when it comes to policy questions of optimal social engineering (see Adams, 1985: 8ff.). The rise of law and economics, therefore, was a major step forward in the debate on liability law (Adams, 1985: 17ff.; Shavell, 1987: 5ff.; Landes & Posner, 1987; Hofstetter, 1990a: 301ff.). Its perception of limited liability as a means of promoting efficiency has created a much more differentiated view of the pros and cons of parent liability (see Hofstetter, 1990a: 306ff.; 1994: Ch. III.3). Parent liability, in this view, appears as a function of three factors:

(a) the relative monitoring cost of alternative risk bearers;
(b) the diversification (relative risk attitudes) of alternative risk bearers; and
(c) the legal (opportunity) cost of developing and operating alternative liability regimes.

Starting from this, I have tried elsewhere to show that the limited liability of subsidiary corporations is above all an efficient solution to the problem of internalizing political risks (Hofstetter, 1994: Ch. III.3; 1990a: 309–12). Such risks are best borne by the political community at large, which is in the better position to prevent those risks from materializing. However, I developed three efficient exceptions to limited liability: first, when a parent abused a subsidiary; second, when the parent diluted or wasted a subsidiary's assets through its interference; and third, where the parent caused direct and separable harm to subsidiary creditors (Hofstetter, 1990a: 315–19).

The traditional remedy of piercing the corporate veil is, thus, a basically efficient instrument within the first category. German corporate group law for "qualified concerns" or French, Swiss and EC *de facto* director laws would belong in the second category. Tort law constructions, like vicarious liability, might have particular efficiency plausibility as instruments of the third category. More specifically, vicarious liability appears to be an efficient tool because it remedies the particular risk-externalization potential in connection with tort creditors of a subsidiary who are not able to renegotiate given corporate group structures or receive compensation for bearing limited liability risks. In addition, if combined with adequate prerequisites and defences for the parent (taking due account of the involvement of the parent at the subsidiary and its relative monitoring cost), the vehicle can be fine-tuned flexibly and restricted to specific situations. Finally, by not being a "broad sweep" rule and by basically covering insurable risks, vicarious liability does not undermine the limited liability of subsidiary corporations in a way that would render them useless as a means of hedging against political risks.

The US courts' parent liability constructions under CERCLA would also seem to have efficiency potential within the above-mentioned third category (see Note, 1989: 986ff.). It is, of course, questionable from an efficiency point of view whether current owners of pieces of land who had no influence on pollution should be jointly and severally liable with the former owners and polluters of the land[26]. The political community at large would possibly be a more efficient bearer of such risks than the parent company of the current owner, no matter how tightly controlled the subsidiary was[27]. Where a parent company controlled a polluter-subsidiary and directed the latter's hazardous waste release and environmental investments, however, a strong efficiency case can be made in favour of parent liability, the major reason being that the monitoring cost for preventing the pollution would basically be lower for the parent than for any part of the political community.

The latter example shows a fundamental trade-off. Uniform rules about parent liability might over or undershoot. Flexible regimes that could adapt to specific situations and different types of risk could hamper predictability. Any parent liability regime does, accordingly, have to aim at some form of optimal simplification (Hofstetter, 1990b: 594ff.).

Beyond economic analysis of law

From a broader social perspective, the economic analysis of law appears reductivist and, hence, arguably inadequate (see Hofstetter, 1994: Ch. III.4). The debate about ecological liability in particular gives rise to doubts about the appropriateness of mere economic thinking, since the ecological problem is in itself an expression of the inability of the economic system to adequately "internalize the environment" (see Luhmann, 1989: 17ff.). In addition, the microeconomic equilibrium model commonly applied by the economic analysis of law has indisputable deficiencies from an epistemological point of view. Features such as methodological individualism, reducing human actors to strategies and opportunistic utility maximizers (see Hofstetter, 1994: Ch. III.4), or the perhaps inherent tendency of economic equilibrium analysis to undervalue the role of innovation and new knowledge (see von Hayek, 1989/91: vol. 1, 10ff., 72ff., vol. 2, 34ff., vol. 3, 97ff.; Ladeur, 1988: 332;

Huber, 1988: 155ff.) are sufficient to warrant that the economic analysis of law should not monopolize any normative legal discussions. Let there be no doubt: economic models greatly help to integrate economic implications into legal thinking. In other words, the economic analysis of law is a most appropriate vehicle for gearing law to the parameters of the economic system. And that is, if no more, at least an economically productive achievement. Yet, where the economic system is part of the problem itself, as is arguably the case with the ecological challenge, it seems at best doubtful whether this challenge can be overcome by merely resorting to economic rationales. The whole "internalization" debate prompted and led by economic thinking in connection with environmental law and liability (e.g. Wagner, 1989: 359ff.; Petitpierre-Sauvain, 1989: 466ff.) could, thus, lead into a vicious circle[28]. For a better ecological future more dramatic changes might be required than just the reallocation of cost within given (assumed) economic conditions.

Is there a superior theoretical framework for assessing and devising ecological liability law? The work of Teubner (1990a: 761ff.; 1991: 204ff.) for example, suggests that systems theory might serve as an appropriate reference. Indeed, systems theory seems particularly well suited to put the ecological problem into a broad evolutionary perspective. There, ecological liability appears as a problem of communication between the legal system and the economic system (Teubner, 1989: 123ff.) where the legal system has taken on the task of channelling signals from the environment into the economy ("risk dialogue"). "Sustainable development", not social efficiency, thus emerges as the ultimate paradigm (Becker, 1991: 19; see also Schmidheiny, 1992: *passim*). Does this make a practical difference? Assuming a natural incongruence between the legal and economic systems, communication between law and the economy is bound to fail if carried out in a one-way linear and hierarchical fashion. Accordingly, communication by legal rules has to be carried out on a two-way basis in that the legal system gears its messages to the parameters of the economy. Yet, this is where law and economics, at least operationally, comes in again. It might, for example, suggest guidelines for promoting the proper application of existing environmental technologies within a given framework of preferences and knowledge.

Stimulating the development of new attitudes, preferences and technology, however, transcends the borders of simple microeconomic equilibrium theory (see Ladeur, 1988: 312ff.). It leads communication between law and the economy to a more subtle level. Under these conditions, law can do no more than work its way intuitively towards the economy and private enterprise (Ladeur: 314ff.; Teubner, 1991: 205; 1989: 81ff.). The task is comparable to one of management trying to stimulate innovation such as in Japanese "knowledge-creating companies". These seem intentionally to refrain from efficiency-oriented, explicit or rational messages. Instead, they subscribe to tacit communication based on symbolism (cryptic (open), simple comments) and redundancy (overlapping messages on various levels and in various forms) (see Nonaka, 1991: 96, 101–102).

Thus, looking beyond simple economic analysis, an appropriate paradigm for ecological liability does have to integrate totally unknown variables into its calculus. It does this at the price of losing conceptual certainty, however. As a consequence, liability law itself becomes a "discovery procedure" (see Hofstetter, 1994: Ch. III.4.; Kramer, 1990: 252; von Hayek, 1969: 249ff.).

The revealed experimental character of liability law suggests openness and perhaps also caution (see Mastronardi, 1991: 449ff.). It calls, among other things, for a diversified approach to ecological problems involving a variety of (soft[29]) instruments within and outside liability law (see Brüggemeier, 1989: 230; Ladeur, 1988: 319ff.). "Decentralized" social arrangements based on such diverse measures as specifically tailored liability laws, liability caps, mandatory insurance, administrative controls, criminal law, ethical norms and education must thereby be given particular plausibility (see Schmid, 1990: 16ff.; see also the concept of the German *Gesetz über die Umwelthaftung* of 1991).

Parent liability law as a means to stimulate self-organization

The open paradigm of sustainable development underscores the autonomy of the economic system and its enterprises *vis-à-vis* environmental demands and liability law. It reveals the necessary lack of an overarching concept that would allow for an all-encompassing internalization of ecological interests into the economic process. What does this mean for parent liability law?

If the optimal ecological conduct by economic actors cannot be deduced from a general theory and translated into unequivocal legal commands, there still remains the possibility of getting closer to such conduct by way of trial and error (that is through falsification, Popper, 1984: 15). The corollary of this is an operational legal concept that is based on piecemeal (Popper, 1969: 157ff.) as opposed to utopian social engineering[30]. In other words, law assumes the more modest role of being "one player amongst many" instead of being the "paramount dictator" overlooking society. This leads, almost by necessity, to some form of labour-sharing between the legal system and private enterprise. Law and the economy observe each other and interact as equals.

This respect of law for the economy and its enterprises calls for legal self-restraint. The task of environmental legal norms is thereby limited to pushing private economic activity in ecologically responsible directions. Law's job, in other words, is to inject environmental messages into the economic system hoping that they will build "structural couplings" (see Teubner, 1990b: 532ff.) towards some form of sustainable development. The targets of such legal norms are accordingly those economic facts (like costs, technology, attitudes or preferences) which can presumably be influenced by being tipped from the outside.

This approach respects the autonomy of economic actors, but it is by no means congruent with *laissez faire*. Law, including liability law, maintains a crucial function in communicating ecological pressures to private enterprise. Yet it does this primarily with the hope of triggering self-organizing processes in the economic system[31]. It issues signals, the consequences of which it can neither foresee nor calculate. However, it can experiment with them, observe their impact upon the economic process, learn from that process and eliminate measures if they fail. An adequate ecological concept, therefore, employs liability law as a tentative measure to generate ecologically responsible self-organization (see Teubner, Chapter 2). Within this strategy, parent liability rules assume both offensive and defensive functions.

On the offensive side parent liability rules amplify the messages of liability law in general. By extending the liability of subsidiaries to parent corporations, parent liability can help stimulate those self-organizing processes that are already aimed at by general liability rules. In other words, parent liability can be a means of carrying environmental liability beyond the basic barriers of corporate law[32]. Yet parent liability can also be a stimulus to self-organization in its own right, that is, it can promote ecologically adequate corporate group structures. In this second variant parent liability does not aim at carrying out the specific purposes of tort-based environmental law, but those of corporate law as such[33]. It focuses on management control and the delegation of responsibilities as between a parent and its subsidiary—those issues which are usually considered to be within the domain of corporate law.

On the defensive side parent liability rules fulfil the task of limiting the liability exposure of corporate groups so as to motivate their continued investment and risk-taking. Parent liability rules, therefore, have a second function of protecting autonomous parts within private firms against the potentially demotivating uncertainties of liability law. Their job is the establishment of a motivational set-up that is conducive to further investment. Its time horizon should ideally be congruent with relevant planning and investment cycles (see Schmidheiny, 1992: 14). The defensive function of parent liability rules accordingly calls for only selective exceptions to the principle of limited liability.

Looking at the results of empirical research the notion of self-organization turns out to be a highly adequate macro-approach to liability law[34]. Eads and Reuter (1983: 72ff.) report that the effect of liability judgments on the conduct of firms is at best ambiguous. The same liability rulings can result in defensive, demotivated and, thus, counterproductive responses as much as they can lead firms to make the desired investments[35]. The outcome seems to hinge primarily on the existence of corporate cultures that are amenable to the productive tackling of the envisaged problems. Yet corporate culture, by its nature, cannot be fine-tuned by law, though it can be influenced by legal signals (see Schmidheiny, 1992: 49ff.). Empirical research, thus, seems to confirm that a very important aspect of liability law is its signalling effects, which influence corporate culture and thus promote self-organizing processes within firms. It is this potential for "scratching enterprises just below the surface" and thereby triggering semi-voluntary changes of attitudes, preferences and motivations which ecological liability can above all hope to exploit[36].

Against this background limited liability emerges in a new light. It is to be perceived as an instrument offering certainty within uncertainty to private enterprise, thereby motivating corporate actors to invest. If exceptions to limited liability remain conceptually narrow, the positive symbolic effects of limited liability may continue. This in turn supports well-tailored parent liability rules. Hence, a multi-layered approach consisting of the principle of limited liability combined with various flexible, yet selective exceptions would seem to deserve preference over broad-based simple parent liability rules (Teubner, 1991: 207–11). It might make sense in this connection to differentiate between organizational risks of corporate group integration and production or transactional risks, whereby the former would be governed by corporate law concepts and the latter by various statutory or civil (common) law constructions[37].

Additional practical conclusions may be drawn from the concept of self-organization. If there is reason to assume that enterprises are in a relatively better position than the legal system to generate new ecological production standards, then there exists an argument for shifting burdens of proof onto companies. This puts them in a position of having to work towards meeting their perceived standards of proof. Law, in turn, could observe this process and "free ride" on it, adjusting the standards of proof as the economy's know-how and attitudes develop. As a consequence it seems important to keep such standards of proof open, in the sense of conceptualizing them procedurally. In other words, if the greater potential for generating sustainable production patterns lies in the economy, then legal standards of proof should focus above all on procedure and organization, and much less on the imposition of substantive targets[38].

Thus, parent liability law could start with a presumption that a certain level of integration with the subsidiary renders the parent *prima facie* liable. But it would then be crucial that the parent be given the right to show the application of appropriate standards of group organization[39]. These standards could be set higher the more industry developed them. Such a process would allow the economy to feed its legally inspired creation of new ecological measures, techniques and attitudes back into the legal system. The legal system, in turn, could screen them and re-signal them into the economy[40]. This self-perpetuating mechanism would seem to be one way of maintaining the differences between an autonomous economy and its dynamic environment at a sustainable minimum (see Luhmann, 1988: 324ff., 348).

CONCLUSION

There seem to be conceptual differences between the USA and Europe with regard to ecological liability of parent corporations. Traditional piercing the corporate veil is still paramount in US corporate law, but statutory constructions like those under CERCLA are partially making up for the lack of development in the corporate law field. The European "action", on the other hand, appears to be concentrated in the area of corporate and corporate group law, even though civil law constructions such as in tort law (vicarious liability of parent corporations) do have some appeal to the European legal community.

Holding these developments in the USA and Europe against the background of economic analysis of law, we find that there is much to be said in favour of both, as long as the respective liability regimes keep focusing on the relative risk bearing advantages of parent corporations (monitoring cost, diversification) and legal cost. An attempt to broaden the perspective beyond efficiency analysis into the realm of systems theory does not change this assessment dramatically. To be sure, it puts significant emphasis on evolution and openness and, thus, on the experimental character of liability law. As a consequence it favours a multi-layered, flexible, diversified and basically cautious approach to ecological parent liability. It also supports the notion of self-organization, that is, it sees liability law primarily as a signal to push economic actors in an ecologically responsible direction. Self-organization calls for selective parent liability rules that might be based on shifting burdens of proof to the parent. It would however seem to be crucial that the parent be offered procedural defences allowing it to escape liability by showing the application of proper standards of group organization.

NOTES

1 *Walkovsky* v. *Carlton* 223 NE 2d 6 (NY 1966); *van Dorn Co.* v. *Future Chem. & Oil Corp.* 753 F 2d 565 (7th Cir. 1985).
2 247 AD 144, 157, 287 NY 62, 76 (1st Dep.) affd., 272 NY 360, 6 NE 2d 56 (1936).
3 In *Re Oil Spill by the "Amoco Cadiz" off the Coast of France on March 10, 1978* MDL Docket No. 3 76 ND Ill. 1984, *American Maritime Cases*, 2123–2199.
4 In *Re Union Carbide Gas Plant Disaster at Bhopal, India, in December 1984* 634 F Supp. 842 (SDNY 1986), affd & mod. 809 F 2d 195 (2nd Dist. 1987).
5 *US* v. *Reilly Tar & Chemical Corp.* 546 F Supp. 1100 (D Minn. 1982).
6 A third alternative is tort law (*respondeat superior*) (Blumberg, 1987: 621).
7 635 F Supp. 665, 670–672 (D Idaho 1986).
8 See the discussion of recent decisions in Hein, 1991: 637–639.
9 724 F Supp. 15 (DRI 1989). The judgement was confirmed by the Court of Appeals of the 1st Cir. (see Hein, 1991: 637, N. 10).
10 724 F Supp., 1991: 22–23.
11 724 F Supp., 1991: 25, n. 5.

12 901 F 2d 1550 (11th Cir. 1990), reh'g gen. en banc, 911 F 2d 7421 (11th Cir. 1990), cert. den. 111 S Ct. 752 (1991).

13 861 F 2d 155 (7th Cir. 1988).

14 See also *Joslyn MfG. Co.* v. *T.L. James & Co.* 893 F 2d 80 (5th Cir. 1990), where the court denied the existence of a more liberal parent liability standard under CERCLA and espoused traditional piercing law.

15 675 F Supp. 22 (D Mass. 1987).

16 724 F Supp. 15 at 19 (DRI 1989).

17 94 NJ 473, 463 A 2d 893 (1983).

18 For details, see Hofstetter, 1990b: 578ff.

19 95 *BGHZ* 330 (1985).

20 107 *BGHZ* 7 (1989).

21 See also the "Video" decision, *Betriebsberater* 1991, 2173ff, and the recent "TBB" decision, *Der Betreib* 1991, 825ff.

22 This is being proposed under Swiss Law by Hofstetter, 1994: Ch. V.3.

23 If a parent has, for example, directed its subsidiary in its hiring decisions or has directly addressed subsidiary employees by giving them work instructions, it automatically assumes joint and several liability to the subsidiary's employees as their "real employer" (see Hofstetter, 1990b: 586).

24 A bank had first refinanced a corporation and acquired its shares before it let it go bankrupt. The bank was declared liable for negligence (see *Der Gesellschafter* 1987: 46–50).

25 For details see Hofstetter, 1994: Ch. I.4.4.

26 There are defences available, for example the "third party defence" and the "innocent landowner defence", which are, however, difficult to meet (see Hein, 1991: 641).

27 If the implementation of CERCLA with its arguably inefficient, that is, too broad liability concept is itself to be seen as an opportunistic political act *vis-à-vis* private enterprise, limited liability could be perceived as a shield against that political risk too.

28 Analogous to the production-consumption spiral, in that the influence of the economic and social environment on attitudes and preferences of the economic actors are neglected.

29 That is not laws which could "kill" an enterprise, like exorbitant punitive damage awards: these could be "wrong" and, thus, irreversible like the "death penalty"; as a consequence, their effect could be overly demotivating and, hence, counterproductive in their effects on private enterprise.

30 Legal engineering along a mechanistic economic efficiency paradigm would arguably belong in this category (see Huber, 1988: 6, 231).

31 See the idea of legally guided self-regulation that is considered as one important road to ecologically responsible conduct by industry itself (Schmidheiny, 1992: 50ff.).

32 See the third efficient parent liability category described above.

33 See the first and second efficient parent liability categories described above.

34 On the operational (micro) level the economic analysis of law may keep its role as an instrument to deduce concrete, yet hypothetical, guidelines for optimal liability rules.

35 See also the critical report on the impact of modern US tort law in Huber, 1988: *passim* and 155ff.

36 It thereby takes advantage of its potential to influence the rationality of the enterprise, something conventional economic analysis of law would usually ignore; on the "bounded rationality" of organizations see Simon, 1981: 114ff.

37 See the attempt made in this direction by Teubner (1990b: 49ff). See also the

Report by the Swiss Tort Law Commission of 1991 which suggests a division
of strict liability for corporations into transactional liability (*Gefährdungshaftung*)
and organizational liability (*Organisationshaftung*). See also Brüggemeier,
Chapter 4.
38 See in this connection the reference in Eads & Reuter, 1983: 79ff., to a study
conducted by Weinstein *et al.*
39 See also Brüggemeier's concept of "proving faultlessness through documentation",
Chapter 4.
40 See Blecher's discussion of group wide quality systems in Chapter 11.

REFERENCES

Adams, Michael (1985) *Ökonomische Analyse der Gefährdungs- und Verschuldenshaftung.* Heidelberg, R. von Decker's: C.F. Müller.
Anderson, Charles B. & Wethmar, Robert P. (1991) US-Umweltrecht: Der Oil Pollu-
tion Act 1990. *Recht der Internationalen Wirtschaft*, 1001.
Becker, Thomas A. (1991) Risikogesellschaft Schweiz? Politische Implikationen einer
segmentierten Arbeitswelt. *Neue Zürcher Zeitung*, **No. 186** (14 August) 19.
Bender, Bernd & Sparwasser, Reinhard (1990) *Umweltrecht, Grundzüge des
öffentlichen Umweltschutzrechts.* Heidelberg: Recht und Wirtschaft.
Blumberg, Phillip (1987) *The Law of Corporate Groups, Substantive Law.*
Boston/Toronto: Little Brown & Co.
Brüggemeier, Gert (1989) Umwelthaftungsrecht, Ein Beitrag zum Recht der Risiko-
gesellschaft. *Kritische Justiz*, **22**, 209–230.
Clark, Robert (1986) *Corporate Law.* Boston/Toronto: Little Brown & Co.
Eads, George & Reuter, Peter (1983) *Designing Safer Products, Corporate Responses
to Product Liability Law and Regulation.* Santa Monica: Rand.
Hayek, Friedrich A. von (1969) Der Wettbewerb als Entdeckungsverfahren. In Hayek
F.A. von. (Ed.) *Freiburger Studien, Gesammelte Aufsätze.* Tübingen: Mohr, 249.
Hayek, Friedrich A. von (1989/81) *Recht, Gesetzebung und Freiheit.* Landsberg a.
Lech.
Hein, Peter (1991) Umweltrechtliche Aspekte gesellschaftsrechtlicher Transaktionen in
den USA. *Recht der Internationalen Wirtschaft*, 636.
Hofstetter, Karl (1990a) Multinational Enterprise Parent Liability: Efficient Legal
Regimes in a World Market Environment. *N.C.J. Int'l L. & Com.Reg*, 299.
Hofstetter, Karl (1990b) Parent Responsibility for Subsidiary Corporations: Evaluating
European Trends. *International and Comparative Law Quarterly*, 576.
Hofstetter, Karl (1994) *Sachgerechte Haftungsregeln für Multinationale Konzerne. Zur
zivilrechtlichen Verantwortlichkeit von Muttergesellschaften im Kontext inter-
nationaler Märkte.* (Forthcoming.)
Huber, Peter W. (1988) *Liability. The Legal Revolution and Its Consequences.* New
York: Basic Books.
Kramer, Ernst A. (1990) Entwicklungstendenzen des Wirtschaftsrechts im ausge-
henden Jahrhundert. *Schweizerische Zeitschrift für Wirtschaftsrecht*, **62**, 249.
Krämer, Ludwig (1991) *Umweltrecht der EG.* Baden-Baden: Nomos.
Ladeur, Karl-Heinz (1988) Umweltrecht und technologische Innovation. *Jahrbuch des
Umwelt- und Technikrechts*, 305–333.
Landes, William M. & Posner, Richard A. (1987) *The Economic Structure of Tort
Law.* Cambridge, Mass: Harvard University Press.
Luhmann, Niklas (1988) *Die Wirtschaft der Gesellschaft.* Frankfurt/M.: Suhrkamp.

Luhmann, Niklas (1989) Ökologie und Kommunikation. In Criblez/Genon, (Eds). *Ist Ökologie lehrbar*. Bern: Zytglogge.

Mastronardi, Philippe (1991) Experimentelle Rechtsetzung im Bund. *Zeitschrift für Schweizerisches Recht*, **NF 110** I, 449.

Nonaka, Ikujiro (1991) The Knowledge-Creating Company. *Harvard Business Review*, Nov/Dec, 96.

Note (1989) Liability of Parent Corporations For Hazardous Waste Cleanup and Damages. *Harvard Law Review*, 99.

Petitpierre-Sauvain, Anne (1989) Le principe pollueur-payeur. *Zeitschrift für Schweizerisches Recht*, **NF 108**, II/4, 433.

Popper, Karl R. (1969) *The Open Society and its Enemies*. London: Routledge/Kegan Paul.

Popper, Karl R. (1984) *Logik der Forschung*. Tubingen: Mohr.

Rehbinder, Eckard (1969) *Konzernaussenrecht und allgemeines Privatrecht. Eine rechtsvergleichende Untersuchung nach deutschem und amenkanischem Recht*. Bad Homburg: Gehlen.

Schmid, Gerhard (1990) Rechtsfragen bei Grossrisiken. *Zeitschrift für Schweizerisches Recht*, **NF 109**, II, 1.

Schmidheiny, Stephan (1992) *Kurswechsel, Globale unternehmerische Perspektiven für Entwicklung und Umwelt*. München: Beck.

Shavell, Steven (1987) *Economic Analysis of Accident Law*. Cambridge, Mass.: Harvard University Press.

Simon, Herbert A. (1981) *Entscheidungsverhalten in Organisationen*. Berlin: Moderne Industrie.

Tercier, Pierre (1990) L'indemnisation des préjudices causés par des catastrophes en droit suisse. *Zeitschrift für Schweizerisches Recht*, **NF 109**, II, 73.

Teubner, Gunther (1989) *Recht als autopoietisches System*, Frankfurt/M.: Suhrkamp.

Teubner, Gunther (1990a) Die 'Politik des Gesetzes' im Recht der Konzernhaftung. In *Festschrift Steindorff*, 261.

Teubner, Gunther (1990b) Steuerung durch plurales Recht. Oder: Wie die Politik den normativen Mehrwert der Geldzirkulation abschöpft. In *Verhandlungen des 25. Deutschen Soziologentages in Frankfurt a.M.* Frankfurt/New York, p. 528.

Teubner, Gunther (1991) Unitas Multiplex—Das Konzernrecht in der neuen Dezentralität der Unternehmensgruppen. *Zeitschrift für Unternehmens- und Gesellschaftsrecht*, 198.

Wagner, Beatrice (1989) Das Verursacherprinzip im Schwcizcrischcn Umwcltschutzrecht. *Zeitschrift für Schweizerisches Recht*, **NF 108**, II/3, 331.

Westbrook, Jay Lawrence (1985) Theories of Parent Company Liability and the Prospects for an International Settlement. *Texas International Law Journal*, **20**, 321–33.

CHAPTER 6

How Law Hides Risk

S. C. Smith
Cambridge

Practising lawyers, who believe themselves to be quite exempt from any intellectual influences, are usually the slaves of some defunct textbook. There are ideas, it seems, which are just too good to be given up, no matter how absurd they may have become. This can scarcely be more apparent than in the law of torts, something which can be readily checked with just a cursory glance at the covers of the "official" texts in the field; those which you can neither afford nor afford not to have. In England, at least, they seem to come resplendent with scenes from some terrible tragedy or other taken from the worst excesses of the nineteenth century. That this is so, just as the structure of the law of torts becomes subject to greater strain than ever before, seems to me a matter of some interest. For it is precisely the problems that we presently consider to be the most pressingly "modern" which seem to send lawyers scurrying for their texts, inventing for themselves a golden age in the history of legal doctrine.

Environmentalists, for instance, have for many years warned us about the disturbing threat of groundwater contamination: "It is not possible," Rachel Carson wrote, "to add pesticides to water anywhere without threatening the purity of water everywhere" (Carson, 1963: 35). So when a Cambridgeshire tannery recently polluted the local water supplies having for years dumped toxic decreasing agents onto its land, there was a great trawling through the law reports for ancient authorities, something which must surely have seemed quite perverse to anyone less than accustomed to what passes for justification in these matters. Apparently, such veneration of the nineteenth century is to be seen as an unfortunate corollary of the absence of any relevant decision in more recent times[1]. However, this deference to authority can hardly be expected from the rest of us, not least because liability law and its

Environmental Law and Ecological Responsibility: The Concept and Practice of Ecological Self-Organization
Edited by G. Teubner, L. Farmer and D. Murphy
© 1994 John Wiley & Sons Ltd

ramifications have now become the stuff of political scandal: "Point of law blocks £ million pollution bill", screamed one national newspaper[2] after the courts had acquitted the tanners, treating a much revered case from 1868 with about as much respect as if it were simply a question of some irritating technicality obstructing the noble aims of legislation, here, the much vaunted "polluter-pays" principle. Such "principles", of course, cannot themselves be relied on as legal authorities; neither, however, can they be simply ignored. On appeal, for example, the court was properly careful not to make the requirement that the polluter should pay a reason for its decision. Nevertheless, it left us in little doubt that it was quite pleased to have dug up an old case that seemed as fashionable as it was binding: "Others must consider whether *Ballard* v. *Tomlinson* [1885] accords with contemporary opinion. Some of them may say that it does" (62g). Others still will be more concerned with the consistency of legal doctrine, since the damage had not even been foreseeable; foreseeability, or so we had come to accept, being a precondition of recovery. As one case-note writer complained: "The polluter pays—regardless"; regardless of fault and, it seems to be implied, regardless of the law.

When finally this decision was itself overturned by the House of Lords, any gesture of surprise would have been almost certainly manufactured, so much had expectations already been frustrated. It will of course be hoped that their unanimously expressed view will help to create greater certainty in the area; always supposing that we are prepared to overlook the equal conviction with which the inferior court had reached the opposite conclusion[3]. In this chapter, I argue that we can expect continued uncertainty for a range of other reasons, reasons which are deeply rooted in problems of what we can call "resonance"[4]. According to this view, problems, especially environmental problems, simply fail to register as such; they leave no social traces, and in this sense resonate too little, so ill-fashioned are our tools of law and politics. On the other hand, and this is something far more difficult to recognize, they also seem to resonate too much, in that we seem to find such limitations both persistently difficult to accept, as well as an increasingly routine source of surprise. Almost as if to compensate, we tend to overreact, producing the sort of grotesque exaggerations of what we are capable of achieving which it is the concern of this chapter to discuss.

EXAGGERATION AND TRANSITION

The usefulness of the concept of resonance in this context is that it draws our attention to the fact that environmental problems are not automatically social problems. There is in this sense an "ecological difference" (Luhmann, 1986; 1989: 147) between the natural conditions which sustain life and the artificial conditions which determine institutional reproduction. In a sense this can be said to have been Hume's problem. Specifically, he was concerned with the lack of symmetry between our desires and the limited capacities to satisfy them which the "cruelty" of nature has afforded us. The rhetorical force of this argument is to expose the real burdens inherent in justifying anything so complex and artificial as capitalism, and to withdraw permanently any persuasive power natural law theory might have been thought to have had in this respect. Instead, stability is derived from convention, that is, "gradually ... and by the repeated experience of the inconveniences of transgressing it". Moreover, this process is self-enhancing: it "assure[s] us still more, that the sense of interest has been common to all our fellows, and gives us a confidence of the future regularity of their conduct" (Hume, 1978: 490). In somewhat clumsier theoretical language, we might have called this, respectively, negative and positive "feedback". In any case, nothing more than this is meant by resonance, the pattern of expectations through which a society modulates its own reaction to its environment. For it is only through changes in such expectations that a society can respond to changes in the ecological climate at all, including whatever changes to that climate it is itself responsible for bringing about. The pattern of expectations becomes the medium through which the possibility of change itself becomes visible, since as soon as it is established it becomes clear what is required in order not to follow it.

In other words, conventions are really a way of handling both stability *and* instability; they carry their own risks as well as whatever benefits they confer. In consequence, Hume could hardly have appealed to conventions alone to justify the *actual* conventions we happen to have; instead the justificatory burden is assumed by an exaggeration of the virtues of property. While stability in expectations is marked out as an artificial, implausible, even a fragile thing, nevertheless property is said by Hume to be "of all circumstances the most necessary to the establishment of human society" after which "there remains little or nothing to be done towards settling a perfect harmony and concord" (Hume, 1978: 491). It is almost as if the very fragility of whatever stability we have managed to achieve is used to justify the actual institutions of property and possession that we now have[5]. Hume's position amounts to the sort of compromise between necessity and contingency which is perhaps characteristic of this transitional period, and one whose consequences we have been struggling with ever since.

This can perhaps be most readily seen in the sociological discussion of functional differentiation, where the argument runs that only a differentiated society can experiment with uncertainty in its structures to any degree. For example, society can tolerate enormous disparities in wealth, just because this is not necessarily converted into inequalities in other areas. And, again, to love means to risk *not* loving because we can be confident that it will nevertheless still be possible to co-operate in other ways, to do business, or even simply to "stay friends". This presupposes a distinction between what we can call *constitutive risk*, that is the risks which differentiation enables us to run, and *reflexive risk*, that is the risks that follow from drawing the boundaries in the way we happen to have drawn them. Exposing us to the former involves the suppression of the latter, at least for the time being; Hume's objectification of property being merely one instance of this. As Luhmann suggests, this sort of exaggeration of the importance of the institutions of private law for society as a whole can be seen in retrospect to have been a condition of their insertion (Luhmann, 1993: 330); what we will have to ask later is whether it might not also be thought of as a presentiment of their demise.

In any case, it is in relation to this discussion that we should understand the problem of over-resonance[6]. For what I have called the suppression of reflexive risk is the same thing as saying that such institutions have bracketed off the question of where their own boundaries lie. Indeed, having exaggerated or objectified them, they come to be seen as co-terminous with the whole of society. For the same reason, such institutions become chronically sensitive to any information which appears on the horizon of their now amply extended field of vision (problems, results, data, events, prejudices) information which they can neither ignore nor properly accept. This in turn requires a "re-problematization" of their boundaries and a rejection of that which is incompatible with their mode of operation. Where this boundary between constitutive and reflexive risk is inadequately managed, the result is over-resonance, the sort of continuing oscillation between near totalitarian ambitions and partial retractions, expectations and disappointments, hopes and despair, all of which seems to be such a characteristic of modern society.

It is this or something like it which accounts for the astonishing combination of vitality and vulnerability with which private law has been invested in recent years (Priest, 1990). For it is in this context that the problem of law's own boundaries gets raised and repressed; the professional task of adjudicators, I want to argue, being to manage what does and does not count as a legal problem. This is so despite or perhaps because private law strategies are being deployed in precisely those areas where they might least be expected to have any chances of success. Problems of environmental protection, for instance, are often provided as clear examples of what is usually described, following Fuller, as polycentric, that is, not easily manageable in a two-party adjudica-tive setting. But this is both true and beside the point. It is better to say that *all* problems are polycentric; only some are more polycentric than others (compare Luhmann, 1984: 84). In any case, we cannot conclude from the fact of polycentricity alone that the problem is non-justiciable, that is unsuitable for adjudication. On the contrary, the whole point about the adjudicative structures which characterize private law is that they represent one of the most successful institutions we have ever devised for breaking such problems up into a manageable form. Far from being dispensed with when things get more than a little complicated, we have become over-reliant on them in the absence of convincing alternatives.

RECIPROCITY AND COMPLEMENTARITY

Perhaps this would go some way to explaining the continuing popularity of ideas such as Garret Hardin's "tragedy" of the commons (Hardin, 1968). At one level, what he is saying is a valuable reminder that property can usefully be made to serve in the management of natural resources. Individual property rights represent one way with which to factor the global environmental problem. However, once again, this is achieved through an exaggeration of their virtues, this time following the well-worn rhetorical strategy of under-mining the attractiveness of alternative regimes. Here land held in common is fictionalized as an area of uncurtailed self-interest and licensed over-production, where clearly this need never have been the case. The commons were in fact maintained by a complex ecology of reciprocal rights which stinted gathering, cutting, herding and other such uses of the land. What is important, and this is beautifully illustrated by E. P. Thompson in his essay "Custom, Law and Common Right", is that such a regime could only be sus-tained in the sort of local and parochial environments from which it derived its meaning (Thompson, 1993: 97), and where custom could be said to "lie with the land". The important distinction, then, is not one between property rights and chaos, but the transformation from what Luhmann calls a regime of reciprocity to one of complementarity (Luhmann, 1970; 1981; 1989). In reciprocity, my rights are mirrored and their exercise conditioned by similar though not necessarily precisely synallagmatic rights of others over the same area. Shifts in population bring about an increased competition for scarce resources which can no longer be regulated by the moral economy of the local inhabitants. Instead, the law becomes mobilized by those landlords who can afford it to protect themselves from third-party interlopers; meanwhile showing little tolerance for the niceties of subordinate servitudes. The modern-izing administrative mind noted the "ruinous effects of a mixture of opposite interests in the same property" (Thompson, 1993: 107), and the juridical notion of individual ownership bundled interests with landlords to the exclu-sion of all others. Their absolute right of ownership is *complemented* by the obligation on an abstract set of putative infringers to respect it. It is this notion of exclusiveness which is novel and, in evolutionary terms, profoundly implausible.

By the time we get to the late nineteenth century, this implausibility is itself obscured, since the development under discussion is more or less taken for granted. For example, in a case of water pollution in 1885 on which the Court of Appeal in the *Cambridge Water* case had placed great reliance, a regime of reciprocal rights in undefined groundwater is rather easily dismissed, and for the same reasons of administrative inconvenience. Instead "every owner has a right to sink a well in his own land and to take from his well all the water he can get, and that as every owner has this right no owner can maintain an action against another for exercising [it] ..."[7]. The landowner's "natural right" to take water is complemented by the obligation on all others to abstain. Looking back, we can see that this solution was already in hand by the end of the seventeenth century at the latest, since it is prefigured in Pufendorf's inversion of the paradox of common ownership. Since Hobbes had spoiled the notion of a positive community of goods in which everyone and therefore no-one can own anything, we are to replace it with the idea of a negative community, one in which because no-one owns the commons everyone has a right to appropriate it provided only that enough and as good is left over for others (Hont & Ignatieff, 1983: 30–36; Luhmann, 1989: 22, 26–27). Indeed, it can be plausibly maintained that the component parts of a theory of "subjective" rights were in hand as far back as the late Middle Ages, and that in this sense it represents what Luhmann would call a "pre-adaptive advance" in the evolution of modernity. So however tempting it may be to see capitalism as somehow necessarily bound up with property rights, we should not lose sight of what is at most only a "structural affinity" (Luhmann, 1981: 79) between the two, and therefore also of the sheer implausibility of their coming together when they did and in the way that they did.

That the development of individual rights has a counterpoint in the rise of administrative power marks it out as one of the most important ideological contributions to the development of modern society, and indeed in a way which we cannot allow talk of formalism in private law to obscure. It was the first person to have enclosed land, saying "this belongs to me", and who was able to find judges simple enough to believe him, who was the real founder of the modern state. For whereas reciprocal rights are mutually conditioning and inevitably localized, there is nothing to condition the seemingly one-sided exercise of individual ownership rights except the power of the state which sanctions them. The vector of complementarity sits orthogonal to that of reciprocity. At one end lie individual rights; at the other lie those against whom they are protected, something which is left inherently abstract and unfocussed, but which gains in concretion each time the right is asserted and protected. It is this combination of abstraction and particularity, constitutive of private law conceived as a regime of complementarity, which lends itself so readily to the very welfarist objectives which the formalist would seek to exclude. In the enclosure cases which Thompson is looking at, this inevitably involves a consideration of which customs it is reasonable for the courts to protect. The lawyer will still want to say that "reasonable" and "unreasonable" are legal "terms of art", and so they are in their way, but as Thompson shows, even "on a very brief view of the case law they were gates through which a large flock of other considerations might come baaing and grunting onto the fields of the common law" (Thompson, 1993: 129–130). It is constitutive of private law at this stage in its development, then, that it turns polycentred, highly-aggregated and multi-valent problems into a binary or dispositive form, one that can be handled in an adjudicative setting. At its very inception private law is indexed to administration; the new regime of individual property rights, for example, was quickly tied, as Thompson shows, to developments in political economy, in particular, notions of agricultural "improvement". Without too much distortion, we might call this the privatization of law enforcement.

To say this is not exactly to defend an instrumentalist conception of law, for if private right is to serve as a vehicle for politics, it always affords in addition at least the opportunity to re-problematize its boundaries[8]. It is better to see it as marking a decisive shift in the legitimation of law in society; whereas jurisprudence had once been indistinguishable from political and social theory, it now has to address more directly the peculiar problems of a positive legal order. For example, how could law justify those risks it was itself responsible for introducing, those associated with enclosure, for example, particularly since they had occasioned such hostility among those adversely affected? Obviously, a detailed answer to this cannot be given here, but the substantial work already done on this period indicates that the generality of the problem would have simply remained invisible for so long as information on the matter remained unpooled[9]. Indeed, that knowledge of this sort does get pooled and distributed at all is largely a result of the very processes of centralization in adjudication and development in legal doctrine which we are discussing; processes which are themselves, of course, a consequence of the new regime of complementarity since this involved rights against the world, rights which would accordingly have to be justified and therefore made available in general terms. In other words, law is self-justifying during this period precisely because it is self-undermining, and history ensures that such a paradox appears for the law only after the makings of a solution are already in the offing.

SYMMETRY AND COMPATIBILITY

It is no part of my argument that law's self-justification should be seen in isolation from more general developments in political theory or more particular developments in other knowledge domains. Rather, the reverse is the case, since the inevitability of self-justification can be seen, at least in retrospect, to have been compelled by the transition from hierarchy to functional differentiation as an organizing principle for the whole of modern society. The complexities of such a transition are still poorly understood, intimidating all but the most expert in intellectual history. Some of these complexities are a consequence of the irreducible particularity of concrete events. Others require consideration of the sort of logical, ontological and epistemological problems which have their roots in Aristotelian metaphysics and cosmology. For as Arthur Lovejoy has shown, whatever other differences there may be among them, there is a more primitive conception of the world as a "great chain of being" which unites societies from the Middle Ages right down to the late eighteenth century (Lovejoy, 1936; 24–66). This conception is not only what motivates people living under early modern conditions of reciprocity, where the question of who they are or what role they have in a wider community is of first importance; a community which is more than just the sum of its parts and can make claims on its participants, whatever their differences in status, in terms of mutual aid. It is also what animates earlier societies organized in equally perfect parts or segments. Here qualitative differences among its parts could not be tolerated except as some form of schism within the whole and therefore as an irritation to reflection and reform (Luhmann & De Giorgi, 1992: 345–353). So, for example, in the conciliar controversy of the late Middle Ages, Jean Gerson could not claim *both* that the General Council was the sole lawful representative of the faithful *and* fail to extend this to all other "perfect" societies, including the secular commonwealth represented in an assembly of its citizens (Skinner, 1978: 116–123). And to take a later example: Michel Foucault has suggested that it is the problem of population, the insight that it develops according to its own configurations, one that is independent of the established sequences in political thought, that finally bankrupts the model of the family as one suitable for society as a whole (Foucault, 1991: 98–102). The details of this can be quite properly left to historians; here, once again, I am only interested in the structure of the argument. What is important is not only that the economy could no longer be conceived along these lines, since this alone would have allowed us to continue with a hierarchical model of political power. Rather, it is because it has failed as a model of any *general application* that it now requires revision not only within political economy but on its own terms and as a model for government. This is perhaps clearest in the writings of Malthus, who retains a commitment to what we would now call "family values" but only instrumentally to the extent that they serve to control population growth (Malthus, 1992: 59–67).

It goes without saying, I hope, that there is an intimate connection between the arguments I have just made and the discussion of over and under-resonance from the previous sections. For so long as society is conceived as a whole composed of parts, then it contains within itself the makings of its own transformation; although this is something we should be careful not to exaggerate, if only because it simultaneously pitches the argument for reform at a level where the risks are highest, that of society conceived in the most global terms. This makes it difficult to accommodate arguments for a form of change which is at once more specific and arguably more radical, something which I argued earlier could be considered the function, as it were, of functional differentiation. Here we are presented with problems of justification of a quite different kind, problems of how the risks of innovation can be contained and which, above all, classical liberal theory is called upon to solve. For example, what appeared in the context of the whole/parts metaphor as a problem of symmetry or even *equivalence* in self-descriptions is now reduced to a problem of *compatibility*. In both its deontological and its utilitarian forms liberalism provides a *matrix* within which legal and economic doctrines can be matched and translations made from one context to another. The task of justification is to suppress the consequences of modernization, including and above all the consequence that modernization makes its consequences more difficult to suppress. Once again liberalism helps out, since it both *reveals* the consequences of our actions, for example, as a net contribution to social welfare even when (or possibly only if) motivated by self-interest; as well as *represses* them, for example by short-circuiting discussion on the matter with appeals to an existing realm of individual freedom which it is necessary for us to protect.

So long as this works, it provides an environment in which the problem of resonance can be managed and contained. As such, it ensured that the consequences of modernization would remain more or less opaque to those who were largely responsible for bringing them about; just as they would have been, it must be added, to those most adversely affected. *If* it is true that the English working classes were present at their own making, they almost certainly could not have known it at the time. Only later does it become possible to see the connection between property and capitalism; between the expropriation of the commoners and the creation of a class of dependent labourers. Again, only later does it become possible to describe law as operating in such a way as to sink risk thresholds and as a subvention to developing industries (e.g. Horwitz, 1977). Or as Michel Serres would have it, property is inherently double-edged: making something mine inevitably involves exclusion of or harm to someone else. The man who spits in his beer before leaving it unattended discovers the true meaning of property in much the same way as does the burglar who defecates in the room he has just emptied of its possessions. No doubt the anthropologist's stercorean account of property is a little forced and one-sided here (Serres, 1980: 183–193). Nevertheless, it serves to dramatize the connection between property and pollution which normative theory was introduced in order to remove. Moreover, it underscores a genuine dilemma for modern liberal thought. For it is perhaps more obvious to us now that individual action nearly always, one might even say constitutively, involves *some* costs on others (see Luhmann, 1992a: 74–79, 111ff.). This has as much to do with the phenomenon of risk as with anything else, that is, the collapsing of the future into the present, presenting such costs as the result of present decisions and activities. While the nineteenth century got by with what Luhmann calls a "temporalization of complexity", a dubious appropriation of history to reflect an always better future, we have to make do with altogether more flimsy metaphors such as "sustainable development". Clearly, teleology still has its charm, but less so the more it gets used.

To say that we have become disenchanted with solutions that use the past to present the future, to say that they have become hollow or have lost their charm, is only another way of saying they have become abandoned to observation by others. If it is true that Hume confronted the contingency of the future with an appeal to the past, that is, to stable expectations, and if it is true that we see this now for the exaggerations it contained, then this can only mean that we see it as a solution given by Hume for his time. However, the more we become conscious of the future in the present, the more the future becomes dependent on decisions we make in the present (including the decision not to make any decisions), how much more fragile, then, becomes the pretence that the future will be the same as the past? The problem lies not so much in the past-dependent present but in the present-dependent future. And this is so despite the fact that although we know that the future will always bring something different, we still do not know quite what it is going to be. Indeed, the problem is not just that we do not know what the future will bring, but that we *know* that we do not know what the future will bring (Luhmann, 1987). And it is *this* which makes the past both more compelling and less reliable.

This probably has consequences for law which have yet to be clearly identified. At the very least, I want to say that it confronts law with a particularly pungent form of double bind. On the one hand, if it is true that the present is increasingly saturated with uncertainty, and this is surely the nub of the environmental problem as we experience it today, then we should expect law to assume a greater significance than is currently fashionable for us to admit. At least, this is the case if you accept that legal technology is specialized in stabilizing expectations for society as a whole. And those who do not must ask themselves how it is at all possible to risk relying on any expectations, but particularly new and improbable ones which have to be introduced in the face of resistance from the past. Law, in this sense, is a licence to run risks. Or rather, to turn to the law is to run a different sort of risk; not the risk that your expectations will be disappointed by the actions of others, but the risk that it will be disappointed by the actions of lawyers. What cannot be eliminated are the risks internal to law, the riskiness of risks, the risk that risk assessments will vary from one time-point to the next. And there's the rub, for it is precisely this reflexivity of risk which, I am arguing, law has found it persistently difficult to recognize. As a result, the law must both accept what it cannot ignore and ignore what it cannot accept.

Perhaps tort offers law a way out. Unable to cope with the riskiness of risks, compensation allows us to postpone commitment until harm has occurred. For where such risk can no longer convincingly be absorbed within liberal political theory, and I have argued that it cannot, then it has to be managed internally using the law's own doctrinal resources. The result is that the ancient competition between utilitarian and deontological conceptions of rights has now intruded more directly into the literature, particularly the Anglo-American literature, on the law of torts, while simultaneously promoting an apparent juridification of political theory itself. For example, as Coleman and Kraus ask in a brilliant exercise in contemporary analytical jurisprudence, does tort law promote a *welfarist* or an *infringement* theory of property (Coleman & Kraus, 1986: 1338)? Specifically, in what sense can it be said that liability rules protect entitlements when they seem to require a transfer of resources based on a sum set not by the holders of these entitlements but by third parties, namely, courts? Infringement theories of property such as those advocated by Richard

Epstein seem to mark out an area of control or autonomy over one's holdings. But if this is so, why is it that tort rules seem to allow, for example, A to pollute B at a price set not by B but by the courts? Using torts, the law tries to have it both ways; to permit A's activities and to penalize them in the event of any harm caused[10]. Tort law seems to make the legality of an action depend on an uncertain future; whereas I said that if law can be made useful at all, particularly when dealing with environmental problems, then it was to immunize action *from* the uncertainty of the future. To resolve this dilemma would certainly be a Herculean task; it is no accident then, that, perhaps more than in any other field of law, tort has exercised the best minds of contemporary jurisprudence. One such is Ronald Dworkin, who has famously described the story of tort law as a kind of "chain-novel". But if such comparisons are to tell us anything about the law, rather than simply the refined literary sensibilities of their author, then it seems to me, or so I argue in what follows, that the law would be far better seen as a rather bad "soap".

ENVIRONMENTAL EPICS

In the first place, of course, law does not resolve itself in the way that a novel does, even a chain-novel. Like a soap opera, it has an open and indefinite future, one that makes law both more mundane and more compelling. For it is not only concerned with judges and "hard cases" but with public opinion and with its audience. The difference is not at all slight; if in a chain-novel it is true that the law has to be constantly rewritten, it is also true that such re-writing is always directed to some purpose. Whether it is the assumed "single right answer" which alone can give such a process meaning, or whether it is the assumption that somehow the law is "working itself pure", it is deemed to be constitutive of writing a chain-novel that its authors "aim jointly to create, so far as they can, a *single unified novel* that is the best it can be" (Dworkin, 1986: 229; my emphasis). Dworkin seems to have noticed the difference here, but only dimly and as a question of degree, for chain-novelists are only "expected to take their responsibilities of continuity *more seriously*" than do the teams of authors working on soaps who have to put together sometimes several episodes a week (Dworkin, 1986; my emphasis).

But the difference is more than one of degree. Soap operas do not merely display "minimal continuity of personality and plot" (Dworkin, 1986); rather, in the notorious complexity and breathtaking revisability of their narratives they show an almost reckless disregard for the sort of continuity which Dworkin might have had in mind. In fact, what is characteristic of soaps is not the continuity in narrative structures but their *repetition*. As Christine Geraghty puts it in her splendid book: "What happens may be predictable but the interest lies in the many variations on how it will happen. A soap's endless future means that an ultimate conclusion can never be reached and soaps are thus based on a premise of continuous disruption." (Geraghty, 1991: 15). In a sense, what actually happens is of less concern to soaps than the question of how the principal characters in the drama cope with events, the choices they have to make. Soap writers might pretend to be "realist" in genre, just as judges claim only to discover the law; but in both cases it is really *how* they see the world that allows them to refract the great issues of the day, turning unstructured complexity into manageable questions in just the way I have discussed it in earlier sections. Moreover, it is this predictability and repetition in soaps which allows their audience to "catch up" even where they have missed several episodes. Unlike the chain-novel, which requires artificial devices, such as "continuity", to connect one episode to another, the very openness of soaps almost effortlessly creates an impression of "real time" through which they become synchronized with the patterns unfolding in everyday life. This is not because we know what is going to happen, or because of any continuity in plot; no soap, no legal system could create continuity in this sense out of an uncertain future and highly turbulent environment; rather, it is because they rely on our understanding of how soaps work. In law, too, the problems of resonance and of synchronization that we have discussed here can only be contained if we have a reflexive awareness of "how the law thinks". This involves its audience as active participants in the drama and the decisions which have to be made, something which is a valuable correction to Dworkin's pre-occupation with judging, and one which, to boot, probably better accommodates his own concerns with community.

This last claim will almost certainly be considered over-ambitious, if only because soaps appear to rely on a notion of community which is unhelpful when it comes to looking at the law. Particularly in the mainstays of British soap opera, it is a community in which everyone seems to know their place as well as everyone else's, where prying is frowned upon, but where we will nevertheless all pull together in a crisis (Geraghty, 1991: 84–90); in short, it is not unlike the model of reciprocity which I discussed earlier. Dworkin, it seems, can at least argue that he takes rights seriously, that in law-as-integrity everyone is entitled to *equal* concern and respect, and that this fits better with the regimes of complementarity I argued were characteristic of modern law. And certainly, if we look at the law of tort, it does seem to be populated with a rather sad litany of stock characters: children who have fallen down manholes; blind men who have tripped over roadworks; the old, the weak and the frail; anyone we should have had in mind, had only we been acting, as Lord Atkin said we should, according to the "neighbours" principle[11]. At the same time, there is an equally stereotypical and one-sided list of defendant characters: profiteering factory-owner, financial shark, borstal-boy attendant, and so on, all readily available pegs on which to hang liability. Certainly, there have been problems with borderline cases: those involving beloved institutions such as cricket clubs; tragi-comic botched rescue attempts; but while controversial these always seemed manageable somehow. Indeed, this is really the point and the appeal of both law and soaps: they confront traditional roles and institutions with new sets of expectations. Many soaps were introduced, of course, in order to introduce professional activity to popular culture; similarly, it could be said that the standard problems in tort law involve both the creation and confrontation of so-called "expectations gaps" in relation to the many well-established professions, such as doctors, lawyers, and the rest. So neither law nor soap opera is exclusively occupied with the familiarities of reciprocity; rather they confront the familiar with the unfamiliarity which is typical of regimes of complementarity. Only in this way do they produce for us such characteristic feelings of both engagement and distance, of commitment combined with a keen sense of artifice. We should go further than this, however, for there are signs that, faced with an increasing pressure to adapt, soap opera's usual combinations of underdeveloped characters and over-intricate plots are ill-suited to testing new themes in the way we have come to expect from it (Geraghty, 1991: 148–166). Is this also true of the law?

There is some evidence for saying that it is. For example, one of the new themes in tort is whether it will be able to protect investors, especially in the wake of financial de-regulation and the scandals of the 1980s. For instance, tort has been used to try and close the expectations gap between the auditors of company accounts and those buying shares on the basis of what they said. The trouble is that according to previous episodes we were supposed to have had little sympathy for the gamblers and the risk-takers in life. Now they are just the sort of people we are being asked to protect, this time as the poor exploited victims of financial deregulation. But if this is so, then surely we are *all* risk-takers now. In which case, who is it that is going to play the necessary role of villain? What, in particular, if no longer risk-taking as such, marks out the defendant as deserving of liability? Alternatively, if we accept that we are all living in a risky environment, then should we not all be expected to take at least some measures to protect ourselves? And does this not indicate that over-protected consumers are really the new villains? Risk gives at most only an ambiguous support to the great "welfare" and "collectivist" themes which had been thought to sustain the development of tort law. It certainly appeals to social welfare, but now in the sense of a regime of complementarity in the purest sense, that is, a vigorous culture of claims and complaints, conditioned only by the expectation that they will be met by others (the state), without this being reciprocated by any specific obligations which the claimant himself is expected to fulfil[12]. Indeed, if anything, tort responds to and reproduces in its turn an increasing individualization of responsibility and risk control (Priest, 1990). This has a levelling effect, eroding the differences on which reciprocity is said to be based. What we have to ask is whether the old stereotypes can be given a more subtle profile here, or will they simply have to be written out of the series?

More likely the courts will simply invent a new character with greater appeal for the mass audience. This seems to be the case with problems of environmental protection. No-one can doubt its popularity; who, after all, does not want to "save the environment"? As a result, it is peculiarly tempting for judges and administrators, confronted as they are with a straightforward competition between otherwise legitimate interests, to appeal to that which most appears to serve the cause of environmental protection. So in the *Cambridge Water* case, the interests of consumers in having cheap water could only be inconclusively weighed against the interests of a much-valued local industry which served as an employer for most of a small town. As I argued earlier, the gravitational pull of the polluter pays principle persuaded the Court of Appeal to accept the naturalization of the plaintiff's right to abstract water from underneath his land. The problem with this is that the value of environmental protection is too abstract to support any one side unambiguously; it can very effectively be appealed to by both participants in a controversy; indeed it is liable itself to generate such conflicts. We can see this in the uncertainties and disagreements which are produced whenever we ask scientists to translate their probabilistic calculations into practical reasoning; it is notoriously both healthy and unhealthy to drink wine, we should both eat and not eat polyunsaturates, and so on. In the *Cambridge Water* case, too, it must have been rather sobering for the Court of Appeal to learn that, just before it gave its decision, the polluting tanners had received a special East Anglian regional award for responsible environmental management[13]!

The real difficulty is not just this difference in evaluation, but its inevitability, since, at least from a sociological perspective, such assessments appear to vary systematically depending on whether we are exposing others to risk or are ourselves exposed to risk from others. There is a wealth of literature on this now which I cannot discuss here (see Luhmann, 1992a: 75–79, 114ff.); it is enough to keep in mind that whatever else we might say about differences among, for example, car drivers, it has always been recognized that the most primitive distinction rests on whether they are situated behind the wheel or in the back seat. Unfortunately, it is just this sort of issue which we seem unable to resist calling upon tort law to tackle; passengers are the very people whose safety we should have in mind when driving, and if we don't, then the threat of liability is supposed to help us "internalize" such risks of harm to others. But if the research on risk is correct, then it is just this sort of issue which we can *never* internalize; risk assessments vary not just between those who decide and those affected, but from one time-point to the next. Perhaps this has been obscured by examples such as that of the car crash, where the gap between action and harm caused is slight. But where it increases, as it does for example in the sort of "long-tail" risks which are characteristic of the environmental problems that concern us here, then it renders recourse to tort as a strategy of regulation just as desperate as it is tempting. As one commentator in this field has recently acknowledged: "It is not 'command and control' but 'compensation' which will capture the imagination of the public and decision-makers in the 1990s—their catch cry: 'the polluter must pay'" (Atkinson,

1992: 105). This is confirmed by the environmental lobby who had hoped that the litigation surrounding *Cambridge Water* would be seen as a test case for the greening of torts, particularly in relation to pollution resulting from the use of land in agriculture. In response, and even before the judgment at first instance, attempts to mobilize agricultural interests were fairly ruthlessly pursued: as one manager of the leather industry complained to the local press, judgment for the water company would involve "every farmer in the country in prosecutions for contamination from nitrates, even if it occurred decades ago"[14]. Because of this, it would be foolish to predict now whether, in the long term, tort law will prove to be environmentally friendly, or whether, instead, the environment will turn out to be unfriendly to the law. Fortunately, however, as in all good soaps, the scriptwriters have been busy working on alternative scenarios in advance.

First scenario: the tragedy of the environment

According to the first scenario, the Court of Appeal's decision generates much enthusiasm for the possibility of extending environmental liability for groundwater contamination. Admittedly, the trial judge had found that the "major pollution" of Cambridgeshire water which had resulted from the tanning had come from what rural economists call "point source incidents", a rather rare phenomenon when compared to the "diffuse" pollution which is said to characterize the dissemination of nitrates (Ward *et al.*, 1993: 16). But the Court of Appeal had said that the right to take water from underneath our land was a "natural right incidental" to ownership (at 60h–61h). Relying on *Ballard*, they designated undefined groundwater as a "common source"; the natural right to take being conditioned by the duty not to contaminate. Moreover, this follows regardless of whether or not the parties are contiguous neighbours, and therefore cuts across what would otherwise be an important limitation to the wider application of the law of nuisance here envisaged. Finally, the Court of Appeal's decision seems to make the polluter pay for the costs of what is now called "historic pollution". An important feature of the *Cambridge Water* case is the fact that when the offending acts took place, the toxins would not have counted as harm relevant for the purposes of tort law. Harm only occurred after the coming into force of legislation implementing the EC Drinking Water Directive which required them to be removed by the water company in order to make their commodity fit for consumption according to standards laid down in the new law. Indeed, prior to the EC directive, we had not even developed the technological know-how to detect such toxins in water to the degree it required. The tanners, then, have been required to meet the cost of removing chemicals from the water supplies where that cost has accrued only as a result of legal changes made after the polluting acts had stopped[15]. In this scenario, then, we can say that the decision gets enthusiastically endorsed by environmentalists for its wide range of applications; this initial euphoria, however, is very soon dissipated.

Already in the Court of Appeal, only "acts of the defendant" had been made subject to liability, something which might easily be read as excluding the more intractable problems of diffuse pollution. Similarly, in their discussion of the case of *Rylands* v. *Fletcher*, it is said that liability "always focuses on *the event* of an escape" (at 59a) from which we should infer that it is *never* to be concerned with a more diffuse process or sequence of many events. Even if we can track anything so diffuse as agricultural pollution down to acts which an individual tortfeasor has done or ought not to have let happen, then it is almost certain that a host of other polluters will soon come into the line of vision, shaking the reliability of our initial judgment. This is especially the case where, the more retrospective tort law becomes, the more potential tortfeasors it will be able to find. It is then possible for someone like Ulrich Beck to argue that the law transforms pollution into *un*pollution, liability into acquittal. The more polluters that can be found, the less likely it is that any *one* of them can be said to have done it (Beck, 1988: 216–220). In a sense, they simply become "invisible" from the point of the view of the law[16]. Then there is the decision of the House of Lords itself. It had been awaited with great excitement. A flurry of broadcasts and newspaper articles voiced the concern of insurers and banks about the implications of retrospective liability[17]. It was hardly a surprise, then, when the requirement that the defendant must have been able to reasonably foresee the harm was reinstated; the application to historic pollution failed. For a while, though, even the environmental lobby was happy, since in all other cases, such as that of nitrate contamination where surely the harm *can* be said to have been reasonably foreseeable, this seemed to constitute clear support from the highest court in the land for plaintiffs to sue in nuisance or under *Rylands*.

But wait! A closer reading of Lord Goff's judgment shows that everything he writes on nuisance and *Rylands*, and that is the vast bulk of his decision, is irrelevant; strictly speaking, it is mere *obiter dicta*. What he says is that if in *Ballard* it was established that a plaintiff had a natural right incidental to ownership, then "[i]n the present context ... this means no more than that the owner of land can, without a grant, lawfully abstract water which percolates beneath his land ... . There is no natural right to percolating water, as there may be to water running in a defined channel" (at 68 g–j). Having established that there is no natural right to the water, he goes on to make it clear that any actions the plaintiff does have depend exclusively on the correct application of the law of nuisance or the rule in *Rylands*. But if the right is only to abstract water, then it is only to abstract water *in whatever condition* it happens to be in. The only other meaningful rule would have been that the plaintiff has a right to abstract uncontaminated water, and it is precisely this which Goff intends to refute. Indeed, on the basis of this reasoning alone, it is possible to argue that there may even be minerals which, while unfit for human consumption, are nevertheless valuable for other purposes, and presumably then the plaintiff would have a right to abstract these too. Of course, Goff is quite correct to say that nuisance and *Rylands* are relevant only where they are relevant. But this can only make the defendant liable for acts or escapes which interfere with the plaintiff's *use and enjoyment* of his *land*, that is, his business of drawing up whatever water happens to be underneath it, whether clean or dirty. If this is the law, it already absolves the tanners, irrespective of whether foreseeability of damage has to be proved in nuisance.

Lord Goff's judgment will no doubt be widely praised; in its way, and we have come to expect it from him, it is quite masterly. In one stroke, however, it also succeeds in reducing centuries of English jurisprudence to a mere academic's quibble[18], since not only "enough" but "as good left over for others" had for long been established as a condition of anyone's right to take from a common resource. This should not be objected to, since it is precisely what the argument of this chapter would have anticipated. Moreover, the decision, though not the judge's reasoning, can be seen to be soundly premised in another logic, one which is almost as ancient, and whose unfolding was long overdue; that of complementarity, intolerant of the paradoxes of inchoate property rights in the same object. They served a different function for another era[19].

Second scenario: the tragedy of the law

The second scenario begins with widespread public acclaim[20] for Lord Goff's admirably clear statement of the fundamental principles governing the complex land torts of nuisance and "the rule in *Rylands*". The "criterion" of reasonable foreseeability might be vague, but it is not so vague as to include retrospective liability for historic pollution. However, once again there is a growing sense that there are subtleties in Lord Goff's judgment which we might have missed.

It had been thought, in the application of *Rylands*, that a defendant would not be liable where what he was doing amounted to a "natural use" of the land. This came to be seen as a licence for courts to engage in an aggregation of social utilities, a balancing of interests. Was the defendant's use of land the accepted practice? How many workers would the decision against him make redundant? The tanners are repeatedly referred to in the judgments as makers of "fine leather" (read: chamois cloths). Indeed, it was the judge's application of this at first instance in favour of the tanners which had caused such outrage in the press; a mere "point of law", they said. But it will perhaps be seen to have been one of the great merits of Lord Goff's firm ruling on foreseeability to have cut a swathe through this rather tired and bogus judicial policy-making. As he puts it; "[i]t may well be that, now that it is recognized that foreseeability of harm of the relevant type is a prerequisite of liability in damages ..., the courts may feel less pressure to extend the concept of natural use to circumstances such as those in the present case" (at 79 h–j). Since the natural user defence only ever operated in favour of defendants, we might ask whether this is not now an environmentalist's charter[21]? In particular, Goff's rejection of the sufficiency of what is common industrial practice, indeed his obvious distaste for undisciplined reasoning of this type at all, suggests that the case may have an application well into and beyond our concerns with the routine use of pesticides and herbicides in farming.

If this is so, it seems questionable whether foreseeability of damage alone can confine the "greening" of torts within legally acceptable boundaries. Indeed, encouraged by their own success the courts will soon be drawn into extending environmental liability into areas where nuisance alone cannot reach. For if farmers are liable in nuisance, why should not banks be liable in negligence for lending money to an environmentally unfriendly concern[22]? Why should regulatory bodies themselves not be liable for having failed to monitor and prevent the polluting acts, or for having licensed the use of a dangerous substance in the first place? In such circumstances, the notion that historic pollution represents an unforeseeable type of damage might begin to appear anomalous and restrictive. Once again, we might like to console ourselves with the thought that it is only in the USA, for example in the New Jersey asbestos cases, that such retroactive liability has been entertained. But already, it is known that the Commission of the European Community is also considering it as a possibility; encouraged by a powerful environmental lobby, it is seen as the fulfilment of the "polluter pays" principle embodied in the amended Treaty of Rome. It is not difficult to imagine a scenario in which a directive is passed introducing liability for historic pollution. Perhaps mobilization of the farmers will delay its implementation in the Member States, and as with the Drinking Water Directive, an intense rearguard action will be pursued seeking to have it repealed. In desperation, the Community will turn to private law for assistance: Member States themselves will become liable in tort for failure to implement (the *Francovich* case), and English courts will be tempted into shifting the burden onto business, enthusiastically holding themselves bound to interpret not just implementing legislation, something which they had only recently been reluctant to concede, but also now the common law itself in accordance with the purposes of European law (*Marleasing*). The problem here is not that there is anything downright wrong with the arguments deployed by lawyers in such cases. On the contary, what is disturbing is just how plausible they can be made to appear. For it is difficult, under these circumstances, to see *which* activities, *if any*, can escape the reach of legal liability[23]. To this the notion of foreseeability perhaps offers no more than a partial and provisional resistance. At the close of this scenario, then, we find out, only too late, that the law itself has played the role of tragic hero; all along, it has been unwittingly contributing to its own downfall.

Those who would defend law-as-integrity might claim that Dworkin's theory is concerned only with problems of legal decision-making. Indeed, in his attempt to steer a course between "conventionalism" and "pragmatism", he can be seen to be offering an account of legal reasoning which is peculiarly sensitive both to the difference between the past and the future as well as the need to overcome it in the present. Thus, law-as-integrity rejects as "unhelpful the ancient question whether judges find or invent law; we understand legal reasoning, it suggests, only by seeing the sense in which they do both and neither" (Dworkin, 1986: 225). Out of the very mutability of traditions and the dynamic nature of interpretative practice, Dworkin infers as a precondition for reaching any judgment that it must be assumed to be the only correct one that there is. The most compelling arguments for the presumed truth of decision-making are drawn from precisely those situations where it is least plausible; out of contingency, Dworkin produces necessity, such exaggerated moves, I argued, being typical of law in transition. Faced with the complexities of decision-making, legal theory has retreated into the most absurd and dog-matic position, a *veritatis splendor* in which law must always be right, even while it routinely changes its mind; indeed, the very presupposition of constant judicial rectitude is precisely what encourages an uninhibited and even reckless pace to such change in the law[24]. For this reason, if for no other, it is unwise, because it is too risky, to bracket off problems of argumentation from an understanding of the operations of the whole legal system. Dworkin might say, correctly to my mind, that his answer is compelled by a two-valued logic, by what he calls "dispositive" concepts which exclude third possibilities, for example, that the decision was an exercise of discretion. But this only means that the decision itself becomes the excluded third, and that principles become a vehicle for the absorption of social values, including that of environmental protection, into the law. Dworkin's theory does not prevent this, it merely attempts to withdraw it from observation. There is no room to explore this further here. It is enough to note that multi-valued logic is already a staple ingredient of soap operas. Here, the audience relishes its role as excluded thirds; observers whose opinions on the development of the soap are actively sought out and reintegrated into the plot. Even the characters in the soap are continually having to resolve seemingly life-or-death questions while openly recognizing that there can be "no right answer" to them (Geraghty, 1991: 44). Legal reasoning cannot simply ignore this; soaps show that the sort of stability which law seeks is achieved through resonance and not merely a concern for right answers. Finally, anyone involved in soap opera is skilled at handling the feelings of engagement and distance which it generates; everyone knows that soap is real life, even if it is all made up. One consequence of this is that soap audiences can, indeed must, relate what a soap character does directly to their background knowledge, for example, that his contract is about to be termi-nated. How skilled are we at relating the developments in tort doctrine to, say, what happens in the insurance economy? For example, is this what it means to say that a case has become "bad law", that it is about to have *its* contract terminated? I explore this briefly in the next section.

THE INSURANCE OF INSURANCE

Of course, the question of compensation is usually not even offered to the court for decision. In a sense, the law compensates insurers for the unpredictability in decision-making by allowing them to purchase the right to control the handling of disputes from insured parties. Where they have control over the decision whether to settle or litigate, then they can select those cases which they will allow the law to decide. To bargain "in the shadow of the law" involves both using and ignoring the relevant legal thresholds; it also darkens the settlement process itself. There is considerable uncertainty about how much a claim is worth, some talk mysteriously about having a "feel" for a case (Genn, 1987: 75–78), and there have even been complaints recently about insurers purchasing the secret and exclusive use of settlement figures. Nevertheless, insurers have not yet been allowed to purchase the whole tort process; they have no direct control over the actual decision where one has to be made, at most only influence. Nor would they themselves have any interest in extinguishing the basis for what is a lively source of income. To some extent, then, insurers have had to come to terms with the realities of torts. Chief among these realities is the gap in time between action and harm, particularly in the area of environmental damage. This gap is partly a creature of tort law, its social, material and temporal dimensions being extended or contracted through changes in doctrinal classification. This is particularly true with the central categories of acts/omissions, causation and harm. Furthermore, once claims are made, the law might also encourage the plaintiff or his backers to resist the offer of settlement, through permitting group actions, markets in unmatured tort claims, and the like. So law can both manage the frequency with which insurers are exposed to claims, as well as influence the decision whether claims are pursued to judgment. Changing that frequency, however, is liable to generate over-resonance in insurance markets.

In what are currently the most notorious of tort cases, for example, those which allow recovery for so-called "long-tail" risks and which are quite typical in environmental claims, liabilities which were priced some 20 or 30 years ago have now matured into awards of damages whose magnitude no actuarial techniques could have anticipated[25]. Moreover, so-called "insurance archaeologists" now specialize in digging up old policies written before the pollution exclusions of the 1980s had been introduced and which had bundled environmental with other risks on a "general and comprehensive" basis. This has led to much litigation in the USA and, while in Europe we like to think that we are not as "claims-conscious" as the Americans, there seems little reason not to expect similar litigation over here, even if not in quite the same quantity. Arguments typically concern the range and depth of coverage, something which amounts to a wholesale redefinition of the controlling categories of acts/omissions, causation and harm. For example, in relation to the *Cambridge Water* case, an insurer might argue that the policy only covered "sudden and accidental" spillages, rather than a protracted period of exposure. He would want to claim that liability covers physical damage only, and not the pure economic costs of cleaning up. And he would want to dispute whether the policy was intended to cover groundwater contamination in the first place. Judges have not always been overwhelmed by such arguments from insurers as to the proper interpretation of their contracts.

However such old disputes are resolved, the law will have to apply different considerations to the new clauses and new policies which began to be introduced in the mid 1980s. For the insurance contract has become a vitally important technology in the insurance industry's response to environmental risk. First, it allows the industry to specialize. In place of the old comprehensive policies, there are now, for example, policies covering only pollution or, somewhat more generally, "environmental impairment". Specialization of this sort has to be linked to a more fundamental development; the reorganization of the insurance industry towards greater centralization. Again the insurance contract is important. For example, one innovation has been the writing of policies on a "claims made" rather than an "occurrence" basis. While in the latter, cover extends to damage which occurred during the policy period, the former allows claims only where the policy is still in effect. At one level, this simply allows insurers to co-ordinate more closely the pricing of risk with their exposure to claims, that is, to review profitability in the light of developments in liability. However, it seems clear that such clauses are also symptomatic of greater centralization in the insurance industry: for example, in the context of several anti-trust suits in the USA, re-insurers are alleged to have tried, in the face of resistance from primary insurers, to coerce the Insurance Services Office into withdrawing from the occurrence form the all-important statistical support on which risk-pooling depends[26]. As I write, it is probably still too early to say whether the "claims made" form and other contractual innovations will receive the recognition they need from the courts[27]. And in the short term, of course, such trends, together with the use of new information technologies[28], are likely only to accelerate the volume of claims to which insurers are exposed. If this is correct, then the insurance industry is presented with seemingly contradictory demands, analogous to those which confronted courts during the period of centralization in administration; in each case, a resolution appears about as unlikely as it is urgent.

For some authors, it appears that this should be considered a necessary and welcome consequence of the extension of environmental liability. Indeed, authors such as Ulrich Beck have elevated such contradictions into a theoretical and political strategy, one in which institutions are purposefully turned against themselves, making them vibrate with the sort of over-resonance which we have been discussing throughout this chapter. His hope is that "organized irresponsibility" can be exposed, and the cloak of anonymity which the welfare state has conferred on those who create danger will finally be removed. Applying this analysis to our problem, we might say that the novelty of environmental liability lies precisely in the way it undermines the risk-packaging and risk-distribution technologies on which institutions such as insurance rely. Because environmental problems cannot be factored in the way insurance practice demands, they get exposed in precisely those areas where it is intended they should be concealed. This leads to what Ulrich Beck might call, loosely translating his excessively rhetorical German, an "immanent erosion" or "institutional destabilization" within the insurance industry. Instead of symbolizing security and progress, society produces uncertainty and self-annihilation. Now if the very peculiar institution of Lloyd's could be said to have had any independent function in an increasingly aggressive and competitive re-insurance market, then surely it would be to have provided the symbolic representation of this illusion of security of which Beck complains. Such illusions are themselves no more than what Luhmann would call a by-product of redundancy, in this case produced by the continual re-insurance of insurance (see Luhmann, 1984: 94, 237–8). But where does it all end? The mythology of Lloyd's insurance market can be said to have provided the symbolic closing of this circle, turning it from a vicious one into a virtuous one. Everything at Lloyd's could be insured, apocryphally, from an actress's legs to a pianist's fingers; and to an unlimited amount, since the liability of its "Names", individuals who provide its capital base, was unlimited. If Beck's image of a society vibrating with environmental tension can be said to have materialized anywhere, then it is surely here in London. With the contraction in traditional lines of insurance covered on a proportional basis, the exclusion clauses written into environmental policies generated a lively market in "excess-of-loss" re-insurance, which were packaged and re-packaged as they were passed around specialist syndicates. The recent unravelling of these "spirals" of re-insurance has inevitably created tensions between those Names who have been asked to meet claims, and those "insiders" who have somehow managed to avoid them (Gunn, 1992). With the recent decision to admit corporate capital to the Lloyd's insurance market bearing only a limited liability, this illusion of infinite coverage has finally been shattered. Beck would then be able to argue that this is not the insurance society, as Ewald claims, but the post-insurance society[29].

UNTHINKING LIABILITY

However swift and entertaining this conclusion might be, it seems to have little else to recommend it. Systems theory presents us with an infinitely more arduous and tedious path, one where the simple has to be understood in terms of what is complex, rather than the other way about, something which I argued earlier was the very attraction of private law, since it presents complex, polycentric problems in a manageable, "justiciable" form. In the absence of any convincing alternatives, we become more rather than less tempted to rely on it, even in the most implausible of situations. As a result, we have tended to resort rather unthinkingly to tort as a strategy of regulation, even in complex areas such as environmental protection. Before going any further, then, it might be helpful to *un*think what tort does in our legal systems and in society. The problem perhaps has its roots in the exaggerated and over-ambitious claims we make for politics; ours is a totalitarian democracy, one in which it is still possible to say "it's all politics". Because we are over-ambitious in this way, we turn to politics first when things go wrong. Some of these claims involve the prevention of damage to the environment. Since we do not yet know whether what will harm the environment harms it less than any available alternatives, this ensures that politics both redistributes these claims elsewhere while depriving itself of any very good reason for doing so. Because tort allows us to postpone the decision over what should be prevented, its use allows us to permit as much as we can on the assumption that any damage will have to be compensated. The persistence of instrumentalism in law guarantees that we cannot ignore demands that tort should be used in this way, even if this drags law into areas which will ultimately undermine it. Accordingly, law too passes the risk problem on to others. It recreates the double bind as a problem for firms; this time in the form of a command which they can neither ignore nor properly obey[30]. With desperate optimism, this is then invoked to support the deterrent value of torts. But if tort becomes so vacuous that businesses risk being made liable no matter how careful they may be, then it would seem rational to take as few precautions as possible, particularly where investment decisions are made within the notoriously short-term perspectives of management[31]. At best it supports a near Luddite hostility in favour of old technologies over what is new (Huber, 1983). Long after managers have moved on, the problem still has to be picked up by someone, usually insurers. If politics can be said to have exported the risk problematic to law, then law can be said to have passed it on to the insurance economy, where it is now translated into the difference between the state of the law which is used to price the risk and the actual award of damages which they are eventually called upon to cover. In this analysis, law "hides" risk by absorbing it from politics before dissipating it in the economy.

But how do we know that this analysis is not simply a consequence of our own theoretical decisions? The theory proceeds from the premise that functional differences are evolved in response to problems of "double contingency", that is, situations where my actions depend on yours and *vice versa*, problems of reflexive risk I called them. Since they have to be suppressed or postponed in order for anything to happen, for modernity to "take off", then it is inevitable that a radical re-problematization of risk, of the future, of environmental danger, will challenge our core institutions, and the whole legacy of enlightenment thought. But if this is an inevitability at all, it could be argued, then it is only one from the point of view of someone who accepts the premises of this theory. If we are to be consistently reflexive, then we have to *un*think functional differentiation too, perhaps seeing it as Luhmann's own particular *Horrorbild* in which we are simply unable to learn how to cope with risk, so entrenched are our institutions in their specific modes of operation (Beck, 1988: 176).

In part this would be easy to endorse, since it follows systems theory's own self-understanding that it should be reflexive about itself. It would simply be an evasion, then, to defend a kind of normative functionalism, and argue that differentiation, all in all, just happens to be the best we can do. Another option would be to see reflexivity as built into the principle of functional orientation, since no matter how hard you try to think of alternatives for functions, they can always be included within the theory as functional alternatives (Luhmann, 1984: 242). This is certainly tempting, but to apply it to functional differentiation as an organizing principle for society as a whole, one which includes all its alternatives, and is therefore now without any, smacks too much of the recent and rather smug restatements of the end-of-history thesis. Typically, it is an option which Luhmann would also appear to have rejected (Luhmann, 1984: 464–5, 1992b: 11). Rather we can say that whatever functions society is prepared at any time to have differentiated is always an open, empirical question; it can never entirely eliminate the fundamental implausibility of its own achievements. This is as true of law as it is of anything else, in particular since it allows us to rely on new and risky expectations in the face of resistance from others. In this sense law is what allows us to learn, since all learning takes place against a background of what we are prepared to hold constant. But so long as we understand law as just one side of the difference between learning and unlearning behaviour, then it is immediately clear that what we hold constant can only ever be provisional. In other words, the whole point about fixing expectations now is precisely in order that we can change them later. That anyone should have ever accepted such a paradox is, of course, fundamentally implausible; so much so that law itself might very well come to be interpreted as a rather eccentric and anomalous development in the evolution of society (Luhmann, 1993: 571–586). We cannot rule this out; at most we can only suggest what might have been lost.

NOTES

1 It seems there had been no authority this century on the question of groundwater contamination. What I will refer to as the "*Cambridge Water* case" (*Cambridge Water Company* v. *Eastern Counties Leather*) is reported in Atkinson, 1992 (the decision at first instance) and in the *All England Law Reports*, [1994] 1 All ER 53 (the decisions of the Court of Appeal and the House of Lords). Page and paragraph references to the latter appear in the text within parentheses.
2 *The Guardian*, 29.8.91.
3 Equal, that is, except numerically; in order to avoid the unpleasant but true

assertion that one court has higher status than another, we are to suppose that the views of five senior judges in the land count for more than three, a proposition which is a scandal in logic, but at least has the merit of promoting a certain residual collegiality.

4 Following Niklas Luhmann (1986); I do not wish to make any claims to originality in this chapter. On the other hand, it is quite unlikely that Luhmann would accept very much of what I say here. I have only tried to allow his always stimulating ideas to "resonate" with my own thoughts on the matter.

5 For arguments in systems theory supporting the same conclusion, see Luhmann, 1990.

6 Concepts such as "interference" are extremely interesting and suggestive; however they should not be introduced in contradistinction to more fundamental concepts in systems theory where in truth they have nothing new to add. It seems to me that everything Teubner wants to say with this idea is already captured by the distinction between the external and internal boundaries to communication, that is, between communication and non-communicative phenomena on the one hand, and the boundaries between communicatively organized systems on the other (see Luhmann, 1986: 40–50, 218–226).

7 Lord Justice Lindley in *Ballard* v. *Tomlinson* 29 Ch. D 115, at 125. And the alternative regime of reciprocity, "too inconvenient to be laid down as law", was to hold that "no-one could take water from his own well to such an extent materially to affect the supply to those whose wells were fed from the same sources as his own" (*ibid.*).

8 This is the reason why Thompson's discussion of the rule of law is always as sympathetic as much as it is sober.

9 Compare Edward Thompson's acknowledgment in *Whigs and Hunters* that his reconstruction of Walpole's administration inevitably involved him in knowing *more* as well as less than its contemporaries (Thompson, 1975: 17).

10 Therapists suggest that the way to deal with a double bind is to shift the problem to another level. Here, tort *is* that level; whether in the form of uncertain *future* damage which might as well be ignored in the present, or as an area of substantive law separate from the law of property, whereas it has never been terribly clear why, on doctrinal grounds alone, this should have been the case, particularly for the sort of land torts which concern me here, for example, nuisance. So does tort law promote an infringement rather than a welfarist conception of property? The answer must be both; depending on whether the plaintiff seeks an injunction (where it does) or only damages (it does not; at least for Winfield & Jolowicz; see Rodgers, 1989: 390). Substantive law v. remedies; the solution is simply to jump from one level within the legal system to another.

11 I apologise for being unable to resist this rather weak joke; it does at least have the virtue of drawing our attention to the way reciprocity's themes, such as neighbourliness, continue to saturate the law long after changes in social structure should have rendered it obsolete.

12 This is a familiar enough theme, one which came to prominence in the risk literature of the mid 1980s (Beck, 1986; Ewald, 1986; and see Luhmann, 1992a: 113).

13 Environmental Data Services Report No. 214, November 1992, p. 7.

14 *Cambridge Weekly News*, 21.9.91.

15 They were found to have stopped in 1976. The Directive (80/778/EEC) had to be complied with by 18.7.85. The implementing regulations are the Water Supply (Water Quality) Regulations 1989 which have effect as if made by the Secretary of State under the Water Industry Act 1991.

16 Note, however, that it is no defence in nuisance to say that the harm was a result

only of the combination of several similar acts where each taken alone would be insufficient. Where the defendants' individual contributions cannot be discretely identified, liability is joint rather than several. At least in this respect, and so far as the common law is concerned, we have no very good reason to accept Beck's "organised irresponsibility" thesis.

17 For example, "Poisoned Profits", *In Business* (BBC Radio 4) 29.9.93.

18 For example, while Lord Goff cites the relevant paragraph from *Halsbury's Laws of England*, 4th. ed., vol. 49, which supports an owner's right to take underground water (para. 392), he fails to cite another paragraph from the same volume which supports the proposition he is labouring to refute, *viz.*; "Since a landowner has a right to take all that he can of underground water that is in no defined channel beneath his land, he has a right of action against any other landowner who pollutes the water so that the plaintiff landowner gets the water in a polluted state" (para. 785).

19 Admittedly, Edward Thompson allows himself to make comparisons between the present-day privatization of water and the enclosure programmes of the eighteenth century; each contained a "logic" which is an "offence to both 'nature' and human community" (Thompson, 1993: 182–183), although he would have been the first to concede that the comparison should not be pushed. There is, after all, probably not that much connecting the romantic poet John Clare with Andrew McIntosh, the Scottish republican who, among other offences, posted a letter bomb to the Cambridgeshire offices of Anglian Water, one which exploded in the hands of "a young clerkess" (*The Scotsman*, 23.12.93). What logic is there to this at all, unless it is to make the somewhat bizarre symbolic connection between the place where opposition is strongest to that where water is dirtiest? Thompson's remarks must be read with caution, then. He is best seen as a rogue historian, anxious to over-come the narrow pointillism of his colleagues while always defending the sound professional sense of sticking to one's period.

20 It has already been welcomed by, among others, the Association of British Insurers, the British Bankers' Association, the Confederation of British Industry, the Royal Institute of Chartered Surveyors; *The Independent*, 10.12.93, *Cambridge Evening News*, 10.12.93.

21 For example, it has been reported that, as a consequence of this decision, firms will now be advised to make sure they have at least the appearance of sound environmental management, by hiring compliance officers, commissioning environmental audits, and the like; see "Manufacturers face new pollution challenge", *The Independent*, 10.12.93.

22 Clearly this has been a matter of general concern for some time; in Britain it seems to be the proposals of the EC Commission's Green Paper on environmental liability which have concerned bankers; see "Pollution liability worries banks", *Financial Times*, 19.10.93.

23 Philosophers, it seems, would not have entertained retrospective liability as a very serious possibility: "did cigarette companies" Steven Lukes once asked, "exercise..., power over the public before it was even supposed that cigarette smoking might be harmful? Surely not" (Lukes, 1974; 52), the resoundingly confident tone in his reply reassuring us that his point in asking was only ever rhetorical. By contrast, it is almost as if tort lawyers have already discovered the fourth dimension to power.

24 This is a development of ideas first expressed in a systematic way by E. H. Levi (see Luhmann, 1993: 347–8, and the notes appended thereto).

25 This can also be a problem for the insured party who may have purchased too little cover. I am grateful to Mr Hamilton of *Eastern Counties Leather* for reminding me of this.

26 An argument made by Ayres & Siegelman (1989) in response to George Priest's influential advocacy of competitiveness in insurance markets. Priest also gave evidence in at least one of the anti-trust cases mentioned in the text; for his response to Ayres & Siegelman, see Priest, 1989.

27 For example, there are signs that it is being rejected in Europe, notably in France where policy-holders have been allowed to recover where a *fait générateur* caused the harm and regardless of the wordings in the insurance contract. See the report "Environmental time-bombs," *Financial Times*, 6.9.93.

28 See the report "Underlying Uncertainty", *Financial Times*, 10.9.90.

29 It may well be that I am over-interpreting Beck's argument here. For his own discussion of Ewald's book, see Beck, 1988: 177–182.

30 "Although product liability exerts a powerful influence on product design decisions, it sends an extremely vague signal. [It] says only: 'Be careful, or you will be sued.' Unfortunately, it does not say *how* to be careful, or more important, *how careful* to be" (Eads & Reuter, 1983: viii–ix).

31 Anyone still unclear about this should read the reports coming from Summitville, Colorado; see "Race to move mountain of waste in the Rockies", *Financial Times*, 8.11.93.

REFERENCES

Atkinson, Nicola (1992) Strict Liability for Environmental Law: The Deficiences of the Common Law. *Journal of Environmental Law*, **4**, 81.

Ayres, Ian & Siegelman, Peter (1989) The Economics of the Insurance Anti-Trust Suits: Towards an Exclusionary Theory. *Tulane Law Review*, **63**, 971.

Beck, Ulrich (1986) *Risikogesellschaft*. Frankfurt/M.: Suhrkamp.

Beck, Ulrich (1988) *Gegengifte: Die organisierte Unverantwortlichkeit*. Frankfurt/M.: Suhrkamp.

Carson, Rachel (1963) *Silent Spring*. London: Hamish Hamilton.

Coleman, Jules L. & Kraus, Jody (1986) Rethinking the Theory of Legal Rights. *The Yale Law Journal*, **95**, 1335.

Dworkin, Ronald (1986) *Law's Empire*. London: Fontana.

Eads, George & Reuter, P. (1983) *Designing Safer Products: Corporate Responses to Product Liability Law and Regulation*. Santa Monica, Cal: The Rand Corporation.

Ewald, François (1986) *L'Etat Providence*. Paris: Editions Grasset.

Foucault, Michel (1991) Governmentality. In Burchell, G. *et al.* (Eds). *The Foucault Effect*. Hemel Hempstead: Harvester Wheatsheaf.

Genn, Hazel (1987) *Hard Bargaining*. Oxford: Oxford University Press.

Geraghty, Christine (1991) *Women and Soap Opera*. Cambridge: Polity Press.

Gunn, C. (1992) *Nightmare on Lime Street*. London: Smyth Gryphon.

Hardin, G. (1968) The Tragedy of the Commons. *Science*, **162**, 1243.

Hont, Istvan & Ignatieff, Michael (1983) Needs and Justice in the *Wealth of Nations: An Introductory Essay*. In Hont, I. & Ignatieff, M. (Eds). *Wealth and Virtue*. Cambridge: Cambridge University Press.

Horwitz, Morton J. (1977) *The Transformation of American Law 1780–1860*. Cambridge, Mass: Harvard University Press.

Huber, Peter (1983) The Old-New Division in Risk Regulation. *Virginia Law Review*, **69**, 1025.

Hume, David (1978) *A Treatise of Human Nature*, 2nd edn. Oxford: Oxford University Press.

Lovejoy, Arthur O. (1936) *The Great Chain of Being*. Cambridge, Mass: Harvard University Press.

Luhmann, Niklas (1970) Zur Funktion des "Subjektiven Rechts". *Jahrbuch für Rechtssoziologie und Rechtstheorie*, 1, 321.

Luhmann, Niklas (1981) Subjektive Rechte: Zum Umbau des Rechtsbewußtseins für die moderne Gesellschaft. In Luhmann, N. *Gesellschaftsstruktur und Semantik 2*. Frankfurt/M.: Surkamp.

Luhmann, Niklas (1983) *Rechtssoziologie*, 3rd edn. Opladen: Westdeutscher Verlag.

Luhmann, Niklas (1984) *Soziale Systeme*. Frankfurt/M.: Suhrkamp.

Luhmann, Niklas (1986) *Ökologische Kommunikation*. Opladen: Westdeutscher Verlag.

Luhmann, Niklas (1987) Brauchen wir einen neuen Mythos?. In Luhmann, N. *Soziologische Aufklärung 4*. Opladen: Westdeutscher Verlag.

Luhmann, Niklas (1989) Am Anfang war kein Unrecht. In Luhmann, N. *Gesellschaftsstruktur und Semantik 3*. Frankfurt/M.: Suhrkamp.

Luhmann, Niklas (1990) Haltlose Komplexität. In Luhmann, N. *Soziologische Aufklärung 5*. Opladen: Westdeutscher Verlag.

Luhmann, Niklas (1992a) *Soziologie des Risikos*. Berlin: de Gruyter.

Luhmann, Niklas (1992b) *Beobachtungen der Moderne*. Opladen: Westdeutscher Verlag.

Luhmann, Niklas (1993) *Das Recht der Gesellschaft*. Frankfurt/M.: Suhrkamp.

Luhmann, Niklas & De Giorgi, R. (1992) *Teoria della Società*. Milan: Franco Angeli.

Lukes, Steven (1974) *Power: A Radical View*. London: Macmillan.

Malthus, Thomas R. (1992) *An Essay on the Principle of Population*. Cambridge: Cambridge University Press.

Priest, George L. (1989) The Anti-trust Suits and the Public Understanding of Insurance. *Tulane Law Review*, 63, 999.

Priest, George L. (1990) The New Legal Structure of Risk Control. *Daedalus*, 119, 207.

Rogers, William V.H. (1989) *Winfield & Jolowicz on Tort*. London: Sweet & Maxwell.

Serres, Michel (1980) *Le Parasite*. Paris: Editions Grasset.

Skinner, Quentin (1978) *The Foundations of Modern Political Thought: Volume Two; The Age of Reformation*. Cambridge: Cambridge University Press.

Thompson, Edward P. (1975) *Whigs and Hunters*. London: Penguin.

Thompson, Edward P. (1993) Custom, Law and Common Right. In Thompson, E.P. *Customs in Common*, London: Penguin.

Ward, N. *et al.* (1993) *Water Pollution from Agricultural Pesticides*. Newcastle: Centre for Rural Economy, University of Newcastle.

SECTION III

ALTERNATIVE INSTITUTIONS OF ECOLOGICAL SELF-ORGANIZATION

CHAPTER 7

Ecological Contracts: Agreements between Polluters and Local Communities

Eckard Rehbinder
Frankfurt/Main

INTRODUCTION

The development of modern environmental law is characterized by an ever increasing legalization. Environmental law has covered the action space of industrial societies with a dense network of laws, regulations and administrative rules. These prohibit certain activities, lay down legal requirements for others, make the exercise of such activities conditional on the fulfilment of certain requirements, such as the granting of a permit or prior notification, and subject prohibited activities to criminal or administrative sanctions. Moreover, the increasing recourse to planning of environmental quality amounts to the establishment of public management of natural resources that allots scarce environmental absorption capacitics to particular polluters according to selection criteria ranging from the principle of priority to equal and fair apportionment. This pattern of legalization is by no means negated or even mitigated by the modern trend towards using so-called economic instruments such as charges, bubbles or pollution licences. The philosophy underlying these latter instruments is that the market be used for economic efficiency purposes in a merely instrumental way in order to achieve environmental policy goals that continue to be set by the state. Seen from this perspective, economic instruments are but a modification of a highly legalized system of environmental protection.

On the other hand, in some countries at least, environmental contracts or, better, environmental agreements, are gaining increasing importance as a supplement to traditional command and control regulation. Environmental

Environmental Law and Ecological Responsibility: The Concept and Practice of Ecological Self-Organization
Edited by G. Teubner, L. Farmer and D. Murphy
© 1994 John Wiley & Sons Ltd

agreements may be concluded between regulators and enforcers on the one hand, and polluters, on the other. This is the most frequent type of environmental contracting. However, there are also environmental agreements between local communities—or the municipalities representing them—and polluters. Our analysis will be limited to this latter type of environmental contracting.

At first glance, one is tempted to seek the reasons for the advance of environmental agreements in the often described implementation and enforcement deficit of environmental law, which, in turn, may be due to over-legalization of society in the field of environmental protection or the lack of acceptance (ineffectivity) of environmental law (Mayntz, 1979; Treiber, 1983: 23ff.; Teubner, 1984: 10ff.; 1992: 139ff.; Ackerman & Stewart, 1985: 1333ff.; Rottleuthner, 1989: 273ff.). As will be seen later, this explains some types of environmental agreements in some countries, but does not represent a comprehensive explanation. Besides, an important role is also played by deficiencies in the normative programme of environmental law that can be attributed to policy failures. In these latter cases, environmental agreements are a means of social self-help in situations of the (relative) inactivity of the state.

The following analysis tries to explore the reasoning behind, the prerequisites for the functioning, and the effects of environmental agreements between polluters and local communities in three countries, namely the USA, Japan and Germany. These countries have been selected because environmental contracting in these countries has markedly different features. A comparison of these countries may supply insights that allow the development of a "theory" of environmental contracting.

ENVIRONMENTAL AGREEMENTS IN THE USA: LEGALIZED DELEGALIZATION

In the USA, negotiation, especially in the form of mediation, has become an important element of environmental regulation in the last two decades. Environmental negotiation ranges from policy development via standard setting and other forms of rule making, to permitting and enforcement. It also ranges from informal *ad hoc* bargaining to mediator-supported negotiation to institutionalized negotiation (Symposium, 1991; Sander, 1990: 14ff.; Rennie, 1989; Mays, 1988; Brunet, 1988; Grad, 1989; Susskind & Cruikshank, 1987; Amy, 1987; Bingham, 1986; Lieberman & Henry, 1986: 424ff.; Susskind, 1985: 101ff.; Susskind & Weinstein, 1980; Cormick, 1976: 215ff.). The discussion has by and large focused on mediation as a means of "alternative" dispute resolution. However, the most interesting texture of this mosaic of different bargaining techniques for us is the movement towards institutionalized negotiation in relation to decisions on the location of hazardous waste disposal sites (Holznagel, 1990: 133ff.; 1986: 352ff.; Bingham & Miller, 1984: 480ff.; Canter, 1982: 482ff.).

In the 1980s, against the background of a growing scarcity of waste deposit and incineration sites, some states such as Massachussetts, Wisconsin, Rhode Island and Connecticut introduced mandatory negotiation into state decision-making procedures for new or significantly modified waste disposal facilities. Other states created optional negotiation procedures.

The relevant laws on mandatory negotiation schemes, notwithstanding differences of detail, all contain certain common elements. The conclusion of a siting contract between the operator of the waste disposal facility and the municipality is a mandatory prerequisite for a positive decision by the state authorities on a siting application. This contract, which is enforceable before the courts, must contain agreements as to the construction, design and operation of the facility, on monitoring, on compensation to the local community for the disadvantages suffered by permitting the facility, on conflict resolution and on renegotiation. It is negotiated by a local negotiation committee, in which the local community, together with members of the local council, is represented. In some states, this negotiation committee can bind the municipality. In other states, the committee only makes proposals which must then be ratified by the organs of the municipality. There are quite far-reaching provisions for public information. First, the environmental impact assessment that must be carried out before approval of the site and the construction of the facility provides the negotiation committee, as well as the general public, with comprehensive information about the project. Second, the local community is to be informed about the results of the negotiation through public hearings that either accompany the whole process, or take place when a preliminary agreement has been reached or the granting of the siting permit is being discussed. In addition, the relevant laws provide for conflict resolution mechanisms in the form of arbitration, and sometimes in the form of preliminary mediation. Where, after a prescribed minimum duration of the negotiations, an agreement between the operator and the negotiation committee cannot be reached, the contents of the siting contract (especially as to the kind and amount of compensation), will be determined by a neutral arbitration tribunal. Finally, the laws provide for the participation and compensation of neighbouring municipalities that are strongly affected by the project. These may either become parties to the siting contract or, at least, be recipients of compensation payments.

It is clear that this model is a long way from being free contracting in the shadow of the law. It is an institutionalized part of the permitting process that contains prescribed substantive law elements, specifically the option for more stringent standards for the operation of the facility and compensation of the municipality for the "sacrifice" in accommodating the facility. Its delegalization effect is, therefore, quite limited. One could term this a form of "legalized delegalization". This model of institutionalized negotiation within the decision-making procedure must be seen against the background of changes in state law that shifted most of the decision-making power as to the siting of

waste disposal facilities from the municipalities to the state authorities in order
to overcome the siting crisis caused by lack of public acceptance of new sites.
Institutionalized negotiation is mainly conceived of (though not exclusively
so), as a compensation for the strengthening of central agency powers.

The model contains some incentives that make the conclusion of a siting
contract attractive for both the municipality and the operator (Bingham, 1986:
8ff.; Susskind & Cruikshank, 1987: 117–8). The conclusion of the contract
is a mandatory requirement for a positive siting decision, but the municipality
is not able to block the taking of this decision because in the case of non-
agreement the conditions of the contract are laid down by an arbitration
tribunal. The municipality would normally be interested in reaching an agreement in order to retain its influence over the conditions of the agreement.
Moreover, the decision to participate in the negotiation exerts some informal
pressures to reach an agreement. The operator also has an interest in
influencing the conditions of the contract, to minimize the resistance of the
municipality or affected citizens and persuade them not to challenge the
legality of the siting decision before the court.

The availability of the results of the environmental impact assessment, in
which the general public may also fully participate, provides an information
base for all participants in the negotiation process that enables them to take
informed decisions as to the contents of the siting contract. By this means
information disequilibria are, to a large extent, removed and the negotiation
committee and other citizens are given an increased chance to influence the
outcome of the process. The procedure also allows the affected citizens to
articulate their fears and anxieties as to the project.

Finally, mandatory arbitration on unresolved issues, apart from providing
incentives to reach an agreement, is a means of bridging gaps between the bargaining positions of the participants, protecting them from a loss of face in
case of acquiescence in a result they had previously opposed, and achieving a
balanced overall result.

Despite all these legislative attempts to make institutionalized negotiations
on siting conflicts attractive, the success of the relevant laws has been mixed
(Holznagel, 1990: 167ff.; Gardner, 1990: 205ff.; see generally Bingham,
1986: Ch. 2). In Massachussetts, the procedure apparently has not greatly
promoted the siting of new waste disposal facilities, mainly because neither the
authorities nor the affected municipalities have ever been willing to go through
the administrative process in such a way that the state of negotiation could be
reached at all. In Wisconsin, by contrast, quite a number of siting contracts
are concluded every year. The differences between the two states may be
explained by differences in population density and the ensuing situations of
sacrifice, but also by different attitudes towards the problem of toxic waste.

It should be noted that the mixed record of institutionalized negotiation on
these politically highly sensitive siting conflicts cannot be generalized to other
forms of negotiation, not even on less politically sensitive issues.

ENVIRONMENTAL AGREEMENTS IN JAPAN: EXPRESSIONS OF SOCIAL CONTROL AND DECENTRALIZED POLICY MAKING

Environmental agreements, in which operators of polluting facilities commit themselves towards prefectures and/or municipalities to comply with requirements which go beyond purely legal ones for the control of pollution or improvement of nature and landscape, are a characteristic feature of Japanese environmental policy (Gresser *et al.*, 1981: 248–9, 346; Weidner *et al.*, 1989: 170ff.; Leane, 1991: 361ff.; Shibata, 1989: 246ff.; Upham, 1987: 248ff.; Yamamouchi & Otsubo, 1987: 221ff.; Young, 1984: 932ff.; Foljanty-Jost, 1988: 45ff., 88ff.; 1990: 36ff. Such agreements are concluded in the case of the construction or significant modification of a major facility.

The total number of these new agreements is between 1500 and 2600, while 200 to 300 agreements being terminated each year. The overall number of environmental agreements is now about 40 000[1]. The areas covered (in order of numerical importance) are industrial effluents, noise, industrial waste and offensive odours. Air pollution which had been rather important in the early 1980s, has recently fallen back.

Along with very general commitments, modern environmental agreements contain concrete and detailed obligations for the control of pollution. These can take the form of emission standards, obligations to install and use particular control equipment, input standards, obligations for the improvement of the natural environment such as the establishment of protection perimeters, obligations to carry out an environmental impact assessment, temporary plant closures in emergency situations, compensation for damages, monitoring of emissions and inspection of the facility. Normally, in heavily polluted areas, these requirements are much more stringent than applicable legal requirements. A major exception seems to be the high-tech industry where environmental agreements are still considered to be unsatisfactory. Emission limits may be more stringent by 50%. In principle, the most recent technology must be used, and the agreements use drafting techniques such as dynamization clauses, short duration of the agreement or renegotiation clauses in order to compel a periodic adjustment by plant operators to the most recent technology.

The use of environmental agreements varies between prefectures (Yamamouchi & Otsubo, 1987: 231ff.; Foljanty-Jost, 1988: 88ff.; Weidner *et al.*, 1989: 189ff.). In some areas and in major municipalities, there is a marked preference for regional or local ordinances that set forth more stringent requirements, the use of environmental agreements being limited to facilities run by the municipality. In others, environmental agreements are the primary instrument of regional and local environmental policy. In others still, environmental agreements are almost exclusively used at the local level.

The legal nature and, in particular the legal force of environmental agreements, are the subject of controversy in legal literature (Yamamouchi & Otsubo, 1987; 226–7; Weidner *et al.*, 1989: 192ff.; Foljanty-Jost, 1988: 90–1). Many authors are of the opinion that these agreements are simple non-binding gentlemen's agreements. However, the lower courts have held that the concrete obligations contained in such agreements are enforceable, although there is a high degree of voluntary compliance, which is fostered by surveillance devices such as rights of inspection that are set forth in the agreements. In cases of non-compliance, the agreements contain a number of sanctions such as the powers of the municipality to limit plant operations or even order a plant closure, the right to cancel a sale or leasing contract connected with the environmental agreement, the right to stop the supply of water, or special contractual fines. Normally, the municipalities enforce the obligations set forth in environmental agreements by recourse to administrative recommendations (*gyosei shido*) rather than use the available sanctions. If the sole parties to an environmental agreement are local communities, they may also turn to the municipality for enforcement of the agreement (Weidner *et al.*, 1989: 194; Environment Agency, 1983: 325).

In most cases, the parties to the agreement are prefectures or municipalities with both often appearing as parties to the agreement. Although the municipalities often consider themselves as the representatives of local communities affected by a new facility, there are also numerous cases of direct citizen participation in environmental agreements (Environment Agency, 1983; 1989; 1990). This may take different forms. Local communities may be co-parties to the agreement or participate as observers; they may also conclude environmental agreements as sole parties. By 1988 there had been 1965 environmental agreements in which local communities participated as co-parties or observers and 4103 agreements had been concluded exclusively by local communities. The figures for new agreements are 149 and 214 respectively in the year 1988, and 127 and 219 in 1989. Roughly speaking, one can say that in about 20% of all environmental agreements there is some form of formal participation by local communities.

Whether or not local communities formally participate in environmental agreements concluded by the prefectures or municipalities, depends on the political configuration. Formal participation is normally designed to make visible to outsiders the success of local communities in compelling the conclusion of an agreement, and their influence over its conditions. Where local communities are the sole party to an environmental agreement, they have often carried through their point of view with the plant operator in opposition to the prefecture or municipality[2]. Moreover, local communities are always the sole party to environmental agreements concluded by municipalities themselves in their capacity as operators of polluting facilities, such as waste incineration or waste water treatment plants.

From the point of view of prefectures, municipalities and local communities, there are a number of good reasons for trying to conclude environmental agreements with local polluters (Yamamouchi & Otsubo, 1987: 228ff.; Weidner et al., 1989: 194ff.; Leane, 1991: 369ff.; Foljanty-Jost, 1988: 46, 49, 89ff.). Municipalities and even (to a certain extent) prefectures have only limited competence to set forth more stringent requirements. While historically the paramount interest was in achieving higher local standards, this has diminished with growing legislation and the stiffening of environmental controls. It nonetheless remains a major force in the quest for environmental agreements, in particular as a means of dynamization of environmental protection, and can also be used as a means of achieving higher local standards in areas where the municipality or prefecture has no competence or where there is legal uncertainty in this regard. In addition, the protracted decision-making process for the making of prefectural or local ordinances can be avoided. Avoidance of public participation plays no role because the agreements are accessible to the public and the negotiations may even be held in public. On the other hand, these negotiations are time consuming and require considerable administrative resources. This is the reason why some prefectures or municipalities either have recourse to standard agreements or prefer ordinances. For local communities, environmental agreements are attractive as an element of democratization of the economy and for the reason that they make the communities' contribution more visible than would be possible in the case of an ordinance. Finally, environmental agreements are seen as devising solutions that are adjusted to the particular case, especially where small and medium-sized enterprises are the parties to the agreement. They redistribute regulatory burdens by resort to a quasi-market model of decision-making.

While the conclusion of environmental agreements can be seen to be attractive to prefectures, municipalities and local communities, one wonders why enterprises should have an interest in entering an agreement that almost invariably results in more stringent requirements for the control of pollution by the relevant facilities than the operator would have to follow under the relevant central regulations (Yamamouchi & Otsubo, 1987: 229ff.; Weidner et al., 1989: 197ff.; Leane, 1991: 370–8).

In many cases, the conclusion of environmental agreements is motivated by the desire of the enterprise to buy or lease extremely scarce municipal land for industrial settlements, especially landfill (the so-called "Yokohama method"). Moreover, the prefectures and municipalities possess a variety of potential means for the exertion of subtle pressure, especially in the field of granting permits. These can be used to give "administrative guidance", but they may also operate as a simple motivating factor.

More important, however, is the interest in establishing an ongoing informal relationship with the administration and the local communities. The sacrifice that the operator makes by concluding an environmental agreement that imposes more stringent requirements is a consideration for the establishment of a special relationship that can be used in the future in the case of disagreement over other matters. One could speak here of a confidence-building strategy. Industry officially considers the conclusion of environmental agreements as an expression of social responsibility. To the extent that the conditions of the agreement are adequate and sufficiently flexible, industry is willing to contribute to the development of regional and local environmental policy and an improvement of regional and local environmental conditions. On the other hand, it would certainly prefer laxer and more uniform requirements. It accepts the conclusion of environmental agreements as a practically indispensable prerequisite for coming to terms with regional and local politicians and local communities. To this extent environmental agreements are an expression of the basic consensus orientation of Japanese society.

Finally, in the past, and to a lesser extent at present, the pressure exerted by local communities and public opinion has played an important role. It has been said that securing the consent of local communities is the primary aim of environmental agreements. In Japanese society plant managers and workers normally live within the local community and, due to the scarcity of land fit for human settlement, cannot simply move away. They are thus dependent on the goodwill of the local community and cannot easily evade social pressure exerted by people adversely affected by the facility. The still vivid memory of the catastrophic health damages caused by industry at the end of the sixties contributes to a weakening of the position of industry in the eyes of the local communities. Environmental agreements, at least in the case of major projects, have a pacifying effect that is a necessary condition for the siting or significant extension of major facilities.

It is difficult to generalize further about the extent to which material advantages, administrative pressures, the interest in establishing ongoing relationships with the administration, social pressure from local communities and social responsibility and consensus orientation motivate the conclusion of environmental agreements.

Environmental agreements are an important element of Japanese environmental policy (Weidner et al., 1989: 200ff.; Foljanty-Jost, 1988: 45ff.; Upham, 1987: 248–9). They are a highly effective local or regional counterweight to weak or inappropriate central regulation that is slow to react to new problems or develop scientific and technical knowledge. Virtually all major polluters, as well as many small and medium-sized ones, have concluded such agreements. Although it is clear that environmental agreements do not comprehensively supplant central regulation, they are the most important source of improvements of environmental conditions in agglomerations and, at least for major polluters, set forth the controlling requirements even outside these areas. The agreements are not merely symbolic. Rather, they set forth concrete and detailed requirements for control of pollution and improvement of the natural environment that are either more stringent than central regulation, or alternatively fill gaps left by central regulation. However, the widespread local and regional differentiation of contract requirements shows that the purpose of the agreements is not pure precaution but, rather, adjustment of protection against pollution and environmental degradation according to local and regional conditions and political preferences.

The political acceptance of environmental agreements is high, and this is not only true for the active participants. The central agencies, industry, environmental groups and local communities all in principle (for their various reasons) accept environmental agreements as a major instrument of Japanese environmental policy.

ENVIRONMENTAL AGREEMENTS IN GERMANY: ISOLATED SETTLEMENT AGREEMENTS IN THE SHADOW OF OVER-LEGALIZATION

In our three country comparison, Germany presents a unique picture of the over-legalization of environmental protection. Not only does it have a very dense coverage by environmental statutes, regulations and administrative rules but also a degree of rigidity and inflexibility of environmental law that makes the conclusion of environmental agreements a highly risky enterprise. In particular, provisions in the Administrative Procedure Act limit the power of public agencies to enter into agreements with private persons.

In the last couple of years, especially under the influence of US experience with negotiation and mediation, there has been a quite extensive discussion on their role in German administration, especially in relation to environmental problems. At best one can talk of a high degree of reservation in relation to the use of contractual instruments for environmental protection while the prevailing view is one of outright rejection (Bohne, 1981: 148ff., 156, 196ff. 244ff; 1984: 343ff.; Bauer, 1987: 241ff.; Breuer, 1990: 231ff.; Kunig, 1990: 43ff.; Winter, 1985: 219ff.; Kunig & Rublack, 1990: 1ff.; Brohm, 1990: 321ff.; Eberle, 1984: 450ff.; more favourable are Holznagel, 1990: 194ff., 228ff.; Hoffmann-Riem, 1989: 36ff., 54ff.; 1990: 19ff.; Arnold, 1989: 125ff.). Generally speaking, it is thought that environmental contracts are normally contrary to the rule of law, violate unalienable fundamental rights or protective duties of the state, encroach upon the planning discretion of authorities, violate the prohibition on unreasonable coupling of agency measures and diminish the judicial protection of adversely affected third parties. This criticism is mainly aimed at environmental agreements concluded by state regulatory agencies with private enterprises, but it also concerns state agencies in their capacity as authors of infrastructure projects *vis-à-vis* private opponents and municipalities which act as representatives of local communities. Moreover, although settlements between the "real" parties to administrative court proceedings on the legality of permits, that is private plaintiffs and intervening private operators of relevant facilities, can in principle be freely concluded, they often have to include the agency as the formal defendant and, therefore, cannot always disregard the legal limits set by the state, in legislation and case law, to the conclusion of such agreements.

In spite of these unfavourable framework conditions, there are quite a number of environmental agreements between central state agencies and the operators of conventional power plants for the speeding up of the mandatory programme for the reduction of SO_2 emissions. There are also agreements of various types that provide for the enforcement of existing source requirements. In addition, mediation has sometimes been used for resolving conflicts between state agencies or municipalities and local communities over the cleaning up of abandoned waste dump sites.

Beyond these types of environmental contracting, which are outside the scope of our analysis, there are isolated examples of environmental agreements between local communities and polluters. These latter agreements can be classified into three categories: agreements that heighten the mandatory requirements for pollution control; those that provide for physical compensation for encroachments on nature and landscape caused by the siting of a major project; and those that award monetary compensation to persons who had instituted administrative court proceedings against a permit, in exchange for withdrawal of their action.

There are two examples of environmental settlement agreements concluded, with the support of environmental groups, by plaintiffs in administrative court proceedings (Weidner et al., 1989: 438ff.; Führ, 1990: 129ff.; Barth & Führ, 1986: 10)[3]. In both cases, the competent agency had granted the operator of a chemical plant a permit whose legality had been challenged by the plaintiffs. The suit had a suspensory effect because the agency had not ordered immediate execution, nor had the lower administrative court reversed the normal suspensory effect of the suit instituted by the plaintiffs. Hence, in view of the possible delay of the investment for some years, the operator of the facility had a genuine interest in achieving a settlement. In the first case, the emission limitations for SO_2 and NO_x envisaged by the competent agency were reduced by roughly 50% and additionally existing facilities were to be included in the emission reduction programme. The parties to the agreement were given access to the results of monitoring undertaken by the operator. In the second case, the settlement prescribed special security measures for the prevention of major accidents in a pesticide plant and once again created inspection rights for the private parties. In addition, the operator promised to compensate for the increased toxic waste caused by the new production facility by reducing the amount of toxic waste originating from other existing facilities in the region.

A third somewhat similar case of a settlement concerned the construction of a third boiler on the site of an incineration plant for toxic waste which had aroused the resistance of a number of municipalities as well as citizens' groups[4]. The settlement as agreed upon by the negotiators for the concerned parties provided for a package of preventive, recycling and treatment measures that were designed to reduce the amount of toxic waste within prescribed time limits, to an extent that one of the two existing boilers that still used out-dated technology could be dismantled. Later on, the second one could also be put into reserve. In exchange, the settlement allowed for a temporary expansion of the capacity of the plant up to 90 000 metric tons. Moreover, the operator had to treat the whole plant as a single facility for the purpose of applying existing emission standards. The municipalities that were parties to the settlement committed themselves to withdraw judicial actions against the permits for the existing boilers and not to institute new court actions against the new permit. However, ultimately the municipalities did not agree to the settlement negotiated by their representatives, mainly because they objected to the negotiated annual capacity of the incineration plant and insisted on retrofitting the existing boilers. They have announced that they will go forward with the court actions against the project. The citizens' groups that had initially participated in the negotiations did not join the settlement agreement either, because they were of the opinion that the temporary increase of the incineration capacity to 90 000 metric tons was not warranted in view of the existing prevention and recycling possibilities. Thus, the negotiations failed because the bargaining positions of the parties were too far apart.

The most prominent example of the second type of agreement, by which physical compensation is provided for encroachments upon nature and the landscape, is a settlement as to the siting of an automobile plant in wetlands near the Rhine in Baden-Wurttemberg. Here, the state agency committed itself to minimize the encroachments caused by the project beyond what was required by the state Protection of Nature Act, construct a new access road in such a way that a sensitive forest was not affected and especially to restore other, degraded wetlands in the Rhine meadows (Holznagel, 1990: 193–194; Benz, 1990). The expenditure involved amounted to between 65 and 105 million Deutschmarks. In exchange for these obligations, the participating environmental associations promised to withdraw all legal actions they had instituted in their capacity as owners of "strategic" property against the siting decision and sell the property to the automobile maker.

Finally, environmental agreements sometimes provide for the sale of the rights of action against a permit for a polluting facility. A good example is the so-called "environmental trade of Bergkamen" (Holznagel, 1990: 191–3; Knödgen, 1981; Duerksen, 1982). In order to avoid a costly delay of the project for a new coal-fired power plant by judicial actions that were about to be instituted by members of an association of affected neighbours, the operator promised the payment of 1.9 million Deutschmarks to 72 members of the citizens' group and other adversely affected citizens in exchange for a withdrawal of administrative objections against the project. Furthermore, the operator committed himself to establish a protective perimeter around the site, renounce the planned construction of residential buildings for the workers of the plant and limit transportation of building materials to certain routes where adverse impacts on the population could be minimized. After the conclusion of the settlement agreement, its legality became quite controversial but was ultimately confirmed by the Federal Supreme Court[5]. Economists have defended the sale of rights of action on efficiency grounds (Endres, 1977: 274ff.).

It is striking that in Germany virtually all environmental agreements have been settlement agreements. This is also true of agreements that were concluded before the formal bringing of an action against the permit granted to the operator of the relevant plant because they were designed to avert imminent litigation. The predominance of litigation or pre-litigation configurations may have to do with the legalization of environmental protection and the more adversarial legal-administrative culture in Germany. Entrepreneurs normally do not voluntarily go beyond environmental standards because of social responsibility for better environmental protection. The prevailing view among industrialists (at least until very recently) has always been that German environmental regulation, in an international comparison, was unduly burdensome and inflexible. Thus, what might induce a plant operator to conclude an environmental agreement would be the negative prospects of protracted administrative court litigation and the financial losses associated with such litigation. Improvement of the corporate image may be a secondary motive which may, in particular, influence the contents of the agreement. From the point of view of local communities, environmental agreements have been attractive mainly because of uncertainties about the extent of environmental rights of plaintiffs and the potential risk of losing the litigation. The toxic waste bubble that was agreed upon in one of the settlement agreements certainly could not have been achieved through court action. In addition, in some cases the fact that the parties were given access to monitoring results and granted inspection rights, thereby ensuring effective control over compliance, was an argument in favour of concluding the agreement. However, local communities tend to consider the results of environmental contracting as merely second best solutions. They would definitely prefer an improvement of the procedural and substantive position of environmental interests.

SOME GENERAL REFLECTIONS

In spite of the considerable differences between the three countries, some generalizations are possible about the reasons for the conclusion of environmental agreements between polluters and local communities, the prerequisites for their functioning and their advantages and disadvantages. Not all commentators welcome the trend towards contractualization, or even "commodification", of implementation and enforcement of environmental laws. The main argument of the critics is that "bartering rationality" tends to be quite different from regulatory decision-making rationality in that it neglects (procedurally fair and sound) fact-finding, preserves inequality, concentrates on divisible and monetarizable goods, prefers private over public values, excludes the general public and prevents structural reform (for the USA see Brunet, 1988:

10515ff.; Edwards, 1986: 668ff.; Schoenbrod, 1983: 1452ff.; for Germany see above, p. 155). However, most of these points are clearly not applicable to the type of environmental agreements that have been dealt with here. They are aimed at environmental agreements between state agencies and polluters by which the former foregoes an administrative order, the denial of a permit or even the making of a regulation or administrative rule. Environmental agreements between polluters and local communities (or municipalities representing them) are but a segment of the whole spectrum of environmental contracting. Many of the objections that have been raised against environmental contracting do not specifically address this type of contracting; equally, some of the advantages of environmental agreements do not concern environmental agreements between polluters and local communities.

Within this framework, the primary problem of evaluation is the development of an adequate system of criteria. In my view, the point of reference cannot be an idealized configuration of traditional regulation, but, rather, the configuration that would have existed if, in a realistic perspective, normal administrative procedures had been followed. In other words, it is not appropriate to compare a realistic contractual configuration with an idealized regulatory one. I have the impression that many critics of contractual techniques in environmental decision-making too easily succumb to this mistaken approach, while I would not deny that advocates of negotiation tend to idealize its benefits and its legitimacy.

Irrespective of their content, environmental agreements between polluters and local communities have one main purpose, namely to overcome acceptance conflicts between a state agency, an author of an infrastructure project or an operator of a private facility on the one hand, and local communities that may be represented by municipalities or selected or self-appointed representatives, on the other. These conflicts have various causes: the project may not in reality be in the public interest, but, rather, bring about net welfare losses; the cost/benefit balance for the people most affected may be uneven; there may be "ideological" opposition to the project and so on. From this perspective, the phenomenon of environmental agreements reflects, and responds to, the crisis of traditional participation as a means of securing acceptance of potentially adversely affected parties by procedural entanglement, presentation of their interests and values, supply of information that supports these interests and values, as well as substantive consideration of these interests in the decision-making process. Arguably, negotiation leads to a broader information input, a participation that accompanies the whole process, the removal of asymmetries of influence and co-operative, problem solving behaviour (Susskind, 1985: 101ff.; Holznagel, 1990: 259ff., 265ff.; Hoffmann-Riem, 1989: 5–6, 32–33).

If this assessment is correct, it is perhaps no longer surprising that negotiation and mediation have gained such momentum in a country like the USA where the administrative process is highly adversarial. Even if many observers stress the advantages of adversarial fact-finding, policy development and adjudication, there is no denying that the phenomenon of environmental negotiation reveals considerable dissatisfaction with the functioning of the adversarial process. On the other hand, environmental agreements are much more frequently used in Japan, a country that is characterized by a consensus-oriented political-administrative culture. To a certain extent, environmental agreements in Japan are a substitute for the lack of, or insufficient, participation. The substantive acceptance-securing function seems to be more important. In a society where compliance by a plant operator with the formal norms is not in itself sufficient to be accepted by the local community, because the community has a primarily extra-legal value orientation, environmental agreements have the function of a local "social contract" that makes the operator of the new facility a member of the community or concretises his obligations towards it.

Beyond the more procedural aspect of securing public acceptance, environmental agreements have a substantive content. Environmental agreements lead to a "fine tuning" of law, considering the particular case, the actual parties, the diverging interests and the actual balance of power (Susskind, 1985: 105–106; Holznagel, 1990: 267ff., 270ff., 283–284; Suhr, 1990: 116–117, 120ff.; Hoffmann-Riem 1985: 64ff.). This is reasonable insofar as the law itself has not taken the complex consequences of administrative decisions and the ensuing impairment of interests adequately into account. Environmental agreements couple topics and actors, which although separated by too narrow or rigid formal law, are nonetheless inseparable politically. In other words, they reverse the reduction of complexity achieved by law where it no longer ensures the "correctness" of the administrative decision.

Environmental agreements between polluters and local communities normally provide for some kind of compensation for disadvantages incurred as the result of administrative decisions such as the toughening of standards, preservation or improvement of the environmental status quo for the future, improvement of the environment at some other place, or some other kind of physical or monetary compensation. There are no substantive criteria for the evaluation of the time horizon, the subject and the kind of compensation granted to adversely affected parties. However, as already stated, the point of reference cannot be the ideal situation but, rather, the situation that would have existed if normal administrative procedures had been followed. From this perspective compensation constitutes an element of an adequate distribution of burdens and advantages between the author of a project or the operator of a facility and the group(s) of adversely affected persons. There may be limits to highly complex coupling of different matters. Also, one may have some doubts on the appropriateness of extreme forms of "cross facility" equity, such as compensation in the form of a swimming pool for additional air pollution or abandonment of a planned prison for a waste incineration plant and so on. But these are extravagant cases that do not falsify the general assessment.

Environmental negotiation requires that the rights and interests at stake be "tradeable" or amenable to compromise. Where there is a sharp conflict of values, for example where members of a local community are ideological opponents of a particular technology or where compensation for the disadvantages associated with the project is not possible, negotiations will not be opened at all or are bound to fail (Cormick, 1976: 217; Susskind & Cruikshank, 1987: 16ff.; Amy, 1987: 17ff.; Holznagel, 1990: 106ff.; Suhr, 1990: 118, 128; Treiber, 1990: 269ff.). In particular, where opponents question the need for a particular project in its entirety, or are firmly against a particular site, there is no room for negotiations. This points to important limits of negotiation as a means of conflict resolution. In a radicalized form, one could argue that the less likely are negotiations to take place, the more they are necessary and *vice versa*.

Within this framework, there are incentives and disincentives for environmental contracting (Susskind & Cruikshank, 1987: 117–118; Amy, 1987: 13ff.; Bingham, 1986: 8ff.; Holznagel, 1990: 272ff.). Considerations such as the ability of the negotiating parties to secure judicial review, their access to information, publicity and public pressure are necessary or at least supportive conditions for contractual arrangements because they diminish unequal bargaining positions. A person who negotiates on rights has a better bargaining position than one whose "unprotected" interests are at stake. The power of opponents to temporarily halt a project constitutes an important incentive to negotiate (Treiber, 1990: 272, 277; Schulze-Fielitz, 1990: 81ff.). While this could diminish the role of negotiations in countries such as Japan and Germany where access to administrative courts is limited, the Japanese example shows that the political culture of a country may provide other incentives for negotiation.

Information is also of crucial importance. Where members of local communities or their representatives are not granted equal access to environmental data by the operator or the agency or are not able to compensate for this deficit in some other way, the chances that negotiations are opened at all, or at least that a balanced result is achieved, are small. From the point of view of the plant operator, an important incentive to negotiate is that through environmental negotiations a long lasting relationship between the parties is often created which can be relied on in later conflicts. As a means of confidence-building, environmental agreements may grant members of local communities access to relevant information.

Theoretically contractual techniques involve the danger of the narrowing of the consideration of affected interests (Amy, 1987: 7–8; Holznagel, 1990: 264, 277–278; Treiber, 1990: 273ff.; Schultze-Fielitz, 1990: 79ff.). Since it is not easy to represent a multitude of different, often very diffuse interests in negotiations, management considerations may lead to an exclusion of certain interests. By limiting the number of negotiating parties the whole breadth of adversely affected interests may not be covered and a decision taken in favour of interests that are present, manifest, easy to organize and ready for conflict.

The merits of this argument are not easy to assess. Once again one should not succumb to the temptation of comparing negotiations with an idealized configuration of regulation and adjudication. In a real world comparison, environmental agreements normally lead to an improvement of the situation of adversely affected parties compared to that which the administration and courts could have achieved in the framework of their legal powers. Where adversely affected parties (or associations representing these parties) participate in the negotiations, an unbalanced result is possible at the expense of non-represented interests (such as persons living far away, persons not interested in environmental protection, future generations and so on). However, on the whole the solution will be superior to the one that would exist without negotiations. In the case of manifest discrepancies between the circle of persons that profit from the agreement and those adversely affected, the latter may sustain resistance against the project. Therefore, even the sale of rights of action against a project is not *per se* immoral even if the sellers had tried to support their case by relying on the public interest (Endres, 1977; Holznagel, 1990: 232ff.; see Knothe, 1983: 18ff.). In essence, it constitutes an attribution of a scarce resource to the more profitable utilization. However it should be noted that monetary compensation may endanger the peace-creating function of negotiations because it may expand the conflict by attracting new conflict parties who strictly oppose such outright "commodification".

The dynamics of a negotiation process and the entanglement in the negotiations always pose the risk that participants may seek quick agreement, be cheated by the other party, underestimate the disadvantages of the bargain, or choose lowest common denominator solutions. In other words they may sell themselves too cheap. Since the discretionary powers of the administration can be used either to the advantage or the detriment of environmental interests, the risk of unfavourable results is, in general, not greater than without negotiations, except for the rare situation where because of a high politicization, the project could have been blocked forever.

Finally, widespread recourse to negotiation may mask a need to reform the regulatory system. Seen from a systemic point of view, contractual techniques amount merely to the curing of symptoms. However, one may question whether, in view of structural conservativism of modern social systems, it would be a responsible, and realistic, position to wait for the aggravation of conflicts and an ultimate collapse of regulation. As long as negotiation is limited to exceptional cases, the systemic problem does not arise. Moreover, the Japanese example shows that even if negotiation goes beyond that, a co-existence between central regulation and local and regional negotiation is possible.

NOTES

1 For these and the following figures see Environment Agency, 1983: 324ff.; 1989: 373; 1990: 460–1; OECD, 1994: 102–3. Part of the figures are accumulated from the various annual reports of the Japanese Environment Agency.
2 Weidner *et al.*, 1989: 194ff.; see Leane, 1991: 39ff.; Young, 1984: 932ff. suggesting the greater role of "administrative guidance".
3 For an agreement between a municipality as operator of a heating power plant and a local community see Führ, 1989: 39ff. For the use of mediation in conflicts between authorities and local communities over the cleaning up of abandoned waste dump sites see Striegnitz, 1990a; 1990b.
4 *Frankfurter Allgemeine Zeitung* of 14 December 1991, p. 51; 26 June 1992, p. 26; 17 July 1992, p. 59; 25 July, p. 45.
5 79 *Bundesgerichtshof in Zivilsachen* 131 (1980).

REFERENCES

Ackerman, Bruce & Stewart, Richard (1985) Reforming Environmental Law. *Stanford Law Review*, **37**, 1333–1365.
Amy, Douglas (1987) *The Politics of Environmental Mediation*. New York: Columbia University Press.
Arnold, Peter (1989) Die Arbeit mit öffentlich-rechtlichen Verträgen im Umweltschutz beim Regierungspräsidium Stuttgart. *Verwaltungs-Archiv*, **80**, 125–142.
Barth, Norbert & Führ, Martin (1986) "Der Fall Basta". *Öko-Mitteilungen* (Issue 41) **9**, 10–12.
Bauer, Hartmut (1987) Informales Verwaltungshandeln im öffentlichen Wirtschaftsrecht. *Verwaltungs-Archiv*, **78**, 241–268.
Benz, Arthur (1990) Verhandlungen ohne Vermittlung—Der Fall "Rastatt". *Informationen zum Umweltrecht*, **1**, 45–6.
Bingham, Gail (1986) *Resolving Environmental Disputes*. Washington DC: Conservation Foundation.
Bingham, Gail & Miller, Daniel (1984) Prospects for Resolving Hazardous Waste Siting Disputes Through Negotiation. *Natural Resources Lawyer*, **17**, 473–89.
Bohne, Eberhard (1981) *Der informale Rechtsstaat*. Berlin: Duncker & Humblot.
Bohne, Eberhard (1984) Informales Verwaltungs- und Regierungshandeln als Instrument des Umweltschutzes. *Verwaltungs-Archiv*, **75**, 343–373.
Breuer, Rüdiger (1990) Verhandlungslösungen aus der Sicht des deutschen Umweltschutzrechts. In Hoffmann-Riem, W. & Schmidt-Aßmann, E. (Eds). *Konfliktbewältigung durch Verhandlungen I*. Baden-Baden: Nomos.
Brohm, Winfried (1990) Verwaltungshandeln mit Hilfe von Konfliktmittlern? *Deutsches Verwaltungsblatt*, **105**, 321–8.
Brunet, Edward (1988) The Costs of Environmental Alternative Dispute Resolution. *Environmental Law Reporter*, **18**, 10515–17.
Canter, Bram (1982) Hazardous Waste Disposal and the New State Siting Program. *Natural Resources Lawyer*, **15**, 421–55.
Cormick, Gerald (1976) Mediating Environmental Controversies: Perspectives and First Experience. *Earth Law Journal*, **2**, 215–224.
Duerksen, Christopher (1982) *Environmental Regulation of Industrial Plant Siting. How to Make it Work Better*. Washington DC: Conservation Foundation.
Eberle, Carl-Eugen (1984) Arrangements im Verwaltungsverfahren. *Die Verwaltung*, **17**, 439–64.

Edwards, Harry (1986) Alternative Dispute Resolution: Panacea or Anathema. *Harvard Law Review*, **99**, 668–84.

Endres, Alfred (1977) Der Umwelthandel von Bergkamen. *Umwelt*, **7**, 274–80.

Environment Agency (1983) *Quality of the Environment in Japan*. Government of Japan.

Environment Agency (1989) *Quality of the Environment in Japan 1989*. Government of Japan.

Environment Agency (1990) *Quality of the Environment in Japan 1990*. Government of Japan.

Foljanty-Jost, Gesine (1988) *Kommunale Umweltpolitik in Japan*. Hamburg: Institut für Asien Kunde.

Foljanty-Jost, Gesine (1990) Kooperation statt Konfrontation: Verhandlungen als Mittel der Beilegung von Umweltkonflikten in Japan. *Zeitschrift für angewandte Umweltforschung*, **3**, 36–50.

Führ, Martin (1989) *Sanierung von Industrieanlagen*. Düsseldorf: Werner.

Führ, Martin (1990) Umweltinformationen im Genehmigungsverfahren. In Winter, G. (Ed). *Öffentlichkeit von Umweltinformationen*. Baden-Baden: Nomos.

Gardner, Joan (1990) *Massachusetts Siting Act and Experience To Date*. In Hoffmann-Riem, W. & Schmidt-Aßmann, E. (Eds). *Konfliktbewältigung durch Verhandlungen I*. Baden-Baden: Nomos.

Grad, Frank (1989) Alternative Dispute Resolution in Environmental Law. *Columbia Journal of Environmental Law*, **14**, 157–85.

Gresser, Julian et al. (1981) *Environmental Law in Japan*. Cambridge, Mass., London: MIT Press.

Hoffman-Riem, Wolfgang (1989) *Konfliktmittler in Verwaltungsverhandlungen*. Baden-Baden: Nomos.

Hoffman-Riem, Wolfgang (1990) Interessenausgleich durch Verhandlungslösungen. *Zeitschrift für angewandte Umweltforschung*, **3**, 19–35.

Holznagel, Berd (1986) Negotiation and Mediation: The Newest Approach to Hazardous Waste Facility Siting. *Boston College Environmental Affairs Law Review*, **13**, 329–78.

Holznagel, Berd (1990) *Konfliktlösung durch Verhandlungen*. Baden-Baden: Nomos.

Knödgen, Gabriele (1981) *The Steag Coal-Fired Power Plant at Voerde or Changing German Clean Air Policy*. IIUG-dp 81/11, Wissenschaftszentrum Berlin.

Knothe, Hans-Georg (1983) Die Rücknahme von Widersprüchen gegen die Errichtung von Kohlekraftwerken gegen Entgelt. *Juristische Schulung*, **23**, 18–23.

Kunig, Philipp (1990) Alternativen zum hoheitlich-einseitigen Verwaltungshandeln. In Hoffmann-Riem, W. & Schmidt-Aßmann, E. (Eds). *Konfliktbewältigung durch Verhandlungen I*. Baden-Baden: Nomos.

Kunig, Phillip & Rublack, Susanne (1990) Aushandeln statt Entscheiden? *Jura*, **12**, 1–11.

Leane, Geoffrey (1991) Environmental Contracts—A Lesson in Democracy from the Japanese. *University of British Columbia Law Review*, **25**, 361–85.

Leberman, Jethro & Henry, James (1986) Lessons from the Alternative Dispute Resolution Movement. *University of Chicago Law Review*, **53**, 424–439.

Mayntz, Renate (1979) Regulative Politik in der Krise. In Matthes, J. (Ed.) *Sozialer Wandel in Westeuropa*. Frankfurt, New York: Springer.

Mays, Richard (1988) Alternative Dispute Resolution and Environmental Enforcement. *Environmental Law Reporter*, **18**, 10087–10097.

OECD (1994) *Environmental Performance Reviews: Japan*. Davis. Paris: Organisation for Economic Co-operation and Development.

Rennie, Sandra (1989) Kindling the Environmental ADR Flame: Use of Mediation and

Arbitration in Federal Planning, Permitting, and Enforcement. *Environmental Law Reporter*, **19**, 10477–484.

Rottleuthner, Hubert (1989) The Limits of Law: The Myth of a Regulatory Crisis. *International Journal of the Sociology of Law*, **17**, 273–85.

Sander, Frank (1990) Alternative Methods of Dispute Resolution: The US Perspective. In Hoffmann-Riem, W. & Schmidt-Aßmann, E. (Eds). *Konfliktbewältigung durch Verhandlungen I*. Baden-Baden: Nomos.

Schoenbrod, David (1983) Limits and Dangers of Environmental Mediation: A Review Essay. *New York University Law Review*, **58**, 1453–76.

Schulze-Fielitz, Helmuth (1990) Der Konfliktmittler als verwaltungsverfahrens-rechtliches Problem. In Hoffman-Riem, W. & Schmidt-Aßmann, E. (Eds). *Konfliktbewältigung durch Verhandlungen II*. Baden-Baden: Nomos.

Shibata, Tokue (1989) Pollution Control Agreements. In Tsuru, S. & Weidner, H. (Eds). *Environmental Policy in Japan*. Cambridge, Mass.: MIT Press.

Striegnitz, Meinfried (1990a) Mediation: Lösung von Umweltkonflikten durch Vermittlung. *Zeitschrift für angewandte Umweltforschung*, **3**, 51–62.

Striegnitz, Meinfried (1990b) Konfliktvermittlung bei der Sanierung von Altlasten. *Informationen zum Umweltrecht*, **1**, 43–5.

Suhr, Dieter (1990) Die Bedeutung von Kompensationen und Entscheidungsverknüpfungen. In Hoffmann-Riem, W. & Schmidt-Aßmann, E. (Eds). *Konfliktbewältigung durch Verhandlungen I*. Baden-Baden: Nomos.

Susskind, Lawrence (1985) The Siting Puzzle: Balancing Economic and Environmental Gains and Losses. *Environmental Impact Assessment Review*, **52**, 101.

Susskind, Lawrence & Weinstein, Alan (1980) Towards a Theory of Environmental Dispute Resolution. *Boston College Environmental Affairs Law Review*, **9**, 143–196, 311–357.

Susskind, Lawrence & Cruikshank, Jeffrey (1987) *Breaking the Impasse. Consensual Approaches to Public Disputes*. New York: Basic Books Inc.

Symposium (1991) Alternative Methods of Resolving Environmental Disputes. *Villanova Environmental Law Journal*, **2**, 1.

Teubner, Gunther (1984) Das regulatorische Trilemma. *Quaderni Fiorentini per la storia del pensiero giuridico moderno*, **13**, 109–149.

Teubner, Gunther (1992) Regulatorisches Recht: Chronik eines angekündigten Todes. In Koller, P. *et al.* (Eds). *Theoretische Grundlagen der Rechtspolitik*. Stuttgart: Steiner.

Treiber, Hubert (1983) Regulative Politik in der Krise? In *Kriminalsoziologische Bibliographie*, 28–54.

Treiber, Hubert (1990) Über mittlerunterstützte Verhandlungen bei Standortentscheidungen. In Hoffman-Riem, W. & Schmidt-Aßmann, E. (Eds). *Konfliktlösungen durch Verhandlungen I*. Baden-Baden: Nomos.

Upham, Frank K. (1987) *Law and Social Change in Postwar Japan*. Cambridge: Harvard University Press.

Weidner, Helmut *et al.* (1989) *Darstellung und Wirkungsanalyse der ökonomischen Instrumente der Umweltpolitik in Japan*. Forschungsvorhaben Nr. 101 03 087 im Auftrag des Umweltbundesamts, IFO-Institut für Wirtschaftsforschung.

Winter, Gerd (1985) Bartering Rationality in Regulation. *Law and Society Review*, **19**, 219–50.

Yamamouchi, Kazuo & Otsubo, Kiyoharu (1987) Agreements on Pollution Prevention: Overview and One Example. In Tsuru, S. & Weidner, H. (Eds). *Environmental Policy in Japan*. Berlin: Sigma.

Young, Michael (1984) Judicial Review of Administrative Guidance. *Columbia Law Review*, **84**, 923–83.

CHAPTER 8

"*Verinnerlijking*": The Limits to a Positive Management Approach

Maarten A. Hajer
Munich

INTRODUCTION

Ecological responsibility and self-regulation are not a natural couple. A conventionally perceived obstacle would be that ecological responsibility requires the foregoing of short-term private gains in order to avoid long-term public degradation. As such it would seem to call for state intervention to force private parties in that direction. Yet even if one rejects this somewhat mechanistic, public choice, way of reasoning, one has to accept that the relationship between ecological responsibility and self-regulation is not self-evident, especially in the context of a highly industrialized capitalist society. On the other hand, self-regulation plays an increasingly important role in present day attempts to come to terms with the ecological crisis. How should we make sense of the contingent relationship between ecological responsibility and self-regulation? Is it bound to fail or can it work? And if it might work, under what circumstances can the dynamics of the process be guaranteed?

This chapter seeks to come to an understanding of this contingent relationship through the analysis of the political discourse in which the terms are embedded and in which the recent coupling of the two has emerged. Subsequently, the questions are raised of how we should understand "self-regulation" under this regime and to what extent ecological responsibility can be secured. We then turn to the analysis of one historical situation in which self-regulation has been employed to further ecological responsibility, namely, the Dutch policy of "*verinnerlijking*" (internalization). It was introduced in 1984 and has been the theoretical backbone of Dutch environmental policy ever since. How did *verinnerlijking* work?

It will be argued that the practice of *verinnerlijking* reflects some of the

Environmental Law and Ecological Responsibility: The Concept and Practice of Ecological Self-Organization
Edited by G. Teubner, L. Farmer and D. Murphy
© 1994 John Wiley & Sons Ltd

weaknesses and contradictions that are characteristic of the new approach to environmental policy making that emerged in Western Europe in the 1980s. The dynamics of this approach, which will be called "ecological moderniza-tion" and the specific Dutch practice of *verinnerlijking* are analysed using the concept of "discourse coalition" as a heuristic device. It is argued that the limits to the success of *verinnerlijking* are partly the result of the confines of the discursive order within which ecological responsibility was being defined. The chapter concludes with an exploration of the possibilities of enhancing ecological responsibility and correcting the bias that *verinnerlijking*, in its present form, logically produces. It is argued that effective internalization of environmental concerns could benefit from the institutionalization of an external mechanism that can help to maintain the dynamics of the discursive order in the search for more effective ways of prevention of pollution and waste.

DISCOURSE COALITIONS

A discourse analysis of environmental politics starts from the presumption that the present concern over environmental change cannot be seen as the "natural" response to physical phenomena. There are many different possible under-standings of "what is the case". The question is why a certain perception of a problem and certain ideas about what constitutes a solution become dominant at a given moment in time.

One way of analysing the process of reification, in which certain under-standings of the world are played down in favour of others, is through the analysis of discourse coalitions in environmental politics (more elaborately see Hajer, 1993 and Haajer, 1995 (forthcoming)). A "discourse" is here defined as an ensemble of ideas, concepts, constructs and categories through which meaning is given to phenomena. Discourse "frames" certain problems by dis-tinguishing some aspects of the situation rather than others. Following this definition we can see law, or physics for that matter, as a discourse. Likewise the arguments of the environmental movement can be analysed as a discourse.

One of the essential characteristics of the political process is the diversity of actors. The participants in environmental politics, for instance, range from various sorts of scientists to policy makers, from local citizens to industrialists, and from all sorts of technical experts to politicians. They all have to find ways to communicate. Indeed, one of the key characteristics of environmental politics is exactly that of finding ways of communicating knowledge from one realm (say tree physiology) to another (say a local community). It is quite obvious that this translation exercise or search for common denominators is accompanied by a loss of meaning. I would even go one step further and argue that coalition formation in politics depends on this loss of meaning and the multi-interpretability of texts. This is why metaphors, analogies, historical examples, clichés, or appeals to collective fears or senses of guilt, can fulfil such a pivotal role in the framing and reframing of problems. These shallow and ambiguous discursive elements are the cement that unites actors with different, or at best overlapping, perceptions and understandings.

Mostly, these metaphors or historical examples do not stand by themselves. Characteristically they feature in the context of what I call "storylines": narratives that combine elements from many different realms into a more or less coherent whole in which actors can recognize a problem and, sometimes, even a perspective for action. The power of a storyline partly stems from the fact that it manipulates many different variables to impose a specific social construct of reality: for example, the perception of the problem, the failure of past regulatory practices, the historical precedents, the potential results of inaction. Yet they are also a way of linking the developments in a given domain (say environmental politics) to wider social commitments. Storylines help to demarcate the boundaries of discursive space in which a problem can be meaningfully discussed and to determine the clustering of knowledge. For instance, in present day environmental politics the concept of "sustainable development" fulfils the role of storyline. It is consciously produced to generate widespread political support but is not necessarily internally consistent (see below). Yet a problem like acid rain also qualifies as a storyline. It imposes somewhat arbitrary boundaries within which a problem like air pollution has to be discussed and clusters knowledge and expertise at the expense of other, often related, problems. The analysis of storylines is of central importance since they are the vehicles through which common sense understandings of environmental problems are created and through which actors exercise power. They impose their view of reality on others, discrediting some institutional arrangements whilst promoting others.

This discursive struggle is not simply a matter of a free exchange of ideas between individuals. The interaction among individual acors takes place in the context of all kinds of institutional practices that are themselves the reified product of past discourses. It is, therefore, paramount to combine the analysis of the discursive production of reality with the analysis of (extra-discursive) social practices from which reality constructs emerge. This is why I suggest using the concept of "discourse coalition".

A "discourse coalition" then can basically be understood as the ensemble of

(a) a set of storylines,
(b) the actors that utter these storylines, and
(c) the social practices that help sustain these storylines.

Discourse coalitions allocate meaning to the vague and ambiguous social world and take care of the structuration of experience through the creation of new practices.

The reconstruction of the complicated discursive relationships in environmental discourse is especially significant because it can clarify how specific terms are to be understood, or, to complicate things further, why it is that different actors sometimes understand the same term differently.

For the purpose of this chapter two further terms should be introduced. If a given discourse is generally used to make sense of the world we speak of "discourse structuration". If this understanding actually leads to a reorganization of the institutional practices we speak of "discourse institutionalization". If these two conditions are met, a discourse coalition can be said to be dominant.

Finally, the fact that many different actors contribute to the reproduction of a discourse coalition also implies that there is a more or less permanent debate on meaning and direction. This can partly be accounted for by the fact that the complexity of the issue results in most actors having only a partial understanding of the whole problem but partly this is also the straightforward result of the political importance of the struggle over competing meanings. In other words, although we can distinguish changing discourse coalitions, there is never a moment at which the struggle over meaning is complete. Empirical research will always have to illuminate the exact character of a given discourse coalition.

ECOLOGICAL MODERNIZATION

One of the most striking events of the 1980s has been the emergence of a widespread recognition of the "ecological crisis" that culminated in the global endorsement of the UN report, *Our Common Future* (WCED, 1987). Elsewhere I have argued that this signified the structuration of a new discourse in environmental policy making in Western Europe that could be termed "ecological modernization" (Hajer, 1995). My argument was that "sustainable development" (the central idea of the UN report, *Our Common Future*) should not be interpreted as a coherent theory of development but is better seen as one of the most powerful storylines of the coalition that developed around ecological modernization (Harborth, 1989; de la Court, 1990; Redclift, 1987). Today this discourse coalition dominates the allocation of meaning in environmental matters and also determines the boundaries of discursive space. If we want to be able to grasp the dynamics and effectiveness of ecological responsibility through self-regulation we first need to have a basic understanding of this discourse coalition since it determines the political meaning of terms like ecological responsibility and self-regulation.

The term "ecological modernization" was first introduced by a group of German political scientists including Joseph Huber and Martin Jänicke. They used it to refer to the attempt to resolve environmental problems using technical means (Jänicke, 1988). Upon reflection, however, this orientation towards technology seems just an element of a much wider paradigmatic change. Ecological modernization, therefore, refers to the political response to the environmental dilemma that starts from the assumption that economic growth and the resolution of ecological problems can be reconciled. Although some supporters might start from moral premises, ecological modernization basically follows a utilitarian logic: at its core is the conviction that pollution prevention pays. This shift to ecological modernization can be observed in at least five different realms.

The primary sphere was that of environmental policy-making. Here market-oriented instruments replaced the judicial and administrative structures of policy-making that had been institutionalized in the late 1960s and early 1970s following the recognition of the need to develop a coherent environmental policy. The state was supposed to control the quality of the separate "compartments", air, water and soil, by granting individual pollution permits. The growing complexity of the system it was supposed to control paved the way for the introduction of more technocratic administrative principles. However, during the 1980s, environmental policy discourse came to be organized around a wholly new set of principles that accounted for the costs of pollution. The last two decades saw the introduction of, more or less in order of appearance, the "polluter pays" principle, the call for an integrated approach to pollution abatement cost-benefit analysis, risk analysis, the bubble concept, the precautionary principle, critical loads, the levy of charges on polluting activities and tradeable pollution rights.

The discourse of ecological modernization also implied a new role for science in environmental policy-making. Here it suffices to say that since ecological modernization started from the acknowledgement that the environment was under severe stress, policy-related science no longer had as its prime task the proof of the existence of damaging effects, but increasingly was given the task of helping to determine the levels of pollution which nature could endure. The policy concept of "critical load", for instance, depended on scientific advice. Closely related to this was the widespread acceptance of the "multiple stress" hypothesis which entailed a shift away from a reductionist epistemology and ontology towards an orientation that took systemic ecological ideas as its starting point.

On the micro-economic level of the firm the discourse of ecological modernization surfaced in the concept of "pollution prevention pays". This idea has long intellectual roots and was taken up following the resource crisis of the early 1970s and led to a number of influential publications (Royston, 1979; Huisingh & Bailey, 1982; Huisingh *et al.*, 1986). It was pioneered in the 1970s by American firms like 3M but really took hold in managerial practice from the mid 1980s onwards.

On the macro-economic level, ecological modernization regarded nature as a public good or resource in contrast to the previous idea of nature as basically a free good which could be used as a "sink". In this respect ecological modernization sought to put an end to the externalization of economic costs to the environment.

Related to these changes, the legislative discourse in environmental politics also changed character. Given the fact that nature could no longer be regarded as a sink and that firms were now supposed to prevent pollution, the strict proof of causality by the damaged party made way for the idea that the burden of proof should be the concern of the suspect of pollution, not the damaged or prosecuting party. Here statistical probability becomes the basis for collective and unlimited liability (see Hofstetter, Chapter 5; Brüggemeier, Chapter 4).

Yet the discourse also drew new boundaries to demarcate the discursive space in which the ecological crisis was to be understood. Here we encounter an essential paradox of convergence. On the one hand ecological moderniza-tion broadened the discursive space, linking environmental policy to economic development. On the other, though, it limited the range of terms in which the ecological crisis could credibly be discussed. Ecological modernization was essentially an efficiency-oriented approach to environmental policy, that sought to remedy degradation through techno-scientific development and technocratic practices. It was committed to market-oriented strategies and accepted the parameters of the capitalist system. As such it ruled out, or at least discredited, the claims of the environmental movement of the 1970s which argued that the ecological crisis might be immanent in the very practices of capitalism (e.g. Enzensberger, 1973), or that saw the core of the problem in the techno-industrial complex and argued for a withdrawal from the prevailing social practices (e.g. *The Ecologist*, 1972; or Schumacher, 1973).

All this supports the contention that the strength of the discourse of eco-logical modernization might also be its principle weakness. Storylines like sus-tainable development suggest that environmentally sound policies are a "positive sum game". This obviously facilitates general endorsement, but re-stricts the degree to which the subscribing actors are committed to radical change. If the ecological problems would be such that more than piecemeal readjustment were required, the discourse would run into difficulty.

The intellectual and social roots of ecological modernization are complex and a full historical account would require a detailed discussion of certain developments in civil society, science and technocracy (cf. Hajer, 1995). Never-theless, it is quite clear that the Organization for Economic Co-operation and Development (OECD) played a key role in the process of discourse structura-tion, that is, in defining and disseminating many of the key concepts and ideas around which the new discourse coalition could be formed. It should also be obvious that ecological modernization was in many respects a timely answer to a debate that was increasingly relating environmental degradation to the overall features of industrial capitalist society. In that respect the discourse of ecological modernization really changed the emphasis of environmental politics. Indeed, it can be argued that many of the solutions to the ecological crisis that were put forward in the name of ecological modernization had, in the decade before, been considered as part of the problem: technocratic management, technoscientific solutions and indeed, economic growth.

This critical analysis should not be taken as a refutation of the general legitimacy, and indeed importance, of ecological modernization as a political response. The discourse of ecological modernization can potentially resolve many of the issues that are perceived to be most pressing. Furthermore, the idea that pollution prevention should be an element of the general conceptualization of ideas had always been a demand of environmental non-governmental organizations (NGOs). Yet with the broadening of both the scope of the debate and the number of actors involved, the environmentalists also lost control over the definition of what the environmental issue really was[1]. For instance, self-organization and decentralization had been central to the "small is beautiful" ideology of the environmental movement, yet in the discourse coalition of ecological modernization these concepts lost their radical edge. Self-organization and decentralization of decision making could now apply to the restructuring of the relationship between the state and a multinational firm such as Shell.

Ecological modernization redefined the boundaries of the discursive space of environmental politics by emphasizing the relationship of environmental care and economic policy. Yet here, the central role of the OECD already indicates the discursive order in which the new strategy should be understood. The OECD provided useful instruments and concepts for environmental policy but embedded these in its preconceived ideas about neo-liberal economic policy and limited state intervention. In other words, it is important to analyse the changing meaning of the terms of environmental discourse by relating them to the context in which they emerge. In the case study below I purport to do just that. My focus will be on the acid rain controversy which was a central part of the environmental debate of the 1980s.

DUTCH ENVIRONMENTAL POLICY IN THE 1980s

In November 1982 the centre-right Cabinet led by Ruud Lubbers came into office. Its top priority was the revitalization of the Dutch economy. Inspired by neo-liberal economic ideas, it aimed to roll back the state, reduce government spending and "deregulate" government policy by simplifying legal rules and procedures, reducing the number of advisory councils, and cutting the permit requirements for business and industry. The new minister for the environment, Dr. Pieter Winsemius, until then a senior consultant at McKinsey and Company[2], was chosen primarily because of his work on deregulation for the employers federation, VNO.

The installation of the Lubbers government coincided with the recognition of the inadequacy of the traditional, predominantly legislative, style of policymaking in the environmental domain. 1982 was the year in which acid rain, the first of a new generation of environmental problems, such as the greenhouse effect and the diminishing ozone layer, became a priority on the governmental agenda. The storyline around the phenomenon of acid pre-cipitation presented acid rain as anomalous to the traditional style of policy-making.

The first result was the *IMP-Milieubeheer 1985–1989*, a White Paper out-lining the environmental programme, published in September 1984[3]. This marked the start of a new phase of environmental policy. Pollution was no longer defined as the inevitable, negative side-effect of production. A clean environment was seen as a precondition for further economic growth. It signi-fied the structuration of the discourse of ecological modernization:

> The challenge confronting environmental policy in 1984 has ... increased rather than decreased. While the first generation of environmental issues has not yet been solved, a new more complex generation has sprung up ... It is important that the environment and the economy not be placed in opposition to one another ... Improving the environment and strengthening a lasting economic development are closely connected and consistent policy objectives. Provided they are managed correctly, environment and economy can be mutually reinforcing[4].

The White Paper also indicates the struggle to set new boundaries to, and to dominate, discursive space. On the one hand the *IMP* gave legitimacy to the call for attention that could be heard from radical environmentalists but, on the other hand, it also immediately distanced itself from these NGOs. The government explicitly followed the ecological modernization storyline promoted by the OECD[5]. While the NGOs blamed the market and called for radical state intervention, Winsemius introduced a market-oriented theory of implementation. Where environmentalists thought in terms of "polluters" and "polluted", Winsemius introduced the term "partnership" into environmental discourse. Furthermore, it was announced that the government was going to put greater stress on economic incentives in the control of pollution than had been done by the traditional order of prohibition. The rationale of the approach was to try to change detrimental social practices by aiming to bring about a mental change within society and further self-regulational practices. It became known as *verinnerlijking* (internalization) after one of its core ideas.

VERINNERLIJKING: THE NEW PARTNERSHIP

The new policy strategy put a great emphasis on integration, effective and effi-ciency oriented priority setting and enforcement. All in all, the new approach emphasized the importance of a new "partnership" both with other authorities (internationally as well as sub-nationally) and with the so-called "target groups" (waste disposal facilities, chemical industry, utility industry, bio-industry, households, refineries, traffic). As part of this target group philos-ophy, the traditional legislative system of granting individual permits was to be replaced by a system of general rules that would apply to specific categories of industrial practice. Hence *verinnerlijking* sought principally to enhance and organize ecological responsibility at the level of target groups.

In this respect one is reminded that the emphasis on partnership was an expression of the philosophy that environment and economic development were not mutually exclusive, provided that the process was well-managed. The old regulatory regime was now labelled as "the hierarchical system of command and control" while Winsemius described environmental renewal (*Milieuvernieuwing*), as the common task for business, households and government to safeguard the physical conditions for a good environment. Self-regulation played a key role in this.

The turn to a more integrated approach to environmental management had, of course, been in the air in the early 1980s[6], yet it is equally true to say that Winsemius added a particular flavour to the Dutch variation on the theme of ecological modernization. Winsemius had strong preconceived ideas about what was to be done. He was a firm believer in "positive management theory" and applied his own management philosophy to the process of reorganization. He saw his job as the changing of cognitive constructions, and in this context drew on the ideas of ecological modernization.

Within his department the struggle against pollution was perceived as an almost hopeless affair. Behind this belief was the conviction that the environment was a zero sum game. The lower ranks of the department and inspectorates were governed by a great mistrust in the willingness of industry to co-operate. Industry had to be forced to comply. However, since this coercive approach to the control of industrial practices and the containing of industrial pollution had not been particularly successful, Winsemius found a department that was literally overwhelmed by fatalism. The ever growing scale of the task ahead and the limited possibilities for change had given rise to what had become known as the "Law of Continued Gloom" (*Wet van behoud van ellende*). Winsemius, however, demanded positive thinking. He believed in the organizational power of shared values and a strong common culture to which end he introduced books like *In Search of Excellence* (Peters & Waterman, 1982) to the department. This, more than anything else, was the driving force of the *verinnerlijking* approach. The central idea of Peters and Waterman is that it is not the discovery of something that is decisive for innovation, but the implementation. This assumption called for important changes, both in management culture and in the structures of environmental policy making.

One of the key ideas of *verinnerlijking* was to broaden the scope of government (see Winsemius, 1986: 63). The Department of Public Health, Physical Planning and the Environment (VROM) should live according to the lessons of Peters and Waterman and change its focus from standard setting to goal achievement and enforcement in society. Here VROM drew on the image of a "chain" of regulatory stages (Winsemius, 1986: 79ff.). Whereas until then the emphasis had been on the first three stages of passing legislation, norm-setting and the granting of permits, Peters and Waterman emphasized the importance of the final two: implementation and enforcement.

Consequently, the Department of VROM aimed to create a much broader "political project" through the construction of a society-wide sense of common purpose. One of the central assumptions of the positive management approach of *verinnerlijking* was that effective environmental policy making required motivated "subjects". VROM in essence tried to create a new discourse coalition. It introduced new storylines on environmental damage, our shared ecological responsibility and on the opportunities for business. Together these had to create a shared view on the reality of environmental politics.

Drawing on the carefully fostered idea that the environment was on the verge of a breakdown, and the prediction that the management of the environmental crisis might, within a couple of years, have to become an almost military operation, Winsemius did not declare a "war on environmental decline" but suggested a much more moderate regulatory regime. Environmental degradation was a serious issue but was still a problem that could be overcome[7]. If one shared in this created image of damage, then most certainly *verinnerlijking* seemed a fair deal.

On the other hand, *verinnerlijking* also was appealing for the Department of VROM. After all, it was hardly the kind of deregulatory move that the Lubbers government had initially announced. If anything, it was much more of a re-regulation with the prospect of a scaling up of environmental policy making from a sector-oriented policy to a central policy of government (see Hanf, 1989). However, Winsemius also argued that VROM had to learn to take the interests of industry into account thereby acknowledging his support and commitment to the OECD proposals that formed the backbone of ecological modernization.

VERINNERLIJKING BETWEEN FEASIBILITY AND ECOLOGICAL NEED

It is easy to see that the positive management of *verinnerlijking* reflected the positive sum game format of the discourse of ecological modernization. Like sustainable development, *verinnerlijking* seemed to lack teeth and claws, and likewise its strength might also be its principal weakness. The fact that meeting the twin goals of economic recovery and ecological sustainability might be problematic was not seriously discussed.

This problem was in fact structural. The introduction of "target groups" was basically the invention of a new level of intermediary actors that had to internalize ecological values. If this had already happened, sectors had to organize themselves and, at that new level, ecological business should be done. The organization of business collectivities would help to solve the problem of disciplining individual industrial projects. But up to what level was this going to work?

Ecological modernization held that ecological responsibility was to be a matter of mutual interest. In other words, as long as pollution prevention paid, no problems were to be expected. Yet if this story line were to lose its heuristic power, or if more stringent measures were required, it would be difficult to see how this newly created order could impose its will on the behaviour of individual enterprises. The target groups were committed to a morally right course of action, but on what basis was the support of individual firms to be obtained?

The key to any understanding of *verinnerlijking* is the insight that its success depended on co-operation and consensus. This obviously imposed limits on what counted as an acceptable solution. Discussing the meaning of the concepts, it came out that for Winsemius, along with the creation of a shared responsibility and the introduction of self-regulation, came the duty to take individual enterprises seriously. *Verinnerlijking* thus became a continuous search for a responsible trade-off between business interests and the public good, between feasibility and ecological responsibility. However, if radical proposals were required to halt physical degradation, this would put the consensus in jeopardy and would make *verinnerlijking* counterproductive. In this sense the social limits to problem closure within the Dutch variation on ecological modernization were contained in the very idea of *verinnerlijking*. Winsemius' positive management philosophy came with implicit limits to action. He was convinced that "nobody should be asked the impossible", meaning that you could raise the targets of your policy but you should beware of a "target overload".

This raises the question of who is to determine when there is a target overload or, in other words, what is feasible. Asked to describe the process in which feasibility of measures for large industrial concerns was determined, key actors invariably suggested that it was the outcome of an intricate process of negotiation. In actual fact, of course, negotiation was not a practice that was introduced under the new partnership of *verinnerlijking* but was common practice before ecological modernization restructured environmental discourse, especially where large firms were involved[8]. Furthermore, the regular procedure of granting permits was primarily a social negotiation during which well informed experts from industry and government tried to reach an agreement on what action should be taken. Yet far less knowledgeable actors such as politicians, general managers or relative outsiders, such as campaigners from environmental NGOs, also participated in this debate.

The structural problem with these discussions was threefold. First, the governmental experts would always be highly dependent on the explanations of the industrial process given by industry's experts[9]. The modern refinery process, for instance, is simply far too complex for outsiders to understand. Second, the market situation of industries is often just as opaque as the technicalities of the plant. Third, the government itself was split since the Ministry of Economic Affairs generally saw its task as protecting industry against irresponsible claims from other government departments. Feasibility, it is easy to see, is an eternally contested notion whose definition depends on the persuasiveness of actors, in a context of structural asymmetry of knowledge.

There are, however, further problems with the relationship between feasibility and ecology. Both ecological modernization and *verinnerlijking* sought to bring about social change by acquiring support through the formulation of their projects in the prevailing discourse of efficiency and money. Pollution prevention would pay. Yet this also made the whole approach vulnerable to other lines of economic reasoning. In management terms, for instance, efficiency could be said to be subordinate to continuity. This seriously restricted the definition of feasibility. As Winsemius argued,

> Of course, you can order an actor to change his practices all at once. But is that fair? Would you not break his competitiveness? With economic feasibility you have to stay within the boundaries of what a society can take, and what the industries in that society can take[10].

Obviously, the state of the environment can be such that it requires tougher policies than industries are willing to take. That is precisely where Winsemius thought *verinnerlijking* touched upon its limits.

For industry, continuity also had a clear claim to priority over sustainability. Indeed, sustainability was described in such as a way as to be logically subordinate: if continuity is at risk, who can pay for the necessary investments? A director of a major industry argued:

> We aspire to curb all harmful emission on a reasonable timescale. We want to have a leading position in this ... [yet] I have been arguing for years to the gentlemen from government: "listen, continuity is what it is all about". Because if a firm loses its continuity, it no longer has the funds to pay the abatement installations. ... Charges, for instance, will always have to be of such nature that a firm does not go broke or become so limited in its profitability that it can no longer pay for the environmental improvements. [Question: "Is this an internal contradiction?"] It is a contradiction but only in time, not in the end result[11].

Interestingly, despite the general commitments to a sustainable solution for the acid rain problem, Dutch environmental discourse hardly ever featured a direct challenge to the argument that sustainability should not be allowed to interfere with continuity. Radical actors mainly challenged the *definition* of feasibility ("they can do much more"), rather than the implicit hierarchical order of continuity and sustainability. Once continuity was accepted as a limit to ecological responsibility, VROM could not do much better given the asymmetrical distribution of relevant knowledge.

CONCLUDING REMARKS: PUTTING SELF-REGULATION TO WORK

One of the *dicta* of Winsemius' theory of positive management was that "a losing side does not have supporters". It was a philosophy that called for feasible goal-setting. In the acid rain controversy this caused problems. The moral outcry (that was partly the result of the public awareness campaigns of the government) called for clear and uncompromising goal-setting. The high visibility of the process of goal-setting forced the government to commit itself to the prevention of the worst damage (i.e. forest dieback). As a result, emissions of harmful substances had to be reduced in the order of 70 to 90%. Yet these goals could not be achieved within the discursive order of *verinnerlijking* The conclusion can only be that it failed to put a stop to the acidification of the environment. The gains that were made through the channels of self-regulation were more than offset by the losses that resulted from the sheer growth in production volumes (especially in traffic and the agricultural industry). In that sense, the acid rain problem highlights the limits of a system of self-regulation relying on discursive principles alone.

The question is, however, whether this approach could have been more productive. It is interesting that whenever new measures were proposed the government failed to present the measures that matched the goals it had set itself, always referring to the need to take account of economic interests. Here the cliché of "international competitiveness" played a central role. Winsemius himself used the competitiveness argument several times in Parliament:

> If measures are taken regarding the refineries this implies an extra burden for these industries that already operate at 60% or 70% of their capacity. ... That extra burden can tip the scales. That cannot be the intention[12].

And in December 1984:

> We have simply thought it expedient ... to mitigate the measures. After all, there is a chance that you chase them away[13].

Yet phrases like "international position of trade" or "international competitiveness" were completely devoid of meaning. Winsemius himself admitted that one really cannot tell to what extent the international position of trade will be affected by certain policy packages saying that, "it is a mixture of analysis and beliefs"[14]. Later he argued that one should not make too much fuss—after all, the Germans, the great competitors of the Dutch refineries, had to cope with at least the same policy package. Clearly, in the 1980s the neoliberal ideology set limits to the meaning of ecological responsibility.

What is even more revealing is the fact that the industries themselves thought that the argument of "international competitiveness" did not hit the core of the problem. According to interviewees, sustainability is much more a matter of long-term investments. A director argued:

> If you want to be sustainable, you have to get away from end-of-pipe technologies. But then you must allow industry five years to develop this technology. You also have to be clear what happens if that period is over and there is no solution. Is that the end or do you get another year?[15]

Yet this was not the way in which the debate on acidification was conducted.

Ironically, in the end it was NGOs like *Natuur & Milieu* and *Milieudefensie* that best understood the lessons of Peters and Waterman. Ecological modernization is not a matter of developing ideas, it is the translation of the discursive principles into concrete policy measures that really matters. Under *verinnerlijking* industry had unjustly got off scot-free and had no real incentive to change their practices. NGOs, therefore, persisted in their campaign that labelled Shell as the biggest acidifier.

Perhaps there is a lesson that can be learned from this. In order to further ecological responsibility through voluntary practices one needs a high overall level of visibility. Bipartite negotiations between state and target groups are not a good basis for this. The dependency of the state that is built into the negotiation process hinders the development of ecologically sound policies at firm level. The environmental officer suggested by Blecher seems to be even more tied to the positive discursive regime (Chapter 11). What seems to be required is the development of a counter-balancing force external to the state, that would for instance capitalize on the critical potential of environmental NGOs. A possible practice that could facilitate this would be the institutionalization of a system of environmental performance review (EPR) (see also Lykke (Ed), 1992). Obviously, the effectivity of this practice would depend heavily on its specific operationalization, nonetheless a combination of the following elements could make EPR into a challenger of the boundaries of conventional politics.

First, a compulsory accounting of certain environmental effects by firms, most fundamentally energy usage, material throughput and recycling effects (for instance expressed in joules). These non-monetized indicators could help to contextualize strict economic performance (mobilize joules against the Dollar).

Secondly, EPR should be run and monitored by an independent agency—after all, the state should also be monitored. Here the dilemma is whether to make EPR a part of an existing authoritative agency or to institutionalize a totally new organization which will have to fight for its own credibility. In any case, one of the key determinants of failure or success would be the possibility of attracting broad-minded top experts. This implies good salaries and, in particular, a commitment on the part of politicians to give any EPR institute real authority.

Third, this EPR agency could stage public meetings on what it, or society, sees as key issues in environmental politics. A topic like acid rain, for instance, which has been high on the agenda, is now fully out of sight. Nevertheless, periodic reports suggest that the problem has not been resolved at all. This leads to a loss of credibility for the *verinnerlijking* approach and does not stimulate the target groups to maximize their efforts in the search for solutions. Dynamism might be brought back into the policy process by reintroducing the issue and evaluating performance. This could take the form of a periodical inquiry such as public meetings in which different "parties" could present their case, question the arguments presented by specific firms, expose the ideology behind the technicalities and show what other courses of action might be taken. Equally, it might also help firms to explain their dilemmas. If the choice would really be laying off workers or caring for the environment, public debate might take a different turn than the care for environmental quality *per se*.

The usefulness of EPR can be illustrated with a small example from the acid rain controversy in the Netherlands. At one point in the controversy two academics published a research paper that suggested that the refineries could fairly easily reduce their SO_2 emissions (Pulles & Wiersma, 1986), but the report failed to attract public attention. Although it was picked up by some MPs, the refineries never had to speak out on it. Environmental performance reviews, organized by an independent EPR agency and led by knowledgeable and independent experts, could provide the stage for such confrontations.

It is interesting in this respect that in the case of the Dutch acid rain controversy every informed actor more or less agreed that the maximum effort had not been made in the 1980s. The problem with *verinnerlijking* was that it relied totally on something as vague as the "social responsibility of enterprises". The meaning of this notion is, by nature, intersubjectively defined. Thus, if, for instance, a firm argues that it will not reduce its emissions appealing to this social responsibility (perhaps because it would put its continuity in jeopardy and result in the laying off of employees) it is up to society to determine which it thinks more important. Social responsibility, after all, takes social priorities as its starting point. Environmental performance reviews could play a useful role in this respect. They could also further the social understanding of what the environmental problem "really" is about: it is not the physical effects but the progressive social reorganization of society.

In this chapter I have consciously restricted myself to the discussion of the discourse coalition and the discursive order in Dutch environmental politics in the 1980s. It is easy to see, however, that the current discussion on the institutionalization of the discourse of ecological modernization in more material practices, such as the levy of charges on energy, will run into the same barriers that we have seen here. Obviously, charges are a practice that can be seen as a system-internal corrective, or a language that is understood by industry. As such it would be a more appropriate play on the autopoietics of enterprises than the moral call to ecological responsibility that lay at the heart of the 1980s conceptualization of *verinnerlijking*. The European Community has recently spoken out in favour of the introduction of a charge on energy thus accepting the need for action. Yet it has made the introduction dependent on the participation of the USA and Japan thereby postponing it for years. The introduction of a unilateral levy of charges is still under discussion in the Netherlands but most likely will meet a similar fate. The arguments that are employed to postpone the levy of charges are precisely the same clichés that prevented the introduction of an offensive against acid rain in the 1980s. Ironically, this resistance has been criticized by some former leading figures from the Dutch business community, one of them an ex-secretary general of the OECD. They maintain that ecological modernization is indeed a positive sum game and that the institutional response to the levy of charges will, if anything, rather increase the competitive advantage *vis à vis* its foreign competitors. Still, the present discussion illustrates once more that many firms that stubbornly resist change are now, nevertheless, happily sailing under the positive and innovative banner of ecological modernization.

Ecological modernization, if it is to maximize its effectiveness, needs to develop its own critical mechanisms to preserve its own systemic dynamics. EPR could play a role here. Recently, Mrs. Brundtland, the Norwegian Prime Minister, reminded a group of representatives from Dutch trade and industry that sustainable development would only work if business itself made it work. The more realistic judgment seems to be that, if left alone, nothing will come from the business community. Instead the consensus around the notion of sustainable development should constantly be renegotiated and for this purpose one will need some system-external mechanism, such as EPR, that can maintain a critical role. This would make it a lot more difficult for the principal architect of the positive management approach to the environment, Winsemius, to admit that though, "the criticism that we are moving too slowly is quite right. Still we cannot speed up"[16].

NOTES

1   Elsewhere I have called this the "discursive paradox" of the new environmentalism. Hajer, 1990: 307.
2   In the Netherlands members of Cabinet can be drawn both from the ranks of parliament and society.
3   *Indicatief Meerjaren Programma Milieubeheer 1985–1989*. This *IMP* prepared the ground for the National Environmental Policy Plan (1989).
4   *Environmental Program of the Netherlands 1985–1989*. Ministry of VROM: Leidschedam, p. 12 (this is the official English version of the *IMP Environmental Management 1985–1989*).
5   In this respect the *IMP-milieubeheer* often referred to the ideas of the OECD Conference on Environment and Economy of June 1984.
6   It is equally obvious that his "target group" philosophy could thrive on the existing

accommodative tradition in Dutch policy-making. Winsemius was not a radical going against the prevailing social bias. In many respects the changes reinforced the historical bias towards co-operation among leaders of politics, government and business. It is also obvious that *verinnerlijking* might thereby also reproduce the inherent limits of the traditional accommodative regulatory regime.

7 This was another case where the influence of Winsemius' positive management philosophy made itself felt. One of his favourite *dicta* was "A losing side does not have supporters. Always make sure people get the idea they are joining a winning team".

8 For instance, during the 1970s the state agreed on more lenient environmental requirements in individual arrangements (*convenanten*) with big firms like Hoechst, Shell Chemical Industries and Mobil Oil.

9 This is very similar to Hawkins' (1984) and Vogel's (1986) findings about the British politics of "best practicable means".

10 Dr. P. Winsemius, interview conducted in the context of my doctoral research.

11 Interview conducted in the context of my doctoral research.

12 *Handelingen Tweede Kamer*, 1983–1984, UCV 95, 14 May 1984, p. 25.

13 *Handelingen Tweede Kamer*, 1984–1985, UCV 32, 10 December 1984, p. 57.

14 Dr. P. Winsemius, in *De Stem*, 26 October 1984.

15 Interview conducted in the context of my doctoral research.

16 Minister Winsemius on the issue of acidification, in *De Stem*, 26 October 1985.

REFERENCES

Court, Thijs de la (1990) *Beyond Brundtland—Green Development in the 1990s*. New York: New Horizons Press.

The Ecologist (1972) *A Blueprint for Survival*. Harmondsworth: Penguin.

Enzensberger, Hans Magnus (1973) Zur Kritik der politischen Okologie. *Kursbuch*, 33 (October 1973).

Hajer, Maarten A. (1990) The Discursive Paradox of the New Environmentalism. *Industrial Crisis Quarterly*, **4**, p. 307.

Hajer, Maarten A. (1992) The Politics of Environmental Performance Review: Choices in Design. In Lykke, Erik (Ed). *Achieving Environmental Goals—the Concept and Practice of Environmental Performance Review*. London: Belhaven Press, pp. 25–42.

Hajer, Maarten A. (1993) Discourse-coalitions and the Institutionalisation of Practice. In Fischer, F. & Forester, J. (Eds). *The Argumentative Turn in Policy and Planning*. Durham: Duke University Press, pp. 43–76.

Hajer, Maarten A. (1995) The Politics of Environmental Discourse. Oxford: Oxford University Press. In press.

Hanf, Kenneth (1989) Deregulation as Regulatory Reform: the Case of Environmental Policy in the Netherlands. *European Journal of Political Research*, **17**, 193–207.

Harborth, Hans-Jürgen (1989) *Dauerhafte Entwicklung—zur Entstehung eines neuen ökologischen Konzepts*. Berlin: WZB.

Hawkins, Keith (1984) *Environment and Enforcement: Regulation and the Social Definition of Pollution*. Oxford: Clarendon Press.

Huisingh, Donald & Bailey, V. (Eds) (1982) *Making Pollution Prevention Pay—Ecology with Economy as Policy*. Elmsford (NY).

Huisingh, Donald *et al.* (1986) *Proven Profits from Pollution Prevention—Case Studies in Resource Conservation and Waste Reduction*. Washington DC: Institute for Local Self-Reliance.

Indicatief Meerjaren Programma Milieubeheer 1985–1989, Tweede Kamer 1984–1985, 18 602, no. 1–2.

Jänicke, Martin (1988) Ökologische Modernisierung. Optionen und Restriktionen präventiver Umweltpolitik. In Simonis, Udo E. (Ed). *Präventive Umweltpolitik.* Frankfurt/M.: Campus, pp. 13–26.

Lykke, Erik (Ed) (1992) *Achieving Environmental Goals—the Concept and Practice of Environmental Performance Review.* London: Belhaven Press.

Peters, Thomas J. & Waterman, Robert H. (1982) *In Search of Excellence—Lessons from America's Best Run Companies.* New York: Harper & Row.

Pulles, M.P.J. & Wiersma, D. (1986) Bestrijdingsmogelijkheden SO_2-emissie door raffinaderijen. *Lucht & Omgeving*, January/February, 14–17.

Redclift, Michael (1987) *Sustainable Development: Exploring the Contradictions.* London: Methuen.

Royston, Michael G. (1979) *Pollution Prevention Pays.* Oxford: Pergamon Press.

Schumacher, E.F. (1973) *Small is Beautiful.* London: Abacus.

Vogel, David (1986) *Styles of Regulation: Environmental Policy in Britain and the United States.* Ithaca: Cornell University Press.

Winsemius, Pieter (1986) *Gast in eien huis: beschouwingen over milieumanagement.* Alphen aan den Rijn: Samson H.D. Tjeenk Willink.

World Commission on Environment and Development (1987) *Our Common Future.* Oxford: Oxford University Press.

CHAPTER 9

Ecological Covenants: Regulatory Informality in Dutch Waste Reduction Policy

Ida Koppen
Florence

The interdependence of a small number of participants in no way guarantees an equilibrium between the parties, but it furthers agreements. When these have become the rule, the hierarchical relation between the state and the "regulated fields" changes into a negotiation system in which order and obedience have been replaced by solutions based on mutual consent (Scharpf, 1988: 70, my translation).

INTRODUCTION

The unilateral imposition of norms is no longer the prevalent model for state intervention in societal processes. In reaction to various critical analyses of "regulatory failure", "legal failure" and the "implementation deficit" (Mayntz, 1983: 61–5; Rehbinder, 1992: 595–8), a model is evolving that is characterized by complex multilateral interactions between the various social actors involved. Interactive network-like relationships emerge around social issues that facilitate the informal exchange of information and typically include some form of negotiated decision-making (Ladeur, 1988: 269–76; Kenis & Schneider, 1991: 26, 42; Scharpf, 1991: 623; Van Vliet, 1992: 209). In this model the state is no longer superimposed on society, but functions as an horizontally juxtaposed entity, a partner, although *primus inter pares* (Willke, 1983: 17, 119ff.; Scharpf, 1991: 622)[1].

One of the areas of government intervention in which this trend can be observed is environmental policy. Various social actors play an increasing role in the formulation and the implementation of environmental policy. Although this influence was originally acquired on the basis of activities in the margin

Environmental Law and Ecological Responsibility: The Concept and Practice of Ecological Self-Organization
Edited by G. Teubner, L. Farmer and D. Murphy
© 1994 John Wiley & Sons Ltd

of the policy-making processes—through protests, boycotts and lobbying of policy-makers—it has gradually been consolidated, at least in some countries, in procedures that are at the core of the policy process.

In this chapter, we will analyse the significance of the institutionalization of informal consultation processes and negotiations between government actors and social actors. Looking in particular at the non-juridical aspects of informal interactions, a first attempt is made to outline the contours of the emerging system of regulatory communication. The theoretical framework within which the analysis will take place is a combination of elements from negotiation theory and the theory of autopoiesis. The empirical data are taken from Dutch environmental policy, in particular waste reduction policy.

INFORMALITY IN CONTEXT

Practice

Recent developments in Dutch environmental policy show the intention on the part of government to enlarge social support for policy measures by involving the affected groups, referred to as "target groups", in an early stage of the decision-making process (Second Chamber, 1988–1989a: 1; Second Chamber, 1988–1989c: 179–81, 188–229). The "target group approach" partly reflects the historical evolution of the Dutch tradition of consultative politics (Lijphart, 1969: 112–5, 122–38; van der Hoeven, 1970), but it is also the expression of a new willingness to face implementation problems in the field of environmental regulation (Winsemius, 1988). The strategic rationale behind the approach is to avoid opposition to measures that are imposed unilaterally without taking adequate account of the interests of the different affected groups in society. The proposed process of negotiations gives each group the opportunity to participate actively in the formulation of policy goals and measures. Essential elements of this process are: to formulate clearly what one's underlying interests are and, hence, to gain insight into the interests of others; to agree about the facts of the issue at stake; and to understand the acceptable solution margins as well as the implications of withdrawal from or obstruction of the negotiations. If parties are able to agree, on the basis of these elements, about what they view as the optimal solution, it is unlikely that the implementation of their decision will meet with any major opposition.

The informality of negotiations and bargaining mechanisms that are utilized to replace or supplement the application of legal norms has often been criticized for violating principles of legality (Bohne, 1981; Winter, 1985; Lambers, 1988; Van Driel 1990; Marin, 1992). The criticism makes the following points. First of all, the legal status of the agreements that result from the negotiations is unclear. Usually such an agreement does not have a binding legal character. Second, the democratic control exercised by elected public bodies is absent in the case that a measure is decided on the basis of informal negotiations. Third, parties not directly involved in the negotiations are in a disadvantaged position since they cannot invoke any of the rights they would have if the measures were adopted according to public law principles. Finally, there is no guarantee that equal cases will be treated equally (Bohne, 1981: 200–23; Winter, 1985: 241ff.; Van Driel, 1990: 419; Marin, 1992: 322–36).

On the other hand, it has been argued that informal bargaining is sometimes the only effective tool when formal legal procedures do not produce the desired effect (Hucke, 1978; Aalders, 1984; Winter, 1985). Hucke found that informal contacts were a necessary fall-back option in regulatory procedures, especially in "situations where the agency can no longer rely on the coercive power of its legal instruments and is forced to choose bargaining to secure a minimum of administrative success in the improvement of environmental quality". He concluded that "a tendency will always exist towards bargaining in policy implementation" (1978: 109; see Damen, 1992: 10).

What we see is that the negative and the positive arguments in fact address different aspects of the same phenomenon. While the criticism of informality primarily addresses its legal connotations, the positive assessments look especially at its communication-enhancing effect. Each conclusion hence reflects only a partial picture. The proposed way out of this dilemma is to make the analytical distinction between the two aspects more explicit. When Bohne criticizes informality, he means informality in the legal sense. Indeed, his definition of informality includes all those government acts that take place outside the realm of application of binding legislative or administrative instruments. Informal acts, according to Bohne, are not provided for by the law, but they are meant to achieve objectives that could have been reached through the application of formal acts. Formal are those acts that follow the prescribed pattern of implementation of legal rules and administrative orders (1984: 344)[2].

The non-legal aspects of informality, on the other hand, refer to the style of communication between the actors. Formal modes of communication often impede the exchange of information that is necessary for a process of joint problem solving. The target group approach in Dutch environmental policy requires communicative informality as a necessary element of consensus building.

The relevance of the distinction between legal and communicative informality, as we will show, is that the very appropriate criticism of the legal aspects of informality can be largely accommodated without compromising the positive impact of informality on communication. In a very straightforward manner, one could say that, "although the juridical component [of the informal act] often poses problems, it is not the main aspect that needs to be solved in the first place", whereas "non-juridical factors are often of decisive significance" (Van Driel, 1990: 432, 433, my translation).

Theory

Once the distinction between juridical and non-juridical aspects of informality is acknowledged, we can explore the non-juridical aspects without having to simultaneously justify the juridical shortcomings. They require different analytical approaches. Focusing, then, on communication we must first of all recognize that it is not a new concept in the field of regulation and policy analysis. Scholars of the second generation of policy instruments have documented extensively the role of enhanced communication and the increased exchange of information. In the consultation and persuasion model presented by Majone, a public dialogue as a social learning process forms a central element. It "shifts the emphasis from legal enforcement to the provision of information and high quality advice" (Majone, 1986: 455). The proposed model is especially pertinent to situations in which the scientific data that underly a regulatory issue are contested, as is often the case in environmental policy (Majone, 1989: 3, 37–41). As a suggested solution, consultation and persuasion are explained on the basis of the rationality and legitimacy of the public policy process. Very recent work by Van Vliet analyses "communicative governance" (*communicatieve besturing*) in the light of negotiation theory (1992: 47–78), showing how negotiation theory can be fruitfully applied to the emerging preference for communicative informality in government intervention. Marin's analysis of "contracting, without contracts" looks at governance beyond law in the context of the autopoietic paradigm of systemic closure (1992: 303). Although he addresses aspects of communication (1992: 345–7), the analysis is dominated by legal considerations.

Missing from the analytic framework that has been developed so far is a theoretical component that analyses the communication process itself in terms of its systemic characteristics. This additional frame can be found if elements of the theory of autopoiesis are combined with elements of negotiation theory. Together these elements provide a broad theoretical "niche" for our notion of communicative informality.

The systemic closure postulated by autopoiesis finds a natural counterweight in the careful exploration of common interests and the process of joint problem solving put forward by negotiation theory. If target groups are viewed as autonomous social subsystems, the informal interactions in which they engage may develop into an autonomous system of communication which might serve as a remedy against autopoietic closure[3]. The direct access to other subsystems, which autopoiesis deems impossible, may be substituted by a negotiation process that searches for the common ground in the interests of the participants. Each of the different groups of participants will have its own rationality, its own perception of the issue at stake. These different rationalities must be reconstructed in the interation process through careful observation and exchange of information. If this process is conducted properly, a common language will emerge, aimed at finding a solution for the problem which is acceptable to all the parties involved. The first common interest recognized by the participating groups is often the desire to find a joint solution. Although this might sound like a truism, it is a significant first phase in the construction of a workable system of communication. Looking at it from a different angle, Marin describes the gradual change from voluntary collaboration to a forced alliance which makes the "open exit" option remote in practice:

> voluntariness turns itself into an iron necessity to go on collaborating, if exit from co-operation becomes ever more costly with ongoing co-operation ... [T]he contracting parties to the policy network find themselves ever more "locked into" an alliance which has become valuable in itself by the sheer amount of "sunk costs" or specific past investment in maintaining the relationship (1992: 338–9).

In autopoietic terms, the fixed binary codes of each system (i.e. actor group) often criticized for over-simplifying social reality (Scharpf, 1988: 64–9), are translated into a new code for the negotiation process in which participants engage. This code, besides being solution-oriented, must be capable of accommodating the interests of each actor group (Koppen, 1991: 148–9; Teubner, 1989c: 418). The negotiations and the ensuing agreements support the thesis that "mutual observation, connection through interference and communication through organization" (Teubner, 1989a: 96, my translation) can lead to the enhanced ability of autonomous subsystems to resolve conflict and agree to joint solutions.

Although Van Vliet presents communicative instruments like negotiation, consultation and co-operation as a third generation of instruments, alongside traditional coercive regulatory instruments and financial instruments (1992: 9–25), most of his examples would qualify as second generation policy instruments in the analytic framework that is proposed here. The first level of instruments is based on the traditional hierarchical regulatory approach and includes the imposition and enforcement of legal norms. The second level comprises those instruments that take some form of negotiated interaction as a starting point (e.g. Willke, 1983: 128ff.). Some financial instruments, like taxes and charges, are first level instruments. Others, like tradeable permission rights, fall into the second category. The third level instruments, that are proposed here, are the "ecological covenants", the structured outcomes of the negotiations, but not the negotiations themselves. Whereas the negotiations are characterized by communicative and juridical informality, a tendency towards quasi-contractual formalization can be observed in the form that is chosen for the outcome of a negotiation process. This outcome is the basis for a new type of social dynamics that has many characteristics of a "relational contract" (MacNeil, 1980)[4]. The aspect of formalization is a crucial element in our definition of third generation instruments. The final step in the analysis is thus a newly formed structure or "organization" among the participating groups in a public policy debate. A kind of joint venture is created that represents a common interest and is aimed at a common goal. The organizational structure reflects the continuous interaction and the long-term orientation of the relationship in which the participants engage. As a policy instrument the "joint venture" has public and private law characteristics, similar to the traditional semi-public bodies that have existed for a long time in the area of economic policy in several West European legal cultures (Willke, 1983: 89; Marin, 1992: 315–7, 319–36). The Dutch trend to intervene in waste policy matters on the basis of covenants will be analysed in these terms.

DUTCH WASTE REDUCTION POLICY: TARGET GROUPS AND STRATEGIC DISCUSSIONS

Dutch environmental policy gives a central role to the active involvement of the "target groups" (Second Chamber, 1988–1989c: 177, 179–81, 188–91). For each component of environmental policy, the Ministry of the Environment identifies the affected social actors and these are invited to participate in the development of policy programmes as well as in the implementation of policies. This approach was already announced in the so-called "Indicative Multi-year Programmes" which were formulated in the early 1980s for the environmental sectors waste, water, air, soil and noise. It was reinforced and elaborated in the National Environmental Policy Plan (NEPP), the first integral environmental programme, which was published in 1989.

The new style of environmental governance introduced by the NEPP is apparent in the opening sentence of the document: "This NEPP contains the strategy for environmental policy in the medium term" (Second Chamber, 1988–1989c: 7). The crucial word is *strategy*. No longer does the government intend to focus on and spell out specific regulatory measures to combat pollution. The emphasis is placed on the design of a strategy, of which regulatory measures will be a part, but which will similarly encompass other kinds of intervention such as research, public information and consultation with the target groups. The idea of environmental management has replaced the notion of environmental regulation, and the traditional sectorial approach has been abandoned in favour of a thematic focus. The themes refer to the flow of substances in the environment the effects of which go beyond the sectorial boundaries that were formerly imposed by legal structures. Acidification, for instance, is now treated as a theme for which an integrated prevention and abatement strategy is to be developed. In the sectorial approach, acidification appeared as a fragmented problem under the sectors air, water and soil pollution. In the area of waste management, the problem of waste processing and waste disposal is integrated with the qualitative and the quantitative reduction of waste.

In this new design, the interaction between the government and the target groups is of central importance. For each of the environmental themes that are identified, the NEPP specifies policy objectives and the date by which they must be reached, it lists the actions the government intends to take and the actions that are expected of the target groups. With respect to waste, the NEPP refers to the Memorandum on Waste Prevention and Recycling which was adopted a year earlier (Second Chamber, 1988–1989b). This Memorandum contains many of the ideas that were later incorporated in the NEPP. When the Minister of the Environment presented the Memorandum in October 1988, he explained the new approach in the following way:

> I regard this Memorandum as the beginning of a process that will cover all aspects of waste policy to discuss with the concerned groups the targets, the measures and the activities needed to achieve a more effective prevention and recycling of waste. It must be obvious that I expect more results from such an approach than from the one-sided use of legislation. Formulating an approach that is not carried by those concerned is at most difficult to uphold and therefore hardly brooking of success (Second Chamber, 1988–1989a: 1).

The government, in other words, views the target group approach as a means to assure the greater success of its waste policy. The involvement of the target groups in the entire policy process, from its inception to its application, is seen as a guarantee for policy effectiveness. The process of consultation is referred to as "strategic discussion" (Second Chamber, 1988b: 3), a form of policy negotiation which the Memorandum describes as follows:

> The introduction of the Indicative Multi-year Programmes on Environment started a process whereby policy and ensuing measures were formulated together with the concerned target groups. Policies and measures should be carried by the persons they most concern and not wordlessly imposed. Fully appreciative of this line of action, it is my intention to embark upon "strategic discussions" with the persons most involved; to provide a guideline for thorough discussion and to arrive at mutual solutions. In the strategic discussions, policy options will have to be formulated as well as the most suitable means to implement the chosen options.

The waste reduction targets that are formulated in the Memorandum have a threefold character. They regard a reduction in the occurrence of a particular waste component, a reduction in the amount of waste to be dumped, or an increase in the amount of waste to be recycled. For car wrecks, for instance, the target set for the year 2000 is to recycle more than 80% of the 500 000 tonnes of wrecks that are discarded each year. For glass the target is to recycle 100% of the 430 000 tonnes generated. For household waste the target is to recycle half and to incinerate half of the total of 5 000 000 tonnes of waste that will be generated in the year 2000. The total quantitative reduction target set in the Memorandum for the year 2000 is an absolute reduction of 5% relative to the total amount of waste produced in 1986 (Second Chamber, 1988b: 19). The NEPP accelerates the 5% reduction target and fixes it for 1994. For 2000, a 10% reduction is foreseen (Second Chamber, 1988c: 148). The NEPP does not make a reservation concerning the uncertainty about the autonomous increase in the amount of waste, as the Memorandum had done. In the NEPP-Plus, intended to update the NEPP, we do find an explicit mention of the factor of autonomous growth (Second Chamber, 1989: 29ff.): a 10% reduction in 2000 is not expected to result in a total net reduction of waste. The different provisions on quantitative reduction targets leave room for diverging interpretations. This was one of the points that was contested by the packaging industry in the negotiations over the Packaging Covenant which we will discuss below.

For each of the 29 waste streams that are identified, the Memorandum lists the target groups that are invited to participate in the strategic discussions. The discussions, it must be emphasized, regard both the targets themselves and the instruments to achieve the targets. The proposed targets are meant to orientate the discussion, they do not have a binding character. In the words of the Minister, the "targets have been formulated to act as guidelines" in the planned discussions and it may very well be that the exchange of views that is to take place in the context of the strategic discussions will lead to "adjustments in the targets" (Second Chamber, 1988–1989a: 1). We will come back to this aspect of the approach, in the analysis of the *primus inter pares* position of the Ministry below.

When a strategic discussion is successful and the parties involved reach agreement about the targets and the instruments, this is often confirmed with the signing of a so-called "covenant", a document which embodies and personifies the formal organization, the quasi-contractual arrangement in which participants wish to engage. The legal status of covenants was originally rather controversial (Second Chamber, 1988–1989c: 181; van Rossum, 1988; Klok, 1989). They were viewed as non-binding declarations of intent, gentlemen's agreements without legal consequences, instruments of government policy to which the various criticisms of legal informality that were mentioned above applied. However, their increased use in recent years and the ensuing discussions of their ambiguous juridical character have led to several adjustments in the use of covenants as public policy instruments. In reaction to the criticisms, recent Dutch covenants have been discussed extensively in Parliament (Second Chamber, 1990–1991) and they contain provisions about the position of third parties. The Covenant on Packaging Waste, moreover, expressly states that the agreement has the legal status of a contract under civil law[5]. Thus we see that legal informality has been replaced by a new type of formality which takes the form of a contractual arrangement, or joint venture.

PACKAGING WASTE: THE TABOO OF DEPOSITS

Packaging waste is identified in the Memorandum on Waste Prevention and Recycling as one of the 29 waste streams. The reduction goal established for packaging waste for 2000 is: no more dumping, 60% of the waste recycled and 40% incinerated. Figure 9.1 shows the graphic illustration of the targets as presented in the Memorandum.

The target groups identified for packaging waste include three Ministries (Environment, Economic Affairs and Agriculture), representatives of the provinces and the municipalities (organized in national associations), the packaging industry (organized nationally in the Association for Packaging and Environment, *Stichting Verpakking en Milieu, SVM*), consumer and environmental groups (represented by the *Stichting Natuur en Milieu*) and the National Institute for Public Health and the Environment (*Rijks Instituut voor Volksgezondheid en Milieuhygiene*). These groups engaged for two years in a strategic discussion about the proposed targets and the best ways to achieve them.

The *SVM* was set up by the packaging industry in the early 1970s in reaction to social concern about the rapidly expanding use of one-way packaging (Peterse, 1992: 202ff.). In the 1970s many products in the Netherlands, especially beverages, were still marketed in reusable packaging. The growth of supermarkets was responsible for a dramatic shift towards one-way packaging which had three clear advantages: less storage requirements, easier to handle and more easily adaptable to changing marketing requirements. Moreover, a separate chain of packaging industries had developed whose interests were also represented by the *SVM*. The *SVM* was responsible for initiating a public

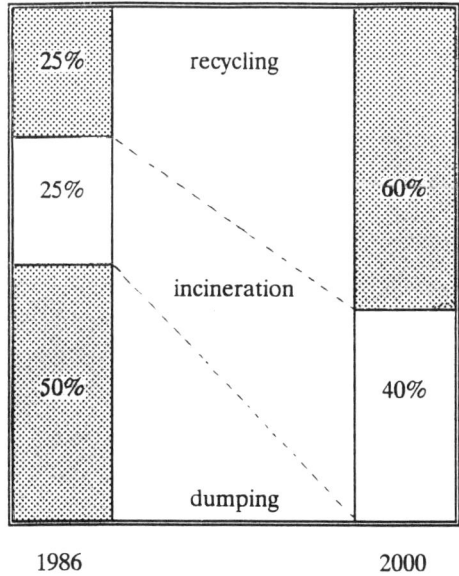

Figure 9.1 Reduction targets packaging waste (source: Second Chamber, 1988–1989b: 50).

dialogue about packaging issues and for collecting data on packaging trends. It supported the initiatives to recycle glass. By organizing itself this way, the packaging industry was from the beginning prepared to enter the debate on waste reduction and to develop alternatives to the use of deposits and reusable packaging.

The self-regulation of the packaging industry in the Netherlands occurred in a social setting that was characterized by the following conditions: a legal framework authorizing the Minister to issue general Decrees on the reduction of packaging waste (Arts. 27–29 of the Waste Act of 1977 and Art. 34 of the Chemical Waste Act of 1981); a very active and well organized environmental and consumer movement; a competitive structure of branch organizations in the different sections of the packaging chain (production, packaging, distribution, retail); and social responsiveness of the dominant firms involved (Peterse, 1992: 207–8). Each of these conditions was crucial for the emergence of the self-regulation. If the regulatory competence of the Minister had not existed, industry would not have been interested in investing so much time in a joint solution. If the public pressure from citizen groups and the willingness to contribute to the effectiveness of waste reduction policies, expressed in powerful boycotts, had not been as substantial as it was, the issue might very well have lain dormant for several more years. The competitive structure of the packaging industry made it necessary for each member of the packaging chain to stay at the forefront of renewal in the branch and hence created an active interest in participation. Finally, the social responsiveness of some of the leading firms was crucial in mobilizing the entire industry for the issue.

After two years of negotiations, the target groups had not reached agreement. The main points of conflict were, first of all, the issue of mandatory deposits on containers. Industry remained opposed to the use of mandatory deposits while environmental groups argued that a deposit system was the only way to assure a high return rate of containers[6]. The issue was connected to the second point of conflict which concerned the target itself: environmental groups, supported by several political factions in Parliament, insisted on an absolute reduction of packaging waste of 10% by the year 2000 (Second Chamber, 1990–1991: 9). As I pointed out earlier, it had not been clear whether the 10% reduction mentioned in the NEPP was to be understood as a net reduction of waste or as a reduction of the total amount including the autonomous increase of waste. In the NEPP-Plus, the Ministry had explicitly stated that the reduction was a reduction of the total, increased amount. But in the negotiations the point remained unclear and in all its correspondence, *Natuur en Milieu* insisted on an absolute reduction of 10%. The Minister himself, in a letter of 17 August 1990 to *SVM* talked about an absolute reduction of 10% in the year 2000 relative to the amount produced in the base year 1986. And in a letter of September to *SVM*, the Minister emphasized that this target was valid for waste materials in general while for packaging waste higher targets were feasible. He even mentioned a figure of 50%[7].

Later that year, the Ministry for Environment and *SVM* decided to break off the plenary negotiations and continue bilaterally. This two-way negotiation resulted in the signing of the Packaging Covenant on 6 June 1991. In the covenant, industry committed itself to a slightly higher objective than the one stated in the Memorandum, but not higher than the targets announced in the letters of the Minister. The amount of packaging waste in the year 2000 shall be *less* than the 1986 level and the quantity recycled shall be *at least* 60%. Other commitments are all phrased in the form of intentions: industry shall make an effort to achieve an absolute reduction of 10% by the year 2000 and to recycle at least 50% by 1995. A commitment is also made to develop new types of environmentally friendly packaging. Furthermore, the covenant spells out a complex process of monitoring and evaluation for the purpose of which several advisory committees have been set up.

The signing of the covenant was reported on the front pages of major national newspapers. The Ministry of Environment and *SVM* hailed the covenant as a far-reaching environmental measure in the spirit of deregulation, leaving the packaging industry room to find the most efficient ways to reach the fixed targets. Environmental groups, supported by a report of the Law Faculty of the University of Amsterdam, criticized the covenant for its lack of binding legal measures and for not introducing mandatory deposit schemes. During the debate in the Parliamentary Committee for Environment, the socialists had insisted that the signing of the covenant should not preclude the adoption of a general Decree on Packaging Waste (Second Chamber, 1990–1991: 6–8).

The issue of linkage between covenants and formal regulations was discussed explicitly in the recent main government documents concerning environmental policy. The NEPP stated that covenants were to be used as instruments of environmental policy, particularly in the following cases (Second Chamber, 1988–1989c: 181):

(a) if the set aim cannot be reached easily or quickly enough by imposing regulation;
(b) where, in the exploratory phase of a particular problem and preceding regulation, covenants can allow preliminary agreements favourable to the environment to be made;
(c) where, prior to regulation, a covenant is made by which the content of the future regulations will be used as far as possible as a basis;
(d) where existing regulations are supplemented or tightened up in a covenant.

It is clear from this citation that the signing of a covenant was not in any way considered an obstacle to the adoption of regulations. On the contrary, it was viewed as an often necessary first step.

In the Memorandum on Policy Instruments, published together with the

NEPP-Plus, this general approach was repeated, now explained over several pages (Second Chamber, 1989–1990b: 12–4). First of all, the general legalist criticism is repeated: covenants suffer from a lack of democratic control, an unclear legal status and they do not contain guarantees for third parties. Covenants that replace regulations, should, according to the document, only be used as temporary arrangements, until the regulatory instrument enters into force. A covenant must indicate clearly which obligations it creates and for whom and how enforcement can be assured. Third parties must be given access to the negotiations and draft agreements must be made public in an early stage of the negotiations.

If covenants are used as predecessors of formal regulations, an ambiguous situation is created that risks being neither fish nor flesh. By fixing the outcome in a regulation, a large part of the additional value of the informal interactions is lost. While communicative informality facilitates the process of consensus forming and responsibilization among the participants, the adoption of a formal regulation would impede the self-organization that is needed to administer and enforce the agreement. The danger exists that the shared sense of responsibility for the success of the agreement disappears when a formal regulation is adopted. This would not be the case if the agreement is formalized in a private arrangement for which participants remained responsible. There is no reason why a private arrangement like this could not fully respect public law principles, such as disclosure of information and democratic control (Konijnenbelt, 1992: 20ff.). The Dutch Covenant on Packaging can be seen as a first attempt to formulate private instruments that accommodate the criticism of legal informality.

INFORMALITY AS A METAPHOR

By dividing informality into legal informality and communicative informality, we have been able to formulate ways of abandoning one without giving up the other. The next step, then, is to unravel the concept of communicative informality itself. Returning to the early positive assessments of informality, we find that the point put forward by Hucke was that informal contacts were a necessary fall-back option, especially in "situations where the agency can no longer rely on the coercive power of its legal instruments and is forced to choose bargaining to secure a minimum of administrative success in the improvement of environmental quality" (1978: 109). Apparently, in the situations described by Hucke, legal instruments were lacking coercive power and, therefore, implementation was unsuccessful. This point coincides with the strategic rationale behind the Dutch target group approach that seeks to enlarge social support for proposed measures rather than relying on the presumed coercive power of legal instruments. That brings us back, however, to the very principle of legality and the danger that an informal approach to state intervention undermines the notion of the *Rechtsstaat*, as was indicated in very clear terms by Bohne in his seminal work *Der informale Rechtsstaat* (1981).

In Germany, the discussion on this point seems to have been temporarily concluded in favour of a formal-legal approach. In the area of packaging waste and beverage containers, two federal regulations were adopted after more than a decade of informal negotiations and voluntary agreements between government and industry[8]. The shift in government approach was presented in rather confrontational terms in the press. "Töpfer pulls the emergency brake", said one newspaper heading (*Süddeutsche Zeitung*, 1990), emphasizing that by relying on informal agreements with industry, the government had never been able to achieve the results that could have been achieved with legal intervention. We have also proposed, however, that the legalist criticism of informality can be accommodated without damaging the virtues of communicative informality.

The description of the Dutch approach to waste reduction shows that the reaction in the Netherlands to the criticism of legal informality differs substantially from the German response. Rather than opting for the immediate imposition of legal norms for waste reduction, Dutch policy shows a shift towards the institutionalization of informality in new forms of decision-making, which differ from existing formal procedures. The essence of these new procedures is not their lack of formal legal context. On the contrary, what we see is that the final form chosen for the concluded agreement is a contract under civil law. The important factor is the particular form chosen for the outcome of the informal process of negotiations which contains a mixture of public and private law elements. At the end of a process of communication, of structured negotiations which give all the participants a chance to listen to each other, to express concerns and ultimately to see to it that one's own interests are accounted for in the final result, a structure has been created for which each participant carries joint responsibility. It is still early to assess in any detail the necessary features of this "joint public/private venture". However, a pattern of internalization can be distinguished that ties in with the notion of self-regulation.

The process of interaction that was described allows for a sensitivization of the actors to each others' needs. A context of reciprocal recognition is created which enables participants to develop mutually beneficial negotiation strategies. The added value of the process does not lie in its legal informality; rather it is the capacity of the process to facilitate the exchange of information which enhances creative problem solving and stimulates a learning process.

The early image of informal interactions between the government and different social actors has changed substantially. Although the actions would still qualify as informal according to Bohne's legal definition, we note that, as a result of the target group approach, interactions have become structured. The example of the Covenant on Packaging illustrates the formalization that has taken place. The image of secretive talks behind closed doors where deals were made that one was not supposed to talk about in public has been replaced by a systematic process of consultations with well-identified interest groups.

The talks are extensively reported in the press and discussed during Parliamentary debates. The final form chosen for the agreement, a binding contract under civil law is a first step towards developing public-private management structures for which participants are jointly responsible.

Although the use of covenants as instruments of public policy is still contested, and the discussion about the legitimacy and the effectiveness, the applicable scope in other words, of agreements as appropriate instruments of environmental policy will continue, we may conclude for the moment that the negotiation processes between the government and the target groups have been institutionalized. What is left of informality is a metaphorical notion of partnership.

COMMUNICATIVE INFORMALITY

In order to better understand the meaning of communicative informality we need to have a closer look at the interaction process as it has developed in Dutch environmental policy. The strategic discussions in the Netherlands have introduced the role of a neutral facilitator who guides the negotiations between the various actors involved. This figure is derived from experiences in the USA where neutrals have played a crucial role in environmental conflict resolution (Susskind & Cruikshank, 1987: 136–85). Similarly, the role of a facilitator or mediator has been introduced in Germany (Hoffmann-Riem, 1989; Weidner, 1993). The neutral functions as an intermediary in a communication process that would otherwise be characterized by prejudices and negative intentions. Participants in a policy debate usually have a very different perception of a problem, they "speak a different language". Industry representatives perceive each suggested solution in terms of its financial consequences. Government representatives typically think in terms of the political consequences. Environment and consumer groups take the amount of protection achieved as a point of reference. Communicative closure can be opened up by creating a common language. As was pointed out earlier, this common language can only evolve if the necessity of a joint solution is recognized as a common interest. This in turn will only be possible if a joint definition of the problem is formulated. Each of these steps in the communication process requires guidance and mediation and it is the role of the neutral to provide this assistance (Koppen, 1991: 150).

Another problem that the neutral needs to mediate is the fact that the participants in policy negotiations often assume they must present themselves as *hard bargainers*: I will bid high in order to be able to give in a little and still be able to achieve my hoped-for outcome. This attitude is very common and often prevents an exchange of information about the underlying interests of the participants. The neutral has to assist the parties in disclosing their interests rather than their fixed positions. This type of information will function as the building blocks to reach agreement: even though the positions that parties take might seem very different, their underlying interests often reveal common points of interest that can be addressed in a joint solution. An integrative approach to negotiating aims to discover the common ground. A neutral facilitator who assists the parties in the negotiation has to help them discover this approach and work together to achieve an optimum joint solution. A very simple example that is often used to illustrate this process is the parable of the two sisters fighting over an orange:

> "It is mine", one of them screams. "I had it first", the other replies, and neither of them is willing to give in. Mother has just come home from work, she is tired and the fighting and screaming disturb her greatly. Angrily she goes over to the two little quarrelers with a knife and is about to cut the orange in half. Giving each of them half an orange seems the most fair solution. But then grandmother walks into the kitchen. "Wait a minute", she tells her daughter, "I will deal with this". Grandmother asks each of the girls to come and talk to her separately. With a smile she takes the knife, peels the orange and gives the peel to one of the girls who goes off to use the peel in the marmalade she had been preparing. The other girl takes the remainder of the orange to press the juice. Each of them has fulfilled her interest at 100%: a maximum joint-gain solution.

The process of guided policy negotiations and the ensuing phases of sensibilization can be interpreted as serving as indirect interferences in the systemic closure proposed by autopoiesis. On the presumption that direct interference is excluded (Teubner, 1989: 96), indirect intervention could occur through mutual observation and structured communication with random outcomes. What we see happening in the practice of policy negotiations is indeed that the outcome of the negotiation process is unpredictable in the sense that the creative learning process leads to the invention of options that could not be foreseen in advance.

The Dutch Covenant on Packaging unfortunately only shows half of the picture of communicative informality since it was concluded between only two of the interest-networks involved in the issue. Consumer and environmental groups did not take part in the agreement. Here we see that the idea of the state as an equal partner in the negotiations is refuted by the facts that show that partnership and self-regulation are still partially imposed.

HORIZONTAL HIERARCHIES

The idea that the role of the state has changed from that of a hierarchical imposer to a consensus seeking negotiator is attractive (Konijnenbelt, 1992: 19). It is tied to a notion of citizenship that values active involvement, transparency and responsibility. The government has become an actor in a complex network of functions that determine the final policy outcomes. In the Dutch case of waste reduction the "equality" of the groups involved was underlined by the statement of the Minister that the proposals were put forward "to act as guidelines", to which adjustments could be made if the outcome of the strategic discussions required such a change (Second Chamber, 1988–1989a: 1). The self-regulatory tendency in this approach is apparent. However, we must note that the context in which the interaction takes place, is obviously not one of complete equality since the state continues to hold an exclusive power unmatched by any of the other players in the steering game. The power to enact formal regulations unilaterally often functions as a stick behind the door for the other actors to participate in a joint process of decision making. The *primus inter pares* role requires sensitivity on the side of the government actors to the subtle balancing of interests necessary to reach agreement. The process of consensus building is intended to create an outcome perceived by all the participants as the optimal joint solution. This requires that the specific power position of each group, including that of the government, is not used to threaten or force outcomes (Van Vliet, 1992: 63). It will play a role in the evaluation by each group of the advantage of a joint solution over withdrawal from the negotiations.

The strategic discussions can be viewed as "negotiations in the shadow of hierarchy" (Scharpf, 1991: 629)[9]. Hierarchy reflects the public context in which the negotiations take place, it refers to the role of law and the role of the state as the guardian of the public interest that determine the contours of the negotiation process.

Whereas the communication process between the involved groups is free, the external effect of their interactions is bound by principles of legality. Even though individual state actors may participate on an equal footing in a negotiation process and in a public-private joint venture, the state as an entity must maintain its public responsibility. It can do so by defining certain general rules to govern public-private arrangements. Thus, the preconditions are created for horizontal negotiations under indirect hierarchy, and the inequality is managed in order to achieve the goal of enlarging social support.

The course of events in the Dutch case of packaging materials illustrates that different alliances may be built. In this case, a contract was signed by government representatives and representatives from the packaging branch. Consumer and environmental groups did not participate in the final negotiations. The evolution of this case denies some of the very interdependencies which the official policies have tried to demonstrate, interdependencies that are moreover necessary to develop trust and confidentiality, essential elements of a process of joint problem solving. It is not yet clear what effect this will have on the success of the agreement. Excluding these groups from the contractual arrangement has created the risk that societal support for the measures will remain limited and hence implementation will be problematic. On the other hand, the organization that is created to implement the contract allows for the inclusion of consumer and environmental groups. Hence, it is the rationality of this organization that functions as a guarantee for the implementation of the contract and the sensitivity that is necessary to achieve the interference into the closed structures of interest representation that have traditionally characterized and obstructed policy implementation.

CONCLUSION

We have made a distinction between three types of policy instruments. If the traditional policy instruments were characterized by hierarchical relationships and the imposition of command and control regulation, the second generation of instruments is typically described in terms of negotiated decision-making and policy networks. In these two scenarios, the role of the state as the guardian of the public interest and the role of law as the guardian of individual interests are distinct. The idea that public entities may engage in negotiations with the private entities they need to regulate is at odds with the image of the state as a non-partisan actor. In this context "informal" regulatory activities have been criticized for not obeying the rule of law.

By dividing the notion of informality into legal and communicative informality, we have tried to demonstrate that the criticism of legal informality could be accommodated without questioning the importance of communicative informality. The third generation of policy instruments that we propose is characterized by communicative informality and new forms of legal formality. As the example of the Dutch ecological covenants shows, negotiated decision-making can take place in a legal context that creates well defined obligations for the participants. The outcome of an informal process of communication takes the form of a contractual arrangement for which parties are jointly responsible. Similar to an organization with members, this "joint venture" resembles the semi-public arrangements that exist in the area of labour law and professional associations.

The process of negotiations, designed to accommodate conflicting interests by defining a common goal and a common perception of the problem, interferes with the systemic closure that is postulated by the theory of autopoiesis. The range of possible solutions is enlarged by a creative process of problem solving, possibly with the assistance of a neutral facilitator. This style of regulatory innovation is exemplified by the strategic discussions that are now an integral part of the target group approach in Dutch environmental policy. As second generation policy instruments they are a well-accepted phenomenon.

The new element introduced here is the explicit recognition of new forms of legal formality that emerge as the outcome of the informal negotiation process. These new forms of quasi-contractual arrangements are private in character, although they are bound by margins set by public law principles. The term "joint venture" is proposed here in order to underline the shared responsibility for the success of the implementation of the agreement. The recent Covenant on Packaging, concluded between the Dutch Ministry for Environment and the packaging industry is a first example of such a private contract. Although not yet accepted by Dutch legal doctrine, this new type of arrangement meets most of the requirements for regulatory effectiveness formulated by negotiation theory and the theory of autopoiesis. The process of negotiation is based on informal communication, characterized by an open dialogue about underlying interests and the formulation of a common objective and a common perception of the problem. The assistance of a neutral facilitator helps to improve the communication and the achievement of the optimal joint solution. The self-referentiality of the different actor groups is transformed into a shared notion of self-regulation in the form of a contract as organization of which all participants are members.

Although the precise contours of these public-private contractual arrangements must still be defined, the Dutch example shows their immediate practicability. It is hoped that other examples will follow and further analyses can be made. One danger must be indicated at the outset. The success of negotiated agreements as third generation policy instruments depends on the process of institutionalization. In this it is essential that communicative informality is preserved if they are not simply to collapse into the traditional forms of regulatory intervention.

NOTES

1 Willke talks about the allocation of new tasks to the state in the context of the formation of a society for which the organizational principles of representation and hierarchy have become too "simple", "*zu einfach*". Scharpf speaks of "*die Enthierarchisierung der Beziehung zwischen Statt und Gesellschaft*", the process of dehierarchicalization of the relation between state and society.

2 Bohne writes (1984: 343–4): "*Informal*" *ist der Komplementärbegriff zu "formal". Das Begriffspaar "formal-informal" bezeichnet alternative Handlungsmodalitäten und bezieht sich auf Entscheidungssituationen, in denen der Staat faktisch—nicht notwendig auch rechtlich—wählen kann, ob er ein bestimmtes Ziel in den von der Rechtsordnung bereitgestellten Handlungsformen oder mit rechtlich nicht geregelten Realakten verwirklichen will.*

   Als "formal" werden also alle rechtlich geregelten tatsächlichen Verfahrenshandlungen und alle Entscheidungen bezeichnet, die auf die Bewirkung von Rechtsfolgen gerichtet sind (sog. Rechtshandlungen).

   Demzufolge erfaßt der Begriff "informal" alle rechtlich nicht geregelten Tathandlungen, die der Staat anstelle von rechtlich geregelten Verfahrenshandlungen oder Rechtsfolgeentscheidungen wählt, die jedoch zur Herbeiflührung des beabsichtigten Erfolges auch in den von der Rechtsordnung bereitgestellten öffentlichrechtlichen oder privatrechtlichen Handlungsformen hätten erfolgen können.

3 In an earlier publication (Koppen, 1991: 149) I argued that actor groups cannot be viewed as closed systems the way that autopoiesis defines these, and I proposed a two-layer analysis, one layer being determined by the interests of the actor groups, the other being cast in terms of the internal coding of the autopoietic system. In this view there was little that autopoiesis could do to resolve a conflict of interests and, *vice versa*, the formulation of common interests did not help to solve the problem of autopoietic closure. In the present analysis, as already proposed by Teubner (1989c: 418), the two layers are integrated and the conflicting interests of the actors appear as just one of the obstacles to effective communication (compare Marin, 1992: 347).

4 The distinction between process and outcome is fluid. The long-term relationships that emerge in the course of the process can also be characterized by a certain amount of formalization (Marin, 1992: 336–43; see Van Vliet, 1992: 66).

5 It must be pointed out that the government is not entirely free to employ civil law instruments for public policy purposes. The complicated question as to the limits of discretion in this matter cannot be addressed here. For the purpose of this chapter it suffices to say that the Dutch legal system places limits on the use of private law if public law instruments are available.

In Dutch case law the so called "*twee wegen leer*" was formulated, the two-track doctrine, which granted government actors a large amount of freedom in choosing between public and private law instruments. In subsequent case law, the *twee wegen leer* was confined to situations in which the guarantees inherent in public law instruments are equally observed by the private law instruments applied. These concern especially the publication of documents, the legal position of the citizens directly involved and of third parties and the mechanisms of democratic control by elected political bodies (decision of the Dutch Supreme Court (*Hoge Raad*) in the *Windmill* case, 26 January 1990, reported in *Milieu en Recht* 1991: 163–7). The extension of public law principles into private law was codified in the new Civil Code (Art. 314, Book 3). A recent discussion of the complex relationship between private and public law can be found in Konijnenbelt (1992: 20ff).

6 Similar discussions have taken place in other European countries, notably Denmark and Germany. The EC Commission has proposed a directive on packaging waste which mentions the possibility of using economic instruments in Art. 11 (EC Commission, 1993).

7 Letters on file with the author. Letter of 17 August 1990, DGM/A nr.1970511; letter of 21 September 1990, DGM/A nr.1290506.

8 The *Pfand Verordnung* of 1988 (*Bundesgesetzblatt* I, 2455) and the *Verpackungs Verordnung* of 1991 (*Bundesgesetzblatt* I, 1234).

9 The phrase refers to the seminal article by Mnookin and Kornhauser "Bargaining in the shadow of the law" (1985), in which they analysed the impact of legal provisions on divorce settlements that take place outside the courtroom.

REFERENCES

Aalders, Marius V.C. (1984) *Industrie, milieu en wetgeving. De Hinderwet. Tussen symboliek en effectiviteit* (Industry, Environment and Legislation. The Nuisance Act. In Between Symbolism and Effectiveness). Amsterdam: Kobra.

Aalders, Marius V.C. and Van Acht, René J. J. (Eds) (1992) *Afspraken in het milieurecht* (Agreements in Environmental Law). Zwolle: W. E. J. Tjeenk Willink.

Bohne, Eberhard (1981) *Der informale Rechtsstaat*. Berlin: Duncker & Humblot.

Bohne, Eberhard (1984) Informales Verwaltungs- und Regierungshandeln als Instrument des Umweltschutzes. Alternativen zu Rechtsnorm, Vertrag, Verwaltungsakt und anderen rechtlich geregelten Handlungsformen? *Verwaltungs-Archiv*, 75, 343.

Damen, Leo J.A. (1992) Samenwerking en afspraken in het milieurecht: doodnormaal? (Co-operation and Agreements in Environmental Law: Dead Normal?). In Aalders & Van Acht (Eds). *Afspraken in het milieurecht* (Agreements in Environmental Law). Zwolle: W. E. J. Tjeenk Willink.

EC Commission (1993) *Proposal for a Council Directive on Packaging and Packaging Waste*. OJ 21.10.93, C285/1.

Hoffmann-Riem, Wolfgang (1989) *Konfliktmittler in Verwaltungsverhandlungen*. Heidelberg: C F Müller Juristischer Verlag.

Hucke, Jochen (1978) Bargaining in Regulative Policy Implementation: the Case of Air and Water Pollution Control. *Environmental Policy and Law*, 4, 109.

In 't Veld, Roeland J. *et al.* (Eds) (1991) *Autopoiesis and Configuration Theory: New Approaches to Societal Steering*. Dordrecht: Kluwer Academic Publishers.

Kenis, Patrick and Schneider, Volker (1991) Policy Networks and Policy Analysis: Scrutinizing a New Analytical Toolbox. In Marin, Bernd & Mayntz, Renate (Eds). *Policy Networks: Empirical Evidence and Theoretical Considerations*. Frankfurt/Boulder: Campus Verlag/Westview Press.

Klok, Pieter Jan (1989) *Convenanten als instrument van milieubeleid* (Covenants as Instruments of Environmental Policy). Enschede: Universiteit Twente.

Konijnenbelt, Willem (1992) *Convenanten met de Gemeente. Fluisteren in het Schemerduister*. (Covenants with Local Government. Whispering in the Twilight). Utrecht: Lemma.

Koppen, Ida J. (1991) Environmental Mediation: an Example of Applied Autopoiesis? In In 't Veld *et al.* (Eds). *Autopoiesis and Configuration Theory: New Approaches to Societal Steering*. Dordrecht: Kluwer Academic Publishers, 143–150.

Ladeur, Karl-Heinz (1988) Perspectives on a Post-Modern Theory of Law: a Critique of Niklas Luhmann, "The Unity of the Legal System". In Teubner (Ed). *Autopoietic Law: a New Approach to Law and Society*. Berlin/New York: de Grutyer, 242–282.

Lambers, C. (1988) Beleidsovereenkomsten in het milieubeheer (Policy-Agreements in Environmental Management). *Milieu en Recht*, 2.

Lijphart, Arend (1967) *The Politics of Accommodation*. Berkeley: University of California Press.

MacNeil, Ian R. (1980) *The New Social Contract: an Inquiry into Modern Contractual Relations*. New Haven: Yale University Press.

Majone, Giandomenico (1986) Mutual Adjustment by Debate and Persuasion. In Kaufmann, F-X *et al.* (Eds). *Guidance, Control, and Evaluation in the Public Sector*. Berlin/New York: de Gruyter.

Majone, Giandomenico (1989) *Evidence, Argument and Persuasion in the Policy Process*. New Haven: Yale University Press.

Marin, Bernd (1992) Contracting without Contracts: Economic Policy Concertation by Autopoietic Regimes beyond Law? In Teubner, Gunther & Febbrajo, Alessio (Eds). *State, Law, Economy as Autopoietic Systems. Regulation and Autonomy in a New Perspective*. Milano: Giuffrè, 303–357.

Mayntz, Renate (1983) Implementation von regulativer Politik. In Mayntz, Renate (Ed). *Implementation politischer Programme II*. Opladen: Westdeutscher Verlag.

Mnookin, Robert H. and Kornhauser, Lewis (1985) Bargaining in the Shadow of the Law: the Case of Divorce. *The Yale Law Journal*, 88, 950.

Peterse, A.H. (1992) Verpakkingsconvenant: element in een alternatieve sturingsstrategie (Packaging Covenant: Part of an Alternative Steering Strategy). In Huls, N.J.H. & Stout, H.D. (Eds). *Reflecties op reflexief recht* (Reflections on Reflexive Law). Zwolle: W.E.J. Tjeenk Willink.

Rehbinder, Eckart (1992) Reflexive Law and Practice. In Teubner, Gunther & Febbrajo, Alessio (Eds). *State, Law, Economy as Autopoietic Systems. Regulation and Autonomy in a New Perspective*. Milano: Giuffrè, 595–624.

Scharpf, Fritz W. (1988) Verhandlungssysteme, Verteilungskonflikte und Pathologien der politischen Steuerung. In Schmidt, Manfred G. (Ed). *Staatstätigkeit. International und historisch vergleichende Analysen*. Opladen: Westdeutscher Verlag.

Scharpf, Fritz W. (1991) Die Handlungsfähigkeit des Staates am Ende des zwanzigsten Jahrhunderts. *Politische Vierteljahresschrift*, 32, 621.

Second Chamber of the States General (1988–1989a) *Letter of the Minister Accompanying Memorandum on the Prevention and Recycling of Waste*. Second Chamber, session 1988–1989, 20 877, no. 1.

Second Chamber of the States General (1988–1989b) *Memorandum on the Prevention and Recycling of Waste*. Second Chamber, session 1988–1989, 20 877, no. 2.

Second Chamber of the States General (1988–1989c) *National Environmental Policy Plan. To Choose or to Lose*, Second Chamber, session 1988–1989, 21 137, nos. 1–2.

Second Chamber of the States General (1989–1990a) *National Environmental Policy Plan-Plus*. Second Chamber, session 1989–1990, 21 137, no. 21.

Second Chamber of the States General (1989–1990b) *National Environmental Policy Plan-Plus. Memorandum on Policy Instruments and Sustainable Construction.* Second Chamber, session 1989–1990, 21 137, no. 22.

Second Chamber of the States General (1990–1991) *Verslag van een mondeling overleg* (Report of an Oral Consultation). Second Chamber, session 1990–1991, 21 137, no. 87.

Süddeutsche Zeitung (1990) *Verpackungsabfall soll drastisch reduziert werden. Töpfer tritt auf die Notbremse.* Süddeutsche Zeitung 18/5/1990.

Susskind, Lawrence & Cruikshank, Jeffrey (1987) *Breaking the Impasse. Consensual Approaches to Resolving Public Disputes.* New York: Basic Books.

Teubner, Gunther (Ed) (1988) *Autopoietic Law: a New Approach to Law and Society.* Berlin/New York: de Gruyter.

Teubner, Gunther (1989a) *Recht als autopoietisches System.* Frankfurt: Suhrkamp Verlag.

Teubner, Gunther (1989b) How the Law Thinks: toward a Constructivist Epistemology of Law. *Law and Society Review*, **23**, 727.

Teubner, Gunther (1989c) "And God laughed...": Indeterminacy, Self-reference and Paradox in Law. In Joerges, Christian & Trubek, David (Eds). *Critical Legal Thought: an American–German Debate.* Baden-Baden: Nomos.

Teubner, Gunther & Febbrajo, Alessio (1992) *State, Law, Economy as Autopoietic Systems. Regulation and Autonomy in a New Perspective.* Milano: Giuffrè.

Van Driel, M. (1990) Zelfregulering als alternatief voor wetgeving. Een werkzaam concept van ruil en compromissen? (Self-regulation as an Alternative for Legislation. A Workable Concept of Barter and Compromises?). Sociaal-Economische Wetgeving, 38, 414.

Van Der Hoeven, J. (1970) De magische lijn. Verkenningen op de grens van publiek- en privaatrecht (The Magic Line. Inquiries on the Margin of Public and Private Law) as cited in Aalders & Van Acht (Eds). 1992: 1.

Van Rossum, E.R.C (1988) *Milieuconvenanten* (Environmental Covenants). Gravenhage: Staatsuitgeverij.

Van Vliet, Leendert M. (1992) *Communicatieve besturing van het milieuhandelen van ondernemingen: mogelijkheden en beprekingen* (Communicative Governance of Corporations' Environmental Behaviour: Options and Constraints. English summary: 208–214). Delft: Eburon.

Weidner, Helmut (1993) *Mediation as a Policy Instrument for Resolving Environmental disputes—with Special Reference to Germany.* Berlin: WZB Publication FS II 93–301.

Willke, Helmut (1983) *Entzauberung des Staates. Überlegungen zu einer sozietalen Steuerungstheorie.* Königstein: Athenäum.

Winsemius, Pieter (1986) *Gast in eigen huis* (Guest in our own Home). Alphen aan den Rijn: H. D. Tjeenk Willink.

Winter, Gerd (1985) Bartering Rationality in Regulation. *Law & Society Review*, **19**, 219.

CHAPTER 10

Environmental Performance Review: Self-Regulation in Environmental Law

Eric Bregman and Arthur Jacobson[1]
New York

INTRODUCTION

Americans have explored self-regulation as a strategy for the legal treatment of environmental issues only slowly and reluctantly. This is surprising. The intellectual roots of environmentalism, after all, are in ecology, the science of self-regulating, autonomous systems. With the notable exception of the National Environmental Policy Act of 1969 (NEPA)[2], the image of the human actor in environmental regulation has been that of a threat to, rather than a participant in, the ecological systems which regulation is designed to protect by controlling human interventions. The role of regulation, from this perspective, is to stop avoidable interventions and to control the harmful effects of unavoidable interventions.

Our thesis is that self-regulation can be an appropriate regime for the legal treatment of environmental issues, if we understand these as problems of co-ordination. By co-ordination we mean the mutual adjustment of individual projects through economic and legal institutions. Co-ordination includes actual collaboration, as in contract, and ordinary duties of civility, as in tort. It includes anonymous allocation through markets and overt government planning by zones, maps and other devices. Co-ordination seeks the adjustment of individual projects in light of a joint plan or state of affairs, and, in turn, derivation of the joint plan or state of affairs from those very projects. Co-ordination is thus the mutual orientation of individual projects through a joint plan or state of affairs drawn from all the projects. In its most developed versions, it includes the evolution of a joint plan through explicit self-regulation.

Environmental Law and Ecological Responsibility: The Concept and Practice of Ecological Self-Organization
Edited by G. Teubner, L. Farmer and D. Murphy
© 1994 John Wiley & Sons Ltd

Co-ordination is thus the extension into law of the idea of the market[3]. It is market exchange with law as a signalling medium, along with money.

Characterization of the environmental problem as the problem of external-ities, or the commons, has distracted attention from the question of co-ordination. All externalities raise issues of co-ordination, but sometimes only in the negative sense of preventing harm to victims. Co-ordination, in general, facilitates the attainment of benefits as well as the avoidance of harm. Good is as much a part of the environment as evil.

Self-regulation, we argue, is the legal regime appropriate to developed versions of co-ordination. Co-ordination is "developed", when it requires individuals to express and take positions on the public interest as part of their projects, and "undeveloped", when, as in contract, it does not force individuals to formulate or pursue a public interest separate from their own interests. In undeveloped versions of co-ordination, the public interest appears only as a constraint on individual interests, not as a component of interests. In the classical theory of contract[4], for example, "public interest" means only the interest expressed by the state—that is, the direct state regulation of con-tract formation or administration. The intersection of interests in an agree-ment can never reasonably be construed as "public". Developed versions of co-ordination put individuals into a position where they are forced, by the very fact of occupying that position, to pursue a public interest of some sort in the course of planning and realizing their individual projects.

Self-regulation is an appropriate regime for developed versions of co-ordination, because self-regulating entities must take positions on the public interest in formulating or enforcing a scheme of regulation. The interests expressed in regulation become a component of the interests of the self-regulating entities, just as the self-regulating entities weave their own interests into the scheme of regulation. Sharing regulation with regulated entities allows them to participate in the co-ordination of their individual interests. This can be to minimize the extent to which regulation interferes with their projects, or to maximize the extent to which it promotes them, but it is always along with the project of promoting a public interest and the not always coincident goals of the regulators. Regulated entities advance self-regulation by pursuing their own projects, which necessarily include, and are transformed by, the project of regulation.

Self-regulation does not wholly identify the public interest with interests expressed by the state. Regulation, like contract, assumes that the state has a monopoly on defining and pursuing the public interest. Self-regulation is con-sistent with the opposite assumption: regulators may participate along with regulated entities in forming the public interest from these individual interests which include the state. Also, regulatory beneficiaries join regulators and self-regulating entities in this project. Each party to self-regulation (regulators, self-regulating entities, and regulatory beneficiaries) expresses its own version of the public interest. Self-regulation thus breaks down the distinction between regulated entity and regulatory beneficiary. The public interest includes the interests of the self-regulating entities, while the burdens the self-regulating entities impose on regulatory beneficiaries cease to be different in kind from the burdens they impose on themselves.

The public interest promoted by regulation is static. It does not change from project to project. When it changes, it does so all at once, all at the same time, for all projects, and only when the regulator, independently of encounters with individual projects, decides to change it. Self-regulation, by contrast, promotes a dynamic public interest, constantly evolving from project to project. The public interest can, in principle, change on each encounter of the regulator with a project.

THE FORMS OF SELF-REGULATION

The forms of self-regulation vary along two axes, according to whether regulated entities and regulatory beneficiaries have a part in the shaping or the enforcement of the regulations governing them. Regulation is self-regulation when government shares with regulated entities and regulatory beneficiaries the power either to set the contents of regulations or to enforce regulations, or both at once. Arrayed against these variables (the power to set contents and the power to enforce) we distinguish four regimes of self-regulation: building codes, markets for rights to intervene in the environment, private enforcement rights, and the environmental review process.

These four regimes form an array of strategies from which a legislator or regulator may choose depending on the objective. Though the order in which we present the regimes is that of increasing degrees of self-regulation, we do not favour self-regulation, *per se*, for political reasons. Self-regulation is appropriate when the objectives the legislator wishes to accomplish can be more effectively accomplished by self-regulation rather than a different regulatory strategy.

In a building code regime, government does not share either the content-setting power or enforcement. Nevertheless, building codes usually permit what we may reasonably identify as a rudimentary form of co-ordination. This is the privilege of adjusting projects in the light of the transcendent public interest expressed in the code. This regime is appropriate where regulation embodies a transcendent public interest, that is to say, a public interest that is general and sufficiently important compared with any possible limiting private interest, to warrant ignoring private interests altogether.

In the market for rights to intervene in the environment, government shares content-setting, but not enforcement, with the regulated entities. This co-ordinates the public interest in containing environmental interventions with the projects of regulated entities, but not with the projects of regulatory beneficiaries except insofar as they have had a political impact on the public interest. Co-ordination allocates benefits to regulated entities rather than deflecting harms away from the regulatory beneficiaries. A market for intervention rights is appropriate where the public has a transcendent interest in the overall level of intervention rights, but not in the allocation of intervention rights amongst individual regulated entities. In other words, the public's transcendent interest tolerates integration with the production functions of the regulated entities. At the same time, this makes no effort to integrate the public interest with the individual interests of regulatory beneficiaries; regulatory beneficiaries can express their interests only by lobbying so as to affect the public interest.

In a regime of private enforcement rights (which includes but is not limited to common law actions), government shares enforcement with both the regulated entities and the regulatory beneficiaries. It does not, in principle, share content-setting with either of them. However, the dynamic of enforcement necessarily leads to an informal sharing of content-setting. This co-ordinates the public interest in limiting interventions with the projects of regulatory beneficiaries, as well as with the projects of regulated entities. Co-ordination allocates harms away from regulatory beneficiaries as well as benefits to regulated entities. This scheme is appropriate where the public interest and the individual interests of regulatory beneficiaries coincide.

Finally, in the environmental review process, government shares content-setting and enforcement with both regulated entities and regulatory beneficiaries. This scheme co-ordinates a series of individual projects with a public interest, drawn at least in part from the projects, in an overall scheme for allocating the benefits and burdens flowing from the projects. The environmental review process serves to co-ordinate the interests of regulated entities and regulatory beneficiaries with a public interest.

APPROACHES TO SELF-REGULATION

Legal theorists have supported self-regulation on three grounds: economic, political, and social. The economic ground is that it is efficient. The political ground is that it encourages citizens, especially regulatory beneficiaries, to participate in government. The social ground is that self-regulation co-ordinates economic goals with social and political goals.

Self-regulation is efficient when it reduces the costs of regulation or increases its impact, or when it accomplishes regulatory goals with the least possible interference in the portions of the market that the regulator does not want to disturb (Ackerman & Stewart, 1985: 1333). Sometimes it only appears to reduce costs, however, by hiding or shifting them from one level of government to another, or from government to regulated entities or regulatory beneficiaries. The model of self-regulation as efficiency assumes that the goals of the regulators are ordinary goals, to be pursued as rational economic actors would pursue them. (Whether rational economic actors would be involved in government in the first place is quite another story!) The goal of regulatory

efficiency is thus at odds with the view that political participation i
in and of itself (Ayres & Braithwaite, 1992: 18), and that citizens sh(
ticipate in government even at the expense of values they pursue in :

Self-regulation may encourage political participation in three ways: by
assigning enforcement to regulatory beneficiaries or to associations of regulated entities; by asking regulated entities or regulatory beneficiaries to set the
contents of regulations; or by providing venues in which the regulators must
at least hear the views of ordinary citizens. Some theorists regard participation
as a virtue in itself, either because it inculcates republican virtue in citizens[5],
or because it supplies a measure of legitimacy to government (Stewart, 1975:
1667). Others value participation when it makes regulation more efficient, by
encouraging the effective expression of under-represented interests or by
reducing the costs of regulation. Apart from wanting participation for its
own sake, however, these are mostly assumptions that have not yet been
empirically proven.

The inspiration for this chapter is the third ground for self-regulation, that
it co-ordinates economic goals with social and political goals. The model of
self-regulation as co-ordination does not assume that the goals of regulators
are ordinary goals to be pursued at the least cost and with the least interference
to markets. Nor does it assume that political participation is a value in itself
or necessarily contributes to efficiency.

The root of the chapter is Gunther Teubner's model of "reflexive law"
(1983: 239). In this model, government best accomplishes self-regulation by
setting up procedures within which regulated entities make decisions. The
procedures themselves push the regulated entities in the directions desired by
the regulator (Teubner, 1983: 254–57). Final decisions, however, are always
the province of regulated entities. Examples of "reflexive law" are corporate
law and labour law, where government does not tell regulated entities what
to do, but how to decide what they are doing (Teubner, 1983: 255). The great
advantage of reflexive legal rationality, according to Teubner, is that it

> requires the legal system to view itself as a system-in-an-environment ... and to
> take account of the limits of its own capacity as it attempts to regulate the
> functions and performances of other social subsystems (Teubner, 1983: 280).

In this model, government officials interact with the participants in autonomous subsystems over issues of co-ordination, but only as guardians of a
procedure for participants to work out substantive issues on their own,
without direct action by the government (Teubner, 1983: 254–57). We would
criticize this model for drawing too sharp or simple a distinction between
procedure and substance. The distinction, while dubious in its application to
corporations and labour unions, loses all its sharpness and simplicity in developed schemes of self-regulation. Government officials conducting an environmental review process, for example, do not just supervise a procedure. They

interact with regulated entities and regulatory beneficiaries to formulate norms that facilitate co-ordination. In environmental review, government officials collaborate with private parties over substantive issues of co-ordination.

NORMATIVE STRUCTURE AND LEGITIMACY

A corollary of our thesis about the relation between self-regulation and co-ordination is a thesis about the relation between legitimacy and what, for want of a better term, we call "normative structure". By this we mean the actual institutions in which people create, transmit and destroy normative material. A normative structure includes relations between people working with norms, but also certain elements in the structure of norms themselves.

Normative structure is useful for understanding the forms legitimacy may take in administrative states that maintain allegiance to the rule of law. Administrative states take responsibility for the outcomes of government action—the economy, education, scientific research, social welfare, and so forth. Legitimacy flows from the legitimacy of the outcomes for which the government takes responsibility, and the legitimacy of actions which the government calculates will achieve those outcomes. In a democracy, only some generally acknowledged version of participation can accord legitimacy to outcomes, and only expertise can accord legitimacy to the calculation of actions. The rule of law, by contrast, takes no responsibility for outcomes. Legitimacy flows either from the procedure by which authoritative officials promulgate a law (positivism) or from the inherent qualities of its contents (naturalism). Administrative legitimation follows, in part, the forms of legitimation in the rule of law: participation resembles procedural legitimation, and expertise, in principle, legitimates the quality of administrative content.

Nevertheless, administrative legitimation requires two alterations in normative structure. Participation allows non-officials to create and destroy norms, whereas positivism confines the creation and destruction of norms to officials. Legitimation by participation requires that parties claiming input to the content of norms conform in different ways. Expertise legitimates content by referring in part to the qualifications of the expert, not just to qualities of the expert's decisions. Naturalism, by contrast, legitimates content solely by reference to the qualities of content. Legitimation by expertise necessarily gives rise to complex relations between procedure and substance.

Administrative legitimation thus focuses attention on four aspects of normative structure:

(a) the behaviour of parties claiming input to the content of norms (e.g. regulated entities, regulatory beneficiaries, government),
(b) the conditions for ousting norms (who may oust them, on what occasions),

(c) the device for creating or transmitting norms (codes, written opinions, form files, and so forth), and
(d) the relations between procedure and substance.

The forms of self-regulation, in effect, mix and transform the methods of legitimation in the rule of law, adapting them to the outcome orientation of administration. Each form of self-regulation thus breaches the divide between substance and procedure in its own, distinctive manner. It may employ its own, characteristic legal devices, make the parties conform in different ways, and set its own conditions for ousting the norms of administration.

Our thesis is that the more elements of regulation government shares with regulated entities and beneficiaries in a scheme of self-regulation, the more the normative structure of that scheme resembles that of legal orders, such as common law, that are especially accommodating to self-regulation. Hence, one can look to these legal orders as alternative sources of legitimacy. The developed forms of self-regulation, we claim, anchor novel normative structures. That is to say, new behaviour of the parties, relations of substance to procedure, conditions for removal of norms and legal devices.

We propose these novel structures as a solution, useful in some circumstances, to what Richard Stewart (1975: 1667) has identified as the crisis of legitimacy in American administrative law: the failure of agencies to exercise their discretion in a manner consistent with ordinary democratic theory while implementing legislative directives (Stewart, 1975: 1673), and the inability of experts to justify distributive decisions in a condition of abundance (Stewart, 1975: 1677–78). He described a new model for achieving administrative legitimacy in a decentralized democratic economy, which he drew, in part, from the then still new environmental review process (Stewart, 1975: 1759–60). The new model, "administrative law as interest representation," expanded the traditional model, "to afford participation rights in the process of agency decision and judicial review to a wide variety of affected interests." (Stewart, 1975: 1760).

Stewart was skeptical that "administrative law as interest representation" could adequately address the problem of administrative legitimacy, because we lack a theory for determining the interests that ought to be represented and the resources and mechanisms for ensuring that they are represented (Stewart, 1975: 1789–90, 1800–05). Nor, for pragmatic economic reasons, is it possible to escape the crisis of legitimacy altogether by deregulation: "dissolv[ing] collective choice into individual exchange and public law into private law, thus limiting the administrative function" (Stewart, 1975: 1807). Administrative law seems condemned to shuttle forever between a powerful administration riding roughshod over regulated entities, and determined regulated entities overwhelming the interests of under-represented regulatory beneficiaries. "The only conceivable way out of the labyrinth," writes Stewart,

> would seem to be a new and comprehensive theory of government and law that
> would successfully reconcile our traditional ideals of formal justice, individual
> autonomy, and responsible mechanisms for collective choice, with the contem-
> porary realities of decentralized, unco-ordinated, discretionary exercises of
> governmental authority and substantial disparities in the cohesiveness and pol-
> itical power of private interests. Such a conception may well be unattainable, and
> in any event will not be achieved in the foreseeable future (Stewart, 1975: 1807).

Like Stewart, we do not believe that a unitary theory of administrative law is
tenable or even desirable. Administrative law, like the common cold, is one
way of speaking about many different viruses.

Nevertheless, the normative structures embedded in developed forms of self-
regulation do suggest models of administrative legitimacy when issues of co-
ordination are prominent. At least one of these models, the environmental
review process, has a normative structure that is not unlike the legitimacy that
backs the common law[6]. Co-ordination in environmental review is characterized, in its best moments, by the ceaseless adjustment of individual interests
against a constantly evolving public interest. Co-ordination of individual
interests with the public interest in the evolution of common law norms, we
submit, bears sufficient similarity to co-ordination in the environmental review
process to make comparison of the two worthwhile.

BUILDING CODES

We use the term "building code" generically, to refer to any regulation
specifying design criteria for a process or project which are deemed necessary
by the government to meet a goal or standard that it has set ("command-and-
control" regulation). The criteria need not specify every aspect of the design.
Nor need the criteria concern the design of physical objects, but may instead
describe the steps in, or parameters of, a process. Thus, requiring "manifests"
for the transportation of hazardous waste is no different conceptually from
specifying the design of air emissions control devices or underground storage
tanks. All use a "building code" scheme.

Building codes put both the content and enforcement of regulation entirely
at the command of government. The assumption behind codes is that the
public interest expressed in the criteria (public safety, aesthetics, or whatever)
transcends the conflicting individual interests of project promoters or process
managers. The further assumption is that the public interest is expressible in
the form of simple and relatively stable criteria. Hence, allowing regulated
entities even indirect influence in determining the content of regulation,
whether individually in "variances" or as a mass in "lobbying", is always
suspect.

Only when individual interest is allowed to modify the public interest expressed in regulation do regulated entities have an interest in aiding enforcement. Because code regulation expresses absolutely no individual interest, regulated entities have absolutely no interest in enforcement. Enforcement is a task of government alone, typically an inspection followed by the grant of the equivalent of a "certificate of occupancy" or a "notice of violation". Because the information needed for enforcement never varies, and because conformity to code criteria is typically physically apparent, inspection costs little in comparison with project costs, and can be charged directly to the promoters.

Building codes are self-regulatory in one, limited, respect: regulated entities may choose to build around code criteria, or incorporate code criteria into their projects in the most advantageous manner. For example, state regulations governing the construction of underground storage tanks typically require diking, tank lining, cathodic alarm systems and so forth, but do not specify tank dimensions or joint and splicing technology for piping. So long as regulated entities satisfy code criteria, they are free to design facilities in whatever way they see fit, for their own advantage[7].

Building codes facilitate only a rudimentary form of co-ordination. The only expression of individual interest they tolerate is adjustment of a project or process consistent with code criteria. As a consequence of the absolute lack of interaction between public and individual interests, the public interest does not change through encounters with individual interests. The public interest is static, precisely because it is transcendent.

Regulations setting performance standards employ a special variety of criteria. Setting performance standards without specifying how they are to be achieved allows more self-regulation than do ordinary design criteria. The content and enforcement of regulation is still at the command of government. However, a performance standard allows regulated entities to determine how they will achieve it. Though performance standards are sometimes set so that there is only one possible or feasible control technology[8], regulated entities in a performance standard regime, at least in principle, can design or install any control technology they wish, so long as they meet the performance standard.

The normative structure of building codes has the following characteristics. First, certain building codes permit only the government, and not regulated entities or regulatory beneficiaries as individuals, to have a direct input into normative content. In such regimes, regulated entities and regulatory beneficiaries can only have an indirect input by lobbying. Other codes allow regulated entities (but not regulatory beneficiaries), to have a direct input into normative content, typically through the "variance" process. Code/variance regimes cultivate a case law structure of decision-making virtually identical to the common law[9]. The major difference is that common law cases evolve against a normatively empty background, while variance cases evolve against a presumptively valid background of code criteria.

Second, codes without a variance process contemplate no normative conditions under which regulated entities can oust criteria. Let us take, for example, codes that prohibit a building inspector from granting a permit to build a house without code level insulation. If the code does not permit application for a variance, the only way to oust the insulation criterion is through amendment of the code. Variance regimes allow ouster. However, the burden on the regulated entity seeking ouster is far greater than the burden on the party in an ordinary common law adjudication, because code criteria are presumptively valid and applicable in all situations. Hence, the regulated entity seeking a variance would have to show that the code criterion would impose hardships or practical difficulties (i.e., economic or technical factors) severe enough to overcome the presumption of normative validity[10].

Third, as a consequence of the rigidity of criteria, codes collapse substance into procedure. The "substantive law" of codes is very much like formalism. The code administrator looks only to whether certain formal requirements of substantive behaviour have been met and no further. Just as formalism looks to whether the overt normalities of a contract are in place ("seal", "offer and acceptance", and so forth) without regard to the contract's substance, a building code looks to whether specific requirements are met (amount of insulation or number of electric outlets) without regard to overall or aesthetic design. Unlike formalities in contract, however, code criteria are, in principle, justified by substance[11]. They are the transformation of substance into procedure in the form of rigid specifications. Codes proceduralize substance.

The legal device characteristic of building codes is the code itself—a single, centralized, lay-administered list of all normative material pertaining to the field of regulation covered by the code. Codes may or may not employ the equivalent of a "building permit", which informs the regulator that regulated activity is taking place, and allows the regulator to determine in advance whether the planned activity conforms to the code. Some code schemes employ the lesser device of "registration", which simply requires regulated entities to notify the regulator of a planned activity. For example, small-scale petroleum storage tanks in New York State are registered, not subject to the regime of permits[12]. The theory of registration is that regulated activities are too numerous, and the dangers of non-conformance too slight, for systematic agency inspection to be efficient. In these circumstances, the impact of the regulation may effectively be at the manufacturing level. The main purpose of registration, however, is often not the enforcement of code criteria, but simply to know where and when tanks, say, were placed.

MARKETS FOR INTERVENTION RIGHTS[13]

Markets for rights to engage in undesirable environmental interventions allow regulated entities to determine for themselves the content of regulation. However, government continues to keep enforcement completely within its power[14]. The instrument of self-regulation is the creation of property rights to engage in harmful interventions and the trade of these rights in an organized market.

The market scheme of self-regulation has been applied primarily to air and water pollution in the form of "transferable pollution rights". Non-transferable intervention rights, or "charges", have been suggested as a device for matters such as airport noise (Harrison, 1983: 41), and have been imposed in lieu of meeting certain zoning requirements, such as provision of parking spaces or public park land. Regulation of hazardous waste disposal sites also effectively sets market conditions for the disposal of hazardous waste.

Government regulates by establishing rights at a desired overall level. Within this context, entities are individually self-regulating in a market by allocating rights to their highest users through trades, and by setting the overall price of rights[15]. The regulated entities "legislate" individual performance standards, which are then enforced by government as if they were criteria of a building code. Self-regulation through markets thus co-ordinates the government's goal of capping interventions with the production functions of regulated entities. Markets for intervention rights express the co-ordination through the price mechanism[16].

Markets for intervention rights facilitate self-regulation by allowing direct expression of the individual interests of regulated entities in a market. Regulated entities balance the costs of intervention rights against prevention costs in order to maximize profits. Nevertheless, the overall level of intervention rights, as in the building code regime, expresses the public interest without reference to the individual interests of regulated entities. By the same token, market regimes ignore the costs of intervention to regulatory beneficiaries as individuals. As in building code regimes, regulatory beneficiaries count only as a mass through their impact on the public interest. Ignoring costs to regulatory beneficiaries as individuals is reasonable only when the costs are low in relation to the administrative costs of ascertaining and compensating them, or when costs are uniform and can be compensated by means other than case-by-case measurement. When neither of these conditions is met, then costs of regulatory beneficiaries ought to be factored into the market for intervention rights as in the private enforcement regime.

Markets for intervention rights also give regulated entities an interest in enforcing intervention limits imposed on other entities in order to protect their own investment in the rights. However, regulated entities have no interest in enforcing intervention limits on themselves, which continues to be the responsibility of the government. Moreover, tracking and permitting intervention rights is an extra cost this regime imposes on either the government or participants in the market.

The normative structure of markets for intervention rights has the following characteristics. Markets for intervention rights are indistinguishable from the building code regime for regulated entities as a mass. Neither regulated entities nor regulatory beneficiaries can influence the overall content of regulation as individuals—the total level of emissions, effluents, and so forth. But a regulated entity, as an individual, can affect the content of regulation as applied to itself by co-ordinating its production function with the price of intervention rights, by purchasing intervention rights on the market or by reducing its need for intervention rights. Regulatory beneficiaries have no similar mechanism as individuals for affecting the impact on them of intervention rights purchased by, say, a neighbouring regulated entity. On the other hand, precisely because a regulated entity has the power to purchase its way out of an intervention limit, it has less motive to join a campaign by regulated entities acting as a group to loosen restrictions on the overall limit. It may even be contrary to the interests of larger firms to join such campaigns, since loosening restrictions on limits lowers barriers to entry for competitors.

Market regimes allow regulated entities to purchase their way out of intervention limits. But these regimes make no provision for regulated entities to oust norms. There is no variance procedure[17]. Normative material, as in building codes, is inaccessible to revaluation through individual effort. All revaluations by individuals are in the currency of money, not norms. Though regulatory beneficiaries may be free, in principle, to purchase intervention rights on the market just like regulated entities, in practice regulatory beneficiaries have no direct influence on intervention rights values.

The application of building codes entails a single procedural decision excluding variances: whether or not a criterion is met. They collapse substance into procedure because the criterion is established by a normative substance. In markets for intervention rights the normative substance establishes only the overall level of interventions, not the allocation or application of normative substance to individual regulated entities. Therefore, the enforcement of an allocative share is entirely procedural—a question of whether the regulated entity complies with its allocation, regardless of how or why it attained its share. The allocation process *is* informed by a substance (money) different from the normative substance informing the determination of total acceptable interventions.

The characteristic legal device of markets for intervention rights is that they treat permits to engage in undesirable environmental interventions as property. The difference between ordinary property and intervention rights is that government needs to protect the use value as well as the exchange value of ordinary property but only the exchange value of intervention rights.

PRIVATE ENFORCEMENT RIGHTS

In this category, governments share enforcement (but not content-setting) with both regulatory beneficiaries and regulated entities through two characteristic devices: private remedies and routine self-audits. Mandatory disclosure links routine self-audits with one version of the private remedy: private rights of action.

Private remedies include the traditional common law remedies relevant to environmental matters (nuisance, ultra-hazardous activities, toxic torts) and, in some instances, private statutory remedies[18]. These remedies apply to regulated, as well as to formally unregulated, entities. However, subjecting formally unregulated entities to private remedies cannot be distinguished, in principle, from other forms of self-regulation. The difference between common law and private statutory actions for damages is that the trigger of liability, hence the measure of damages, in private statutory actions is often more definite than in common law remedies. Precisely because they may be more definite, private statutory damage remedies tend to focus on harm to the regulatory beneficiary to the exclusion of benefit to the regulated entity, whereas common law remedies (nuisance, for example) are more likely to encourage discussion of both benefits and harms.

According to a typology developed by Richard Stewart and Cass Sunstein, modern American administrative law includes four distinct private remedies, three of which are relevant to self-enforcement (Stewart & Sunstein, 1982: 1198)[19]. The right of defence, which is the traditional private administrative remedy in American law, allows a regulated entity to seek a "judicial determination whether the legislature has authorized [an agency to commit] what would otherwise be a common law wrong" (Stewart & Sunstein, 1982: 1203). The right of defence enables regulated entities to participate in enforcement. The modern right of initiation allows regulatory beneficiaries to seek "review of agency inaction or of action that is assertedly inadequate" (Stewart & Sunstein, 1982: 1205). The right of initiation enables regulatory beneficiaries to participate in enforcement. Private rights of action, either expressly provided by statute or implied by courts, give regulatory beneficiaries the right to sue regulated entities in damages for statutory wrongs (Stewart & Sunstein, 1982: 1206).

Private rights of action enable regulatory beneficiaries and regulated entities to participate in enforcement together under the ultimate supervision of a court. They are no different, in this respect, from traditional common law remedies. In both cases, however, the absence of formal supervision by an administrative agency allows regulated entities and regulatory beneficiaries to use statutory or non-statutory norms to negotiate settlement of their differences according to their special situations, and thus to formulate subordinate norms with greater freedom than parties participating in an exclusively administrative process.

Private rights of action and common law remedies form a version of the market for intervention rights. Instead of setting the overall level of interventions by formulating the public interest in a political process, private rights of action and common law remedies allow regulatory beneficiaries, acting singly against individual regulated entities, to influence the overall level of interventions by pursuing their private interests[20]. Private rights of action and common law remedies thus factor costs to regulatory beneficiaries, as well as benefits to regulated entities, into the equations of value. (In markets for intervention rights, as we have seen, only benefits to regulated entities are included in the equations of value, once the overall level of interventions has been politically determined). As a result, the cost of intervention rights to one regulated entity may differ starkly from the cost to another. No market for intervention rights socializes value in the form of a single price, because regulatory beneficiaries introduce non-generalizable elements of cost.

Private rights of action and common law remedies implement a form of self-regulation in which regulatory beneficiaries play a significant role, along with regulated entities and the judiciary. In one sense, the regulatory beneficiaries play a leading role, since they, not an administrative agency, have the exclusive power to decide when to initiate enforcement. Consequently, at least in the US institutional framework, government participates in enforcement through a "passive" judiciary, not through an "active" administrative prosecutor. Because regulatory beneficiaries play a leading role, government may even lose control over enforcement altogether, when regulatory beneficiaries and regulated entities negotiate settlements outside the doors of the court. Indeed, such settlements may involve benefits to the regulated beneficiaries unrelated to the regulatory goals, in return for the benefit to the regulated entity of being left free of private (or any) enforcement.

Private rights of action and especially common law remedies co-ordinate the projects of regulated entities with those of regulatory beneficiaries. Markets for intervention rights, by contrast, co-ordinate the projects of regulated entities with a transcendent public interest expressed in the overall level of tolerated interventions. In a private enforcement regime, the public interest is intertwined with the working out of the interests of regulatory beneficiaries in numerous separate battles with regulated entities. A transcendent public interest never appears separately, as it does in building codes and markets for intervention rights. Nonetheless, a form of public interest does appear indirectly as normative standards incorporated in the rights of action.

Routine self-audits implement a version of building code inspection. Self-audits serve one or more of three functions:

(a) informal self-enforcement,
(b) formal self-enforcement by certification, and
(c) disclosure.

Even agencies jealous of their enforcement powers must rely in large measure on informal self-enforcement (business routine, ethical practice, and the like) for the efficacy of administration. In the securities industry, for example, internal compliance officers are prevalent[21]. Also, within the last three years, US industry has become aware that it is only by installing internal environmental control systems that corporations can stop employees from incurring catastrophic civil and criminal environmental liabilities[22].

The US Environmental Protection Agency (EPA) has not, so far as we know, seriously considered replacing direct reporting to the EPA with certification based on mandatory self-audits. Apparently, it believes the current public enforcement regime to be sufficient (Herz, 1991: 1249; Van Cleve, 1991: 1224). The European Community, by contrast, has proposed that industrial companies engage external licensed auditors to conduct voluntary audits, leading to certification of compliance, along the lines of audits in securities regulation[23].

The need for routine audits by corporations of their compliance with environmental regulations has been balanced by the inevitable demand of litigants and regulators for disclosure of the information developed in the audits. Since 1971 the Securities and Exchange Commission has required publicly held firms to disclose information concerning environmental activities, and has consistently interpreted these requirements rather broadly (Herz, 1991: 1254–57; Van Cleve, 1991: 1235–38). Congress enacted a limited audit/disclosure scheme in the Emergency Planning and Community Right-to-Know Act of 1986[24], which requires facilities to determine whether they use more than minor quantities of certain hazardous substances and to communicate this information for planning purposes to state and local emergency response authorities. Though the EPA has used auditing provisions in enforcement settlements and decrees, it has declined to encourage the use of voluntary audits by limiting disclosure requirements (Van Cleve, 1991: 1223–24).

Routine self-audits are instruments of self-regulation in that regulated entities assume primary responsibility for developing or certifying compliance information. Routine audits effectively subcontract enforcement to those burdened by regulation. Whether an enforcement agency can achieve the same level of compliance by monitoring self-enforcement as by direct enforcement is an empirical question, depending on such factors as the efficiency of selective *in terrorem* enforcement, the growth of a profession of outside auditors whose reputations or licences depend on reliable audits, the sensitivity of capital markets to compliance levels, and so on. There is no doubt, however, that routine audits linked to effective internal control systems reduce the cost, and enhance the effect, of regulation (Ayres & Braithwaite, 1992: 103).

Self-audits facilitate co-ordination in two ways. If disclosed, self-audits provide regulatory beneficiaries with more information about the actions of regulated entities. If undisclosed, self-audits provide the regulated entities with more information about their own actions than they would otherwise have, and allow them to control actions to enforce compliance.

The normative structure of private remedies has the following characteristics. When the standard of liability for a right of action is drawn from statute, government continues to have a building code input, but at a remove from actual enforcement. Government's input is even more diffuse in common law actions. Regulated entities and regulatory beneficiaries end by controlling standards through their domination of enforcement. Self-audits only accentuate the dominance of private parties *vis-à-vis* the government. Government can exercise direct control over self-audits by setting up the parameters of the audit or by sampling audits for validity, and it can exercise indirect control (through the judiciary), by compelling disclosure of audits or the methodology used to conduct them.

Government can mandate self-audits or create conditions in which prudent managers want to conduct self-audits voluntarily, to avoid liability-creating incidents. In the latter case, the regulated entities effectively control the content of self-audits to the exclusion of government and regulatory beneficiaries.

When the standard of liability for a right of action is drawn from statute, courts and the private parties have only a limited role in ousting the statutory norm in favour of a common law normative development. Because the enforcement power is exclusively in their hands, however, regulatory beneficiaries can effectively oust norms through private deals with regulated entities. The motive for making such deals is strongest when the regulatory beneficiary has a business relationship with the regulated entity. But even where there is no such relationship, enforcement costs may drive the regulatory beneficiary to a settlement, effectively ousting the statutory standard.

A statutory standard can also provide an umbrella for the creation of subsidiary norms amongst regulated entities sharing common liability. For example, regulated entities allocate CERCLA clean-up liability in settlement conferences, in which the government has only the *in terrorem* role of threatening to impose its own allocation should the settlement conference fail. The group sharing liability may also include regulatory beneficiaries, and the subsidiary norms may effectively create new classes of regulatory beneficiaries who are protected by these norms. The regime of private remedies thus furthers the breakdown of distinctions between regulated entity and regulatory beneficiary. This breakdown is especially apparent in common law actions, where the regulatory burdens are, at least in theory, always reciprocal.

Privatization of enforcement transforms procedure entirely into content. The private remedies regime is thus the opposite of the building code regime, which treats content formally, as if it were procedure. The procedural role of government in adjudication stays in the background, when regulatory beneficiaries work out their relations with regulated entities outside the doors of the court.

Precisely because its normative substance is shorn of any procedural aspect, the private remedies regime discourages regulated entities and regulatory beneficiaries from devising any general scheme of value: each circumstance is different, and the regulated entities and regulatory beneficiaries have the opportunity to tailor the normative standards to fit their own, unique circumstances.

In such situations insurance is often the agent for generalizing costs. However, the difficulties faced by the insurance industry in the USA in administering environmental liabilities under the standard Commercial General Liability Policy, or even under policies designed to insure against environmental hazards, proves the resistance of a private remedies regime to even this form of generalization. Only when courts or private parties turn some part of substance into procedure can insurance perform its function of classifying typical risk situations.

The characteristic legal device of private enforcement is the ordinary private damage action, sometimes coupled with a self-audit. To the degree that legislation provides the standard of liability for the private action, it takes on the character of building code enforcement. Nevertheless, the primary device of private enforcement, the private damage action, is the classic common law enforcement device. The private remedies regime achieves the level of self-enforcement of common law jurisprudence. To the degree, however, that government requires self-audits, the private enforcement regime transcends self-enforcement in common law.

ENVIRONMENTAL REVIEW PROCESS

Description

We use the term "environmental review process" to mean the review of the potential impacts on the environment of a particular proposed discretionary action by government, ranging from a small variance for a sideyard zoning setback in order to construct a patio to major actions such as area-wide rezoning, construction of a superhighway, or the condemnation and redevelopment of city centre areas. The models are NEPA and the New York State Environmental Quality Review Act (SEQRA)[25], together with their respective regulations[26]. The purpose of these environmental review procedures is to inject environmental factors into the government's decision-making process, but at least in principle, the agency's decision is not substantively determined by the question of environmental impact[27].

The review is undertaken in connection with discretionary actions[28]. Thus, where building codes (as we use the term), apply and the regulated entity seeks government approval (a permit to build, operate, transport, or occupy), "as of right" because specific criteria are met, there is no environmental review. The same is true for transactions in intervention rights.

Where discretionary action is involved, the first step is the determination of who, or what part of government, is responsible for the review—which agency is the "lead agency" and which other agencies are "involved agencies"[29]. This can be complicated by politics and turf wars among agencies, but usually is a simple decision as to which agency has jurisdiction over the most important approvals for the project.

The lead agency's first decision is whether the project may have a significant impact on the environment[30]. The determination of significance may be a subjective analysis of general standards or a mechanistic application of specific criteria or a combination of the two[31]. If there are not considered to be any, the agency makes a "negative declaration" or a "conditional negative declaration" (CND); that is to say, that it considers that there will not be any significant impacts if specified conditions are incorporated into the project[32]. This initial determination is premised on information presented to the agency in an "environmental assessment form" (EAF), the content of which is established by regulation and is often the subject of vigorous debate amongst regulated entities, regulatory beneficiaries and the agencies that oversee the environmental review process[33]. The negative declaration or CND is generally issued simultaneously with the agency's substantive ruling. The EAF is usually prepared by the project proponent (actually, the proponent's environmental consultants and lawyers), as where a variance is requested. Sometimes, the lead agency prepares the EAF (actually, the agency's own consultants, usually another agency or bureaucratic division, and their lawyers), as when a wide-scale re-zoning or land use plan is proposed.

The public as such does not participate in these phases of the environmental review process, although records and files of negative declarations must be kept and CND's must be published. However, public participation is frequently mandated by the agency's general rules, as with zoning variances or condemnation, in the course of which environmental factors can be, and often are, raised[34].

If the agency makes a "positive declaration" that the proposed project may have a significant impact on the environment, a much more elaborate environmental review proceeds. The core of this process is the environmental impact statement (EIS)[35]. The agency may prepare the EIS or delegate its preparation to the project proponent[36]. (Again, the real work is done by environmental consultants, or their bureaucratic equivalents, and lawyers.)

The first step in the EIS process is "scoping": defining the scope of the EIS, or the issues that it will address. Under SEQRA, formal scoping is optional for the lead agency. The agencies involved "should provide input ... reflecting their agency's concerns", and the lead agency then determines whether or not to invite "interested agencies" and the public to participate[37]. The methods used to obtain scoping information "should reflect the complexity of the project, the degree of public concern and the significance of the environmental impacts"[38].

The impact analysis is simple in theory, complex in application. The EIS must contain the following key elements: a description of the proposed action, including its purpose, public need and benefits; a description of the environmental setting of the areas to be affected (the substantive areas of concern having been identified in scoping); a statement and evaluation of the environmental impacts of the proposed action, derived from comparing a description of the locations involved in the future without the project (no-build) with a description of the same locations with the project (build); identification and discussion of any environmental impacts which cannot be avoided or adequately mitigated; a description and evaluation of "the range of reasonable alternatives to the action which are feasible, considering the objectives and capabilities of the project sponsor", including alternative sites, technology, scale, design, timing, use, and so forth; and a description of mitigation measures[39].

After scoping, the lead agency or the applicant prepares a draft EIS (DEIS)[40]. If prepared by an applicant, a preliminary DEIS (PDEIS) is submitted to the lead agency, which must then determine whether it is "satisfactory with respect to its scope, content and adequacy for purposes of commencing public review"[41]. This can be one of the most important stages in the process. The adequacy of the methodology identified at scoping is reassessed by its ability to reflect accurately present or future conditions. The results may disclose limitations of methodology as applied, leading to modifications or even changes in baseline methods. Entirely new methods may be needed when application of prior methods proves impractical. Impacts disclosed may lead to modification of the project so that they are avoided or the development of "mitigation" to counteract impacts. In short, this period between scoping and determination by the lead agency that the DEIS is complete is the stage where the regulated entity/applicant has the greatest involvement and participation in the normative and substantive definition and application aspects of the regulatory process. It is the high point, or period, of the regulated entity's self-regulation.

We present three examples of self-regulation at the PDEIS stage of environmental review, drawn from typical situations in practice:

(a) Initial investigation disclosed the possibility that portions of the site of a proposed subdivision contained significant prehistoric artifacts. Before submitting the PDEIS, the applicant redesigned the layout to leave some of these areas in an open space reserved area so that they would not be disturbed by construction. The applicant and the applicant's special archaeological consultant devised a protocol for testing to determine whether or not those areas which could not practically be avoided were archaeologically important, that is, whether the artifacts found would significantly advance understanding of prehistoric peoples. Initial testing indicated that the areas were not significant. The agency's review of the methodology and results of that testing led to a spirited debate over how much testing was needed. The consultants for the applicant and the agency each cited literature and anecdotal support for requiring more and less testing. The lawyers and agency heads mediated a compromise.

(b) A proposed urban renewal residential development was to be located near a major airport, an elevated rapid transit line and a six-lane boulevard.

How were worst case noise impacts to be analysed? Actual noise monitoring was impractical. The municipal agency rejected the noise contours analysis of the airport previously prepared by the Federal Aviation Administration as being too general. The applicant called on an expert in aircraft noise to develop theoretical aircraft noise levels throughout the project based on the actual, published airline schedules and the specific aircraft used. This was finally accepted by the agency, but only after field tests of noise generated by several specific aircraft were found to be consistent with the theoretical analysis. This left the further question of how to integrate aircraft, traffic and rapid transit noise to get a worst-hour noise level. Each noise-generator caused the greatest noise at a different time. Eventually, the applicant and agency agreed on a methology and changes were made to the project where necessary to assure an acceptable interior noise level. This *ad hoc* methodology was thereafter required by the agency for other projects subject to significant aircraft noise.

(c) In scoping, the study area for various impacts is defined. During preparation of the DEIS for a major urban centre project, it became apparent that there was a potential for significant traffic and mobile-source air quality impacts beyond the defined study area. Detailed, intersection-by-intersection application of the agreed methodology for assessing impacts was virtually impossible. The applicant's consultants and the agency's technical advisors developed a multi-level screening procedure to identify potential impact locations. This methodology was further refined and used later for another large project in the same municipality.

Public review follows the lead agency's completion of its own DEIS or acceptance of the applicant's DEIS[42]. The DEIS is "published" by notification of its completion in appropriate publications and by distribution of copies to all involved and interested agencies and to all persons, entities and groups which have an interest in the matter[43].

A period for public comment follows, which under SEQRA must be at least 30 days[44]. The agency may, but need not, conduct a public hearing for comments on the DEIS[45]. Where a public hearing on the proposed action is required under the agency's non-environmental regulations, the two hearings are often held as one. This is also the case where review is required under both SEQRA and NEPA. One EIS and one public hearing serves for both.

The public comment period is when the regulatory beneficiaries participate most directly in the review and promote their interests in the process. If they fail to bring forward their concerns at this point, they may well be barred from raising those concerns, however weighty, in a later judicial review of the lead agency's action[46].

The public's comments must be answered. This is done in the final EIS (FEIS), which consists of the DEIS, copies or a summary of all comments on the DEIS (whether or not a hearing was held), and the lead agency's response to the comments[47]. Where the EIS is prepared by an applicant, the responses to comments are initially prepared by the applicant and its EIS consultants. They must be approved by the lead agency, of course, since it is the agency's EIS even when prepared by the applicant. Therefore, the time from receipt of public comments to publication of the FEIS is another period of intense interaction between the applicant/regulated entity and the regulating agency. All the issues of methodology, project revision, mitigation and the like may be revisited. The agency is likely to be in contact with some of the public commentators as well, during this period. It is not uncommon that additional data collection and analyses are ordered as a result of information obtained during the public comment period. Indeed, the failure to follow up and investigate issues raised during this period can invalidate the entire review process and project approval[48].

When the government takes action, having been informed of environmental effects, it must prepare a Record of Decision (NEPA)[49] or Findings (SEQRA)[50]. Although there is some debate about whether there is a difference in the two models, the net effect for present purposes is that the government must describe what has been reviewed and set forth a rationale for its decision.

Analysis

In agency action governed by the environmental review process, determining the content of regulation is not distinguished from enforcement. The regulator builds regulations into the proposed project by planning the project along with the regulated entity. In building codes, the determination of content precedes project planning. In private rights of action, the determination of content follows project planning. Only in the environmental review process do the determination of content and project planning occur together. By the same token, the regulated entity may modify the project as agencies formulate regulations in the course of the process.

The environmental review process constitutes self-regulation in that the regulated entity has input into and participation in determining the scope, methodology, and standards for defining impacts, and mitigation. It is of utmost importance that the regulated entity initiates project proposals, and typically it is also responsible for environmental data gathering, analysis and presentation. As in self-audits, the regulated entity is responsible for providing information to the government. Trusting data-gathering to regulated entities alters the balance of power between regulatory and regulated entities in favour of the regulated entities. Involvement of the regulator "on the ground floor" of a project also alters the balance of power, this time in favour of the regulator.

Environmental review clarifies the fact that regulatory beneficiaries play a role in self-regulation too. The only difference between regulatory beneficiaries and regulated entities is that regulation, in formal terms, does not purport to regulate the behaviour of regulatory beneficiaries, even though their behaviour may be affected, often by legal compulsion. Regulatory beneficiaries may play as active a role in environmental review as regulated entities, sometimes even sharing the task of formulation project proposals.

The environmental review process extends co-ordination to allocation of benefits and harms among regulated entities and regulatory beneficiaries, unlike markets for intervention rights and private damage actions. Markets for intervention rights co-ordinate benefits to regulated entities with an overall public interest in avoiding harmful consequences flowing from the production of those benefits. The interests of individual regulatory beneficiaries are irrelevant. Also, the government sets the supply of intervention rights so as to achieve the desired level of interventions, rather than to shift benefits from regulated entities to regulatory beneficiaries. The government may, however, tax holders of intervention rights, and allocate portions of the revenue to regulatory beneficiaries. Such schemes do begin to achieve a broader kind of co-ordination, but to that extent depart from a pure market for intervention rights. Private damage actions do not, in principle, allow regulatory beneficiaries to capture any part of the benefit to regulated entities, since the measure of liability is always harm to the regulatory beneficiary. Only environmental review allows regulated entities and regulatory beneficiaries to co-ordinate the allocation of benefits flowing from a project as well as harms caused by the project.

The normative structure of environmental review has the following characteristics. It involves the concerned agencies, the regulated entity and regulatory beneficiaries. As a result of the process, the government formulates an explicit public interest. Private rights of action, by contrast, suppress the role of agencies in formulating a public interest. Individual regulatory beneficiaries do not control the environmental review process, however, as they may control private rights of action.

In the building code regime and in markets for intervention rights, regulatory beneficiaries have no formal role other than lobbying to form the public interest. In private rights of action, the formal role of regulatory beneficiaries is limited to collecting damages for injuries already caused by regulated entities or the agency's failure to act effectively. In environmental review, regulatory beneficiaries can influence agency action before damage occurs.

The environmental review process also raises the issue of co-ordination among multiple agencies with jurisdiction over a project. In principle, the lead agency is supposed to handle the intra-governmental co-ordination function. However, the degree to which agencies can assemble a coherent position on fundamental issues of policy distributed among them is problematic. Involved agencies sometimes even take a "second bite at the apple" by requiring the regulated entity to undertake additional information-gathering and analysis which could, and in principle should, have been part of the EIS process.

Some standards entering the environmental review process are predetermined as in building codes, and cannot be ousted as a result of the individual efforts of either regulated entities or regulatory beneficiaries[51]. Other standards are *ad hoc*, and are produced in the course of, and confined to, the process[52]. A third category are standards which pre-exist, but are modified in the course of the process, either in their application to the particular project or more generally[53].

The only lawful responses to building code criteria are compliance or avoidance. The environmental review process adds mitigation in certain circumstances. Avoidance of impacts allows a regulated entity to redesign a project to avoid significant impacts. Mitigation allows a regulated entity to add items to a project to remove or compensate for impacts. For example, a subdivision could be redesigned to avoid disturbing an area of potential archaeological significance, or the potential significant impacts of construction in the sensitive area could be mitigated by undertaking a full-scale archaeological dig. A project could design its buildings to avoid the impacts on them and surrounding areas created by the buildings' disturbance of emissions patterns from a neighbouring stationary source, or it could mitigate the impacts by taking steps to reduce the emissions from that source. A builder may construct new wetlands to replace the benefits lost when a project requires the destruction of pre-existing wetlands.

Building codes proceduralize substance in the form of rigid criteria—once these are determined, the regulated entities and the code administrator take a totally external posture towards them, as if they were formalities without substance. The environmental review process takes just the opposite tack. It "substantiates" procedure by requiring agencies to adopt procedures for enhancing the probability of maximizing the achievement of co-ordination of the various interests associated with the impact of governmental action on the environment.

Environmental review can be looked at as forcing the co-ordination of interests: the government, regulated entities, and regulatory beneficiaries must go through the stipulated processes which force them to interact and co-ordinate their interests with respect to a proposed action. While courts are very reluctant to substitute their judgment for the results of that co-ordination, they are not at all slow to intervene where the co-ordination process is flawed or incomplete. The environmental review process is a matrix, the purpose of which is to co-ordinate these interests.

The process is futile, however, if the government does not pay attention to the input it receives, or fails to look at concerns which are otherwise brought to its attention. In consequence, the courts have evolved the "hard look doctrine", which holds that a necessary part of the process is that government takes a "hard look" at the relevant environmental issues before taking action[54]. Sloppily applied, this can blur the line between courts imposing their own substantive views as to the desirability of a particular government action and the enforcing of the process of co-ordinating the various interests concerned. In principle, though, it should not.

The environmental assessment form and the EIS appear to be completely novel legal devices. Nowhere else, in our understanding, does the regulated entity tell the regulator what the problem is and how to solve it, with the regulatory beneficiaries looking over everyone's shoulder and also giving their opinion as to the nature of the problem and appropriate solutions.

The EIS provides the material for a case law development, reminiscent of the role played by written opinions in common law. Agency decisions that would otherwise be protected by "informality" are subject to review by the courts of whether their actions were arbitrary and capricious. The process goes further than common law, which focusses on adjusting rights between individual private parties, and co-ordinates the interests of government together with the interests of private parties as regulated entities and as regulatory beneficiaries. It needs also to be said that it shares with common law the fact that the process itself can be burdensome, time-consuming and expensive, and sometimes out of proportion to the ultimate benefits.

CO-ORDINATION THROUGH SELF-REGULATION: A NEW COMMON LAW

Environmental review attempts to institutionalize a form of co-ordination, in which regulated entities, regulatory beneficiaries and the government interact to formulate a public interest, that is informed by and is an expression of private interests, which are in turn transformed by the process. For all its flaws, accusations of agency capture on the one side, agency rigidity on the other side, and crushing expense on all sides, the environmental review process does exhibit a legitimate vision of co-ordination.

Two other visions would co-ordinate interests by relying entirely on the market or entirely on building codes or city plans promulgated by the government. Market co-ordination is efficient, innovative and tolerant of diversity, particularly in an advanced industrial society which is so complex that traditional government regulation has broken down. But markets are always short-term, and environmental survival requires co-ordination of long-term concerns, which markets ignore whenever possible.

Government co-ordination is efficient in that regulated entities and regulatory beneficiaries always know where they stand and can adjust their actions accordingly. However, it is extremely intolerant of innovation and diversity, except through the political process which acts only in gross and from which agencies are often insulated.

The advantage of environmental review is that it brings more information about environmental impacts into the decision-making process of government, and it does so project-by-project. Market co-ordination keeps that information "secret", so that the government cannot easily "learn" what regulations may be necessary to correct the flaws of the market. Government does learn what regulations may be necessary in the case of pure government co-ordination, but cuts itself off from information that can be generated in planning of individual projects. In the environmental review of individual projects government sets out to learn, and to apply what it has learned, project-by-project. Environmental review thus embeds the process of formulating a public interest in the self-regulating system of the market.

The values expressed in the environmental review process certainly strike a defensible balance between the needs of the whole and the energy, creativity and vision of individuals. However, in reality, agencies can be just as obdurate and arbitrary from the point of view of either regulated entities or regulatory beneficiaries, particularly as their sophistication in handling the environmental review process grows. Nevertheless, environmental review does impose certain constraints on agency behaviour, and the practical effect of NEPA and SEQRA is that agencies are indeed more responsive to the balancing of environmental factors "with social, economic and other essential considerations"[55].

Whatever values we seek to promote in co-ordination, and however effectively we are able to promote these in our actual administrative institutions, there should be no doubt that each of the four regimes of self-regulation we have described possesses a normative structure, each with a claim to legitimacy within its sphere. The normatively and technically unconstrained decision-making that critics of the US administrative process point to in other spheres does not afflict the regimes of environmental self-regulation.

The tendency of US lawyers and commentators to brand the pull-and-tug of environmental review as "not law" is a result, we believe, of their ideological commitment to, and admiration of, the normative structure of common law. The discourse of environmental review is a recognizably "legal" discourse, with its own peculiar, and still little understood, normative structure. The processes of environmental review are no more or less legitimate than the processes of common law adjudication.

Common law is our own universalizing jurisdiction, against whose normative structure US lawyers judge the legitimacy of all subsidiary legal regimes[56]. Administrative law harbours subsidiary regimes, such as labour law, that pass the common law test with flying colours. However, to the degree that a legal regime competes with common law as a universalizing legal order, the peculiar needs served by the competing regime will be neglected or scorned by the colonial agents of the universalizing jurisdiction. As devotees of "law office law" know very well, the concept of discretion, even within common law precincts, is not always a licence for caprice, but often a commission to work out predictable, legitimate solutions in terms of a different normative structure[57].

Just such a commission has been granted to agencies to co-ordinate environmental benefits and harms in projects presented to agencies for approval. Any legal regime in which regulated entities swear that government is ignorant, rigid and high-handed, in the same moment that regulatory beneficiaries swear that government has been captured by regulated entities, and government officials swear that both regulated entities and regulatory beneficiaries are excessively demanding, is probably doing its job as well as any other.

NOTES

1 The authors wish to thank Eva Hanks, Michael Herz, and Stewart Sterk of the Cardozo School of Law for their valuable comments on earlier drafts. Subsequent drafts benefited greatly from discussions at the Conference on Ecological Responsibility and Self-Organization, 13–15 April 1992, sponsored by the European University Institute.

2 42 USCA §§ 4321ff.

3 Philip Heymann (1973: 805–13) defines co-ordination from the perspective of individual rational choice. From this perspective, the public interest can be an instrument for attaining individual interests only. People co-ordinate behaviour only if by doing so they maximize their interests, but no one in Heymann's model defines their interest as necessarily including, contributing to, or working on a public interest. Heymann's is a private law view of co-ordination. Our view spans both public and private law. In the sphere with which we deal in this chapter, people include, contribute to, or work on a version of a public interest, not because they are beings with a higher consciousness, but by virtue of the situations into which regimes of self-regulation put them.

4 By "classical theory of contract," we mean the work of theorists such as Williston, who constructed contract doctrine entirely from individualist premises (Gilmore, 1974: 6).

5 As do, for example, Ayres & Braithwaite, 1992: 17–18.

6 It is well to remember, for this purpose, that the origin of common law in medieval England was not the heavens, but the Chancery, the highest administrative office of a centralizing, bureaucratically rational power: Plucknett, 1956: 139–40.

7 Regulation in the tax code is analogous.

8 See, e.g., Clean Air Act, 42 USC 7501ff.

9 Applications for variances often trigger an environmental review process. Hence, analysis of codes with a variance procedure properly belongs under that heading.

10 See Anderson (1986) vol. 3, §§ 20.02 and 20.72.

11 The corporate form insulates shareholders from laws which would otherwise impose personal liability; code insulation criteria protect homeowners from the cold. Also, the reason shareholders choose the corporate form is totally self-regarding. The reason codes require insulation may have nothing to do with the immediate self-interest of the regulated entity.

12 6 NYCRR § 612.

13 By "intervention rights" we do not mean rights to intervene in the administrative process, but rights to engage in undesirable environmental interventions.

14 This first explicit step towards a legal regime of self-regulation by humans as participants in ecological systems was taken by policy-makers quite early on in the USA, in the mid 1960s, and by commentators soon after (Ackerman et al., 1974).

15 In July of 1991, the Chicago Board of Trade voted to create a private market for trading rights to emit sulfur dioxide. See Peter Passell, "A New Commodity to be Traded", *New York Times*, 17 July 1991, at A1.

16 The literature on self-regulation sometimes describes "law-making by private groups" as self-regulation. See, e.g., Ayres & Braithwaite, 1992: 101–06. Even where groups self-legislate, however, agencies seldom abandon legislation entirely. The literature refers to government back-stopping industry self-regulation as "co-regulation." See Ayres & Braithwaite, 1992: 102. Agencies may impose guidelines, or supervise private legislators informally. On self-regulatory organizations in securities law, see Ayres & Braithwaite, 1992: 104. See also s. 19 of the Securities Exchange Act of 1934, 15 USC s. 78s, as amended, and rr. 19b-4, 19c-1, 19c-3 to 5, 19d-1 to 3, 19g2-1 and 19h-1, promulgated thereunder. On self-regulation by trade associations, see Ayres & Braithwaite, 1992: 128. Government at least tacitly adopts the private legislation as its own. Though government establishes the overall level of intervention rights it allots to a market, it is indifferent to the allocation, or changes in allocation, of rights among regulated entities. It adopts the allocation accomplished by the market only for enforcement purposes. In private legislation the government does not establish a baseline of acceptable interventions, but adopts the baseline established by the private legislator.

   Whether agencies exercise control or not, private legislation simply shifts the political issues of regulation, well described by Stewart (1975: 1790–99) to the private legislator. The private legislator "stands in" for the government. Individual regulated entities regulate themselves only to the extent they participate effectively in the decisions of either government agencies or a private legislator. Even if the private legislator is democratic (often it is not), the extent to which self-regulation may be equated with ordinary democratic theory is as much a question for private legislators as for agencies. See *ALA Schecter Poultry Corp.* v. *United States*, 295 US 495 (1935).

17 Building codes may also set an overall level of interventions, e.g., housing density in a zone. The variance process allows individual regulated entities to affect the overall level of interventions—a variance allowing greater housing density for a particular project. Individual regulated entities cannot change the overall level of interventions in a market for intervention rights; that can be accomplished only through legislation.

18 The US Supreme Court has sharply limited the application by courts of private damage actions for statutory wrongs (Stewart & Sunstein, 1982: 1304–05). Nevertheless, some federal statutes do provide for private damage actions in limited circumstances. For example, the Comprehensive Environmental Response, Compensation, and Liability Act, 42 USC §§ 9601ff., provides a private action for the recovery of response costs consistent with the national contingency plan, 42 USC § 9607(a)(4)(B). States too have statutes providing for private damage actions, and also may be more generous than federal law in implying private damage actions for statutory wrongs. See, e.g., the private right of action implied under Art. 12 of the New York Navigation Law for clean-up, lost income, and other damages caused by oil spills: *NY Navigation Law* § 181 (McKinney, 1992).

19 The fourth remedy, "new-property hearing rights," gives a trial-type hearing to the recipients of certain individual statutory benefits, such as welfare payments (Stewart & Sunstein, 1982: 1207). It is not relevant, so far as we know, to environmental law.

20 The private interest of regulatory beneficiaries may include the prospect that they will also be regulated entities in other proceedings.

21 The permanent presence of Internal Revenue Service agents in large industrial concerns conducting almost constant audit supervision is unusual.

22 The Department of Justice has promulgated sentencing guidelines for environmental crimes. They include, indeed highlight, whether the convicted corporation

voluntarily implemented a self-audit plan and reported the violation in question when it was discovered in the self-audit process. The guidelines suggest, but do not require, that this voluntary self-disclosure of criminal activity should eliminate or reduce fines and jail terms. See Wallance & Kornreich (1991). Van Cleve (1991: 1224–27) quotes a letter from Richard B. Stewart, Assistant Attorney General, Land & Natural Resources Division, Department of Justice, to Admiral James D. Watkins, Secretary of Energy and William K. Reilly, Administrator of EPA (14 September 1989), expressing a similar policy. What a Hobson's choice!

In the fall of 1990, the Cardozo Law Review and Cardozo's Samuel and Ronnie Heyman Center for Corporate Governance held a symposium, "Corporate Governance and the Environment: Beyond the Transactional Audit," in which representatives of industry emphasized the control function of audits. See, e.g., Friedman, 1991: 1320–23.

23 "Draft Elements on a Proposal for a Council Regulation Establishing a Community Environmental Auditing Scheme for Certain Industrial Activities" issued by the Commission of the European Communities (DG XI), 5 February 1991. See also Freeman & Cunningham, 1991.

24 42 USC, §§ 11001ff.

25 *NY Environmental Conservation Law* §§8-0101ff. (McKinney, 1992) (hereinafter ECL).

26 40 CFR §§ 1500ff., and 6 NYCRR Part 617. With respect to the federal regulations, the majority of agencies have issued their own regulations. See, e.g., 24 CFR Part 50 (Department of Housing and Urban Development); 32 CFR Parts 650 and 651 (Department of the Army).

27 Courts have interpreted NEPA as requiring only that federal agencies follow certain procedures for developing information about the environmental impact of "major Federal actions significantly affecting the quality of the human environment," 42 USC 4332(2)(C). See *Vermont Yankee Nuclear Power Corp.* v. *The Natural Resources Defence Council, Inc.*, 435 US 519 (1978). At least in principle, SEQRA requires New York state agencies to "use all practicable means to realize the policies and goals set forth in this article, and to act and choose alternatives which, consistent with social, economic and other essential considerations, to the maximum extent practicable, minimize or avoid adverse environmental effects, including effects revealed in the environmental impact statement process." ECL § 8-0109(1). However, the statutory command is only as good as the willingness of a reviewing court to second-guess the decisions of agencies as to practicability, and so forth. Given the standards of review employed by New York courts over SEQRA determinations, its substantive cast may be somewhat illusory. See, e.g., *Town of Henrietta* v. *Department of Environmental Conservation* 76 AD 2d 215, 222, 430 NYS 2d 440, 447 (4th Dept. 1980). See also Gerrard *et al.*, 1990 § 6.03[5].

28 For NEPA, see *Dale Robertson, Director of US Forest Service* v. *Methow Valley Citizens' Council*, 490 US 332 (1989). For SEQRA, see ECL § 8-0105(5)(ii).

29 For NEPA, see 40 CFR §§ 1501.5 and 1501.6. For SEQRA, see ECL § 8-0111(6) and 6 NYCRR § 617.2(t) and (v).

30 For NEPA, see 42 USC § 4332(2)(C). For SEQRA, see ECL § 8-0109(2).

31 For example, 6 NYCRR § 617.11 lists very broad standards for determining significance, such as: "(1) a substantial adverse change in existing air quality, ground or surface water quality or quantity, traffic or noise levels; a substantial increase in solid waste production; a substantial increase in potential for erosion, flooding, leaching or drainage problems". The New York City Department of Environmental Protection has evolved specific impact criteria for noise (an increase of 3 dB is significant) and traffic congestion (a one-tenth of 1% increase in the volume/capacity

ratio), but has evolved no such criteria for increases in sewage or solid waste flows, secondary displacement of residents through gentrification, changes in neighbourhood character, visual impact and the like. Some of these areas are not readily amenable to hard criteria because of their inherently subjective character, such as impacts on neighbourhood character or visual impacts, but they are all subject to an evolutionary process of normative transformation similar to that in the common law.

32 For SEQRA, see 6 NYCRR § 617.6(g)–(h). NEPA has no formal CND device. The same result is accomplished informally through a Finding of No Significant Impact, CFR §§ 1501.4(e) and 1508.13.

33 For NEPA, see 40 CFR § 1501.3. For SEQRA, see 6 NYCRR § 617.6(g).

34 The authorization for such rules is in 40 CFR § 1501.4(e). See, e.g., 10 CFR § 51.33 (Nuclear Regulatory Commission).

35 For NEPA, see 42 USC § 4332(2)(C). For SEQRA, see ECL § 8-0109.

36 For NEPA, see 40 CFR § 1506.5(a) and (c). For SEQRA, see ECL § 8-0109(2) and (3).

37 6 NYCRR § 617.7(b).

38 Ibid.

39 Ibid. § 617.14(f).

40 For NEPA, see 40 CFR § 1502.9(a). For SEQRA, see ECL § 8-0109(4), and 6 NYCRR § 617.8.

41 6 NYCRR § 617.8(b)(1).

42 For NEPA, see 40 CFR § 1503.1. For SEQRA, see ECL § 8-0109(5), and 6 NYCRR § 617.8(c).

43 For NEPA, see 40 CFR §§ 1502.19 and 1503.1. For SEQRA, see ECL § 8-0109(4) and 6 NYCRR § 617.8(c) and (d).

44 6 NYCRR § 617.8(c).

45 Ibid. § 617.8(d).

46 See Vermont Yankee Nuclear Power Corp. 435 US at 553–54; Jackson v. New York State Urban Development Corp. 67 NY 2d 400, 430, 503 NYS 2d 298, 313 (1986).

47 For NEPA, see 40 CFR §§ 1502.9(b) and 1503.4. For SEQRA, see ECL § 8-0109(4) and 6 NYCRR § 617.14(i).

48 See, e.g., West Branch Assoc. Inc. v. Planning Board, Town of Ramapo 177 AD 2d 917, 576 NYS 2d 675 (3d Dept. 1991), where the Appellate Division overturned a town planning board's subdivision approval made 2-1, 2 years earlier, because at a public hearing a qualified expert had identified three threatened or endangered plant species on the property and the board had not undertaken any investigation.

49 40 CFR § 1505.2.

50 ECL § 8-0109(8) and 6 NYCRR § 617.9.

51 For example, the one-tenth of 1% change in volume/capacity ratio as the definition of significance for traffic impacts.

52 For example, visual impact or impact on neighbourhood character.

53 For example, the evolution of noise impact analysis described above.

54 See Jackson case cited above, n. 46.

55 6 NYCRR § 617.9(c)(3) and (4).

56 Teubner (1991: 127) explores the limits and advantages of pluralist norm production, using the relationship between British colonial masters and indigenous law as a model.

57 Yablon (1990: 231) states that judicial discretion may be justified by the exercise of skill, the result of expediency, or the promotion of creativity.

REFERENCES

Ackerman, Bruce A. *et al.* (1974) *The Uncertain Search for Environmental Quality.* New York: The Free Press.

Ackerman, Bruce A. & Stewart Richard B. (1985) Reforming Environmental Law. *Stanford Law Review*, 37, 1333.

Anderson, Robert M. (1986) *The American Law of Zoning 3d.* Rochester: The Lawyers Co-operative Publishing Company.

Ayres, Ian & Braithwaite John (1992) *Responsive Regulation: Transcending the Deregulation Debate.* New York: Oxford University Press.

Freeman, D.J. & Cunningham, C.C. (1991) The Environmental Audit: Management Tool or Government Weapon? *New York Law Journal*, 30 December 1991.

Friedman, F.B. (1991) Environmental Management for the Future: Environmental Auditing Is Not Enough. *Cardozo Law Review*, 12, 1315.

Gerrard, Michael B. *et al.* (1990) *Environmental Impact Review in New York.* New York: Matthew Bender.

Gilmore, Grant (1974) *The Death of Contract.* Columbus: Ohio State University Press.

Harrison, D. Jr. (1983) The Regulation of Aircraft Noise. In Schelling, Thomas C. (Ed). *Incentives for Environmental Protection.* Cambridge: MIT Press.

Herz, M. (1991) Environmental Auditing and Environmental Management: The Implicit and Explicit Federal Regulatory Mandate. *Cardozo Law Review*, 12, 1241.

Heymann, P.B. (1973) The Problem of Co-ordination: Bargaining and Rules. *Harvard Law Review*, 86, 797.

Plucknett, Theodore F.T. (1956) *A Concise History of the Common Law*, 5th edn. Boston: Little, Brown and Company.

Stewart, Richard B. (1975) The Reformation of American Administrative Law. *Harvard Law Review*, 88, 1667.

Stewart, Richard B. & Sunstein Cass R. (1982) Public Programs and Private Rights. *Harvard Law Review*, 95, 1193.

Teubner, Gunther (1983) Substantive and Reflexive Elements in Modern Law. *Law & Society Review*, 17, 239.

Teubner, Gunther (1991) Autopoiesis and Steering: How Politics Profit from the Normative Surplus of Capital. In In't Veld, R.J. *et al.* (Eds). *Autopoiesis and Configuration Theory. New Approaches to Societal Steering* 127. Dordrecht: Kluwer Academic Publishers.

Van Cleve, G. (1991) The Changing Intersection of Environmental Auditing, Environmental Law and Enforcement Policy. *Cardozo Law Review*, 12, 1215.

Wallance, G.J. & Kornreich, M.R. (1991) Department of Justice Guidelines on Prosecuting Environmental Offenders. *New York Law Journal*, 31 December 1991.

Yablon, Charles M. (1990) Justifying the Judge's Hunch: An Essay on Discretion. *Hastings Law Journal*, 41, 231.

CHAPTER 11

Environmental Officers: Management in an Ecological Quality Organization

Michael Blecher

Bremen/Vicenza/Tirana

> Standards are only temporary ... standards exist only to be revised. Standards are a never-ending challenge (Kaizen) (Bockerem, 1991: 244).

INTRODUCTION

Who is to blame if enterprises do not fulfil social expectations of care in relation to the ecological effects of their operations?

From the (prevalently political or sociological) point of view of "suspicion" it is obvious that enterprises act irresponsibly by trying to escape their ecological duties and thrive on the realization of the profit principle. Nothing can be done about this other than calling for a strong legislator who would be able to install efficient means of environmental regulation, control and liability in order to cope with the devil in disguise.

From the point of view of "free entrepreneurial discovery" it is just this totalizing tendency of socio-political values which threatens the self and other healing mechanisms of markets by excessive regulation.

Having some experience of "third positions", the law at least might stop blaming others and run the risk of starting to take the responsibility for their failures as a part of its own failures. To say it or sing it with Marilyn Monroe: "Put the blame on me".

In attributing rights and duties, legal awareness would, paradoxically, look

Environmental Law and Ecological Responsibility: The Concept and Practice of Ecological Self-Organization
Edited by G. Teubner, L. Farmer and D. Murphy
© 1994 John Wiley & Sons Ltd

at law itself and at other systems *sine ira et studio*, that is without hasty negative or positive judgments. Even without having reached the state of "enlightenment", one can (and should) observe one's continuously limited
reality-constructions (one's ignorance) in order to see things clearly as they are,
as Buddhists say. What is thus necessary is the understanding that the continuous opportunity for both personal and social improvement (and freedom
from ignorance) lies in the changing nature of those reality-constructions.

The introduction of environmental officers by various German environmental laws (e.g. Federal Emission Protection Act 1974; Water Resources Act
1976; Waste Disposal Act 1976) may be regarded as an early result of such
legal reflections on regulatory self-limits and on the limits of economic enterprise in developing environmental responsibility (Rehbinder *et al.*, 1972). The
model of the environmental officer, thus, appears to be an instrument of
indirect regulation or structural intervention, or perhaps a chance for
entrepreneurial self-control; all in all, it seems to be an example of reflexive
law trying to cope with the so-called regulatory crisis (Rehbinder, 1988:
115ff.).

Until now, the environmental officers' general legal functions have been the
following: surveillance of compliance with statutory requirements; innovation
of procedures and products with regard to their environmental impact; communication with and instruction of employees; and representation in relations
with agencies (Rehbinder, 1988: 118; 1989: 327; 1991: 609ff.). They are
employed by enterprises without any participation by agencies, works councils
or environmental organizations. The relevant agency and works council must
only be informed about changes in their employment and duties. They are protected against discharge, as well as against discrimination, but only with
respect to their role. Management is, in any case, free to give them another
position in the enterprise. Management must also ask them to state their point
of view regarding environmentally relevant investments and justify the rejection of measures proposed by the officer. The management is under a general
obligation to ensure that the officers may adequately perform their functions.

Experience with environmental officers has shown that their role has been
"successfully absorbed" by enterprise structure. In spite of this organizational
integration, the principal legal target of introducing new means of environmental protection and an "environmental conscience" of enterprises has not
been reached. Their decisional competences have proved "too weak"
(Rehbinder, 1989: 344ff.). These failures have been described as failures of the
"reflexive law" in question (Rehbinder, 1988; 1989; 1990a; 1991).

By dealing with the limits of reflexive law, as well as with the limits of its
critique, I would like to show that more recent changes in law and economic
enterprises will, and indeed must, lead to a change of the legal and economic
estimation of the role of the environmental officer. Even without actual empirical studies a "logical" (theoretical and methodological) and "historical" observation of the enterprise, environment, and related law can show the way to

improvement of the environmental officers' role. I shall put forward the following arguments:

(a) the necessary framework of a "theory of ecology" shows that there is no (legal) reflection without the procedural definition of "ecologically sustainable quality (value) standards" deriving from, and always necessarily referring to, what I call "the quality of uncertainty";

(b) the corresponding regulatory model is "regulation by coupling" or "ecological regulation";

(c) enterprise organization is moving from quantities to qualities, or from Fordist/Taylorist means of mass production to total quality management (TQM);

(d) law can successfully regulate TQM as ecological quality management (EQM);

(e) consequently, the role of environmental officers is going to be legally re-evaluated and improved as a part of EQM.

THEORETICAL ASPECTS: THE QUALITY OF UNCERTAIN BUT SUSTAINABLE DECISION-MAKING

Paradoxically, we can approach the present role of the environmental officer by gaining a theoretical distance from it through the good old Marxian idea of "ascent to concretion".

In recent years, various reports have dealt with the trouble areas of global ecological and demographic development. These have attempted to outline a "sustainable society" and its "sustainable development" in order to divert the psychosocial lifestyles of global society towards ecological survival standards[1]. Likewise, in preparation for the 1992 UN Rio de Janeiro Conference for Environment and Development (UNCED), 48 top managers from all over the world set up a "Business Council for Sustainable Development" whose object was to develop concepts for ecological entrepreneurship[2]. What *they* understand by "sustainable development" is the "true" unfolding of a market society that does not seem to be sufficiently unfolded. What has hitherto contributed to the degradation of the environment should now contribute to its salvation.

In any case, and whether "the threat" is faced with fear or "composure", there is agreement on the tendency to serious and irreversible changes in the global natural "households". Furthermore, there is the global "insight" expressed in these reports, summits and World Bank projects that different parameters for world society's "house-keeping" will have to be provided if we are to cope adequately with "the needs" of the natural environment, which appear also to be the needs of society itself. To be sure, the "difference" may consist in the above-mentioned proposal of a versatile, dynamic, reactive and profitable economic system[3]. Society's own household (i.e. its economy), however, is obviously not the natural one and all observations/descriptions of the latter are made in and by society itself. Because of this and due to the absence of a central observing authority, the search for justifiable, tenable, tolerable or sustainable social development leads to dissent about the sustainability of competing models or, in other words, to a controversial balancing of the appropriateness of social self-descriptions and decisions. Nevertheless, *all* rival theories (see Wiethölter, 1991: 8ff.) are looking for changes while no longer trusting in concepts of linear regulation and planning, which have proved to lack the necessary complexity to deal with the changing independence/dependence constellations (i.e. the ecology) of modern societies and their subsystems. Furthermore, there is tacit agreement on the fact that the descriptive/constructive horizon has changed "from subject to self-reference" (Frey, 1989)[4]. The experience of highly interdependent systems which determine the use of their materials and informations (rationalities) regardless of each other and their common natural environment has led (and has to lead) to temporalized, contextual, compatibilizations of systems' recursive (economic, political, legal, scientific as well as "subjective") *eigenvalue* constructions which take the protection of the natural environment as one of their various criteria of relevance. In other words, the threat to the natural environment ("the ecological risk" in a stricter sense) becomes the reason for a ("polyvalent") continuous total revision (recombination) of psychosocial reality constructions, and only this gives us the chance of coping with the "needs" of the natural environment[5]. Such a "proceduralized" and, in this sense, "balanced" decision-finding continuously requires[6]:

(a) collision standards/criteria for the appropriateness of decisions;
(b) a corresponding definition of areas of competence (*fora*);
(c) a corresponding creation of procedures; and
(d) the awareness of the remaining irreducible inappropriateness of these attempts at appropriate decision-making.

In terms of the theory of self-reference, that means a reflexive "re-entry" of the distinction of appropriateness/inappropriateness into the decision-maker's self-referential definition of appropriateness. Then the self-reference of the procedure takes the form of "self-constructions as self-revisions" (as the *eigenvalue* of a "second order proceduralization" or an "ecological recursivity")[7].

We shall call this attitude towards social rules and social knowledge the "ethics of self-reference" and declare that its social realization is the particular task of law. It requires the realization and maintenance of the legal dilemma (paradox) to fulfil immunizing functions for society and its subsystems (Luhmann, 1984: 509ff.) and simultaneously immunizing functions against those immunizing functions. This is nothing else but the reconstruction[8] of the 300 year old tension between law's protection-of-rights aspect and its protection-of-institutions aspect in terms of the "coupling" of functionally differentiated systems. In our translation, it means that the constructive and organizational development of social and psychic systems provided by law must be simultaneously judged by criteria of "ecological sustainability" in the sense of an equal consideration of *all* environments. This is the case because, according to "the logic of self-reference", all inequalities of systems (and environments) have proved to be contingent constructions, that is to say, they are neither necessary nor impossible but always possible in a different way. There are always other possibilities of construction and this is why their selection demands justification. The discovery of the contingency of systemic decisions/distinctions (including system and environment) reveals the (normally blind-spotted) "paradoxical equivalence" ("paradoxical symmetry") of all actual and possible forms of reality in systems and between systems (and environments). There is thus no basic prevalence of the actual in comparison with the possible. The necessarily "asymmetric" and, therefore, non-equivalent systemic construction of normality must refer to the continuous presence (present) of this "ecological equivalence" as the binding condition of its own possibility. Otherwise systemic constructions will, sooner or later, come across this equivalence in terms of "stalemate experiences". This might be in the form of environmental catastrophes but can also be seen in the once so-called "regulatory trilemma" of "juridification" (Teubner, 1984).

The neglect of the fact that the self (one's own "system") can only operate as one side of a distinction from another side ("environment"), and the resulting promotion of self-interests or self-portraits of the other, lead to the surprising reappearance of the excluded other (for example, the industrial waste of "the north" exported to "the south") and to the (momentous or final) "failure" of the distinction as such in the course of time. More precisely, it leads to the failure of the historical structures which define or determine the "coupling" of the above-mentioned basically equivalent but undetermined psychosocial spheres[9]. The effects are: "reparadoxifications of deparadoxifications" or, in legal terms, "deregulations of regulations" and "re-regulations of deregulations"! These effects may be disastrous or beneficial or a mixture of both. In any case, one discovers the "risk" of (the contingency of) those couplings. As a result, one may try to redefine the old distinction by a new one, for example by the dissent of those deciding and those affected by the decisions (Luhmann, 1991a: 111ff.). In spite of the adequacy of this description it only defines the starting point of "risk-management" (the "management of ignorance") (Zimmerli, 1991) which has to deal with the reduction of this split by new means of structural coupling. These would have to overlay the distinction with a new one, say "risk and rationality" (Japp, 1992: 36ff.). This tries to cope with the dilemma of risky, irreversible and self-binding decisions by requiring the introduction of a kind of "reflexive interval", defining criteria for the quality of the "relation" of psychosocial spheres in the light of contingency, that is to say, maintaining a maximum of reversibility and flexibility with respect to the remaining irreducible "lack of ecological adequacy" of decisions.

This means, first, that (negative) surprise reducing behaviour ("prevention") has to accept and cultivate "the other side" of self-definitions as one of the possibilities of the self that represent the self's necessary condition. Therefore, the understanding of contingency does not (only) lead to a post-modern devaluation of psychosocial historical development but, on the contrary and against all expectations, to the "intrinsic" binding quality of modern uncertainty and its constructive processing. In other words, having moved from Musil's "man (society) without qualities" to the post-modern "man (society) without certainties", we have now reached the "man (society) without certainties *and* its qualities" (Prandstraller, 1992).

Second, in terms of psychosocial reality constructions, this means that any (legal) programmatic decision must be realized as a necessarily temporalized and continuously revised attempt to approach this undetermined binding equivalence ("justice"), even though it can never completely achieve this "impartiality", which simultaneously moves away from it by producing its own contingent (!) "constraints", being necessarily a determined, asymmetrical, and "partial" decision.

From this point of view, contingency becomes the key concept of social and legal theory[10]. As all systems regulate everything according to their own rules of functional reproduction there is no "voice" left for their "ecological coherence"; "unity" is a paradise lost to be in fact replaced by "contingency". The ultimate impossibility of obtaining the ecological equivalence of systems and environments through psychosocial constructions (that is by decisions based on distinction and indication) translates itself into the fact that the recursive self-referential "life-span" of any constructive distinction and decision remains unstable, unsatisfying and uncontrollable with regard to its "reciprocity" with other possibilities—including other systems and their possibilities. This complete(!) awareness of contingency leads to the logical and ethical conclusion that any (legal) attempt at construction must shift from experience-led boundary-orientation, the attachment to "states" of knowledge and the abstract-universal overcoming of contingency to temporary contextual decisional procedures that incorporate "self-change" and "self-rejection". This is a shift to procedural moments of creation and preservation of variety and flexibility of self and alter-observation in economic, political, legal, scientific and "personal" operations (Ladeur, 1988; 1991)[11].

A system can, therefore, only be responsible if it is not attached to certain identifications, that is only if it realizes and implements contingency. Responsibility, then, is the "insight" into the binding "reciprocity-content" of contingency, recognizing it as the criterion for self-referential development (as criterion for criteria) and, therefore, recognizing the reciprocal responsibility for its "unfolding" through insufficient, but inevitably, systemic approaches and their coupling. Then, any decision can be "criticized" from the point of view of not having achieved the maximum possible reciprocity and its "normative surplus-value" in a certain historical context[12]. Consequently, one must admit that any psychosocial organization is, as such, a psychosocial limitation of responsibility ("organized irresponsibility") (Beck, 1988). Therefore, organizational attribution of responsibility is neither sufficient nor socially avoidable. This insight, in fact, leads to queries regarding the social conditions of limited liability which appears to be admissible only if the effort is made to reach ecological responsibility in the above-mentioned sense. This means that the remaining irreducible ecological inadequacy of decisions is translated into the responsibility for continuous improvements or an increase of quality while, at the same time, uncertainty itself increases.

ECOLOGICAL REGULATION

If the above expresses our available "knowledge" in spite, and because, of the "ecology of ignorance" (Luhmann, 1992), we can now be certain that consensus on truth, objectivity, certainty, safety and justice, as well as the ecological appropriateness of systemic distinctions and decisions, is constructively unattainable. Nevertheless, these cannot be neglected and there are continuous attempts to realize them as "representative models" of the quality of ecological reciprocity that contingency itself stands for. It is only by aiming at this quality that the attempt can be "sustainable": sustainability is nothing but a definition of the current(!) state of the relative quality of the "structural coupling" of systems (and environments), thriving on the total quality of their binding ecological reciprocity which is always present but has to be continuously determined.

Being responsible for the "quality" of the functions and performances of social systems, law takes advantage of the vehicle of structure coupling to implement criteria for the increase of the ecological quality of system decisions and organization. In fact, the idea of functional differentiation, which has produced the problem of how to combine autonomy with regulation, only seems to function because it is, at the same time, "ecologically" invalidated. Recently, however, there has been an observational shift in legal systems theory from an "autonomy orientation", with its emphasis on "reflexive" regulatory elements, to an "ecology orientation", with an emphasis on mutual "perturbation", "a-causal synchronization" and the "reciprocal misreading" of legal and economic processes bound by "ecological recursivity" (Luhmann, 1990: 29ff., 163ff.; Teubner, 1991b, 1991c; Santos, 1987; Blecher, 1991: 184ff.).

Teubner's recent reflections on (law's) reflexiveness try to point out how "the new legal pluralism" is reconstructing different social processes as a production of legal norms[13]. The reason for legal accomplishments in society consists in the "specific entanglements of these legal norms with their social environment" by "ultra-cyclic bindings of structural couplings". The visible regulatory strategy corresponding to this new definition of the "invisible hand" is the "instrumentalization of societal law" for regulatory political purposes.

He proposes three criteria and corresponding key-questions for monitoring the functioning of the legal design of society:

(a) Establishing actual bindings: which real social process can be constructively reinterpreted or "misunderstood" by law in order to enable the real social process to reread the results?
(b) Observing the limits of coupling: do the political manipulations of pluralist law observe the motivational limits of social subsystems?
(c) Maintaining the institutional division of couplings: is the differentiation of binding cycles (social process—legal misunderstanding; legal process—political misunderstanding) maintained?

As a fourth criterion I would add that the "unity" of the other three is established by law's sensitivity in taking up the tendency of social systems to change their structuring attitudes. The concentration on "actual bindings" is above all a concentration on the place and the way in which signals of change occur in the client's system. So, on the one hand, "to skim off its normative surplus value" (Teubner, 1991c) law observes the autonomous functional logics of the social sphere in question together with its "contingency-sensitive" areas which will serve for legal couplings. While, on the other, as the above-mentioned legal strategies are naturally valid for law itself, it is important to find out how the relevant "local" legal regulation can become "responsive" in the sense of being disposed to be overbid by the never-ending increase of quality.

TOTAL QUALITY ORGANIZATION OF ENTERPRISE

Legal observation often likes to interpret improvements in product and environmental safety as the result of product and environmental liability laws. Undoubtedly this is true to a certain extent. Enterprise has to reduce costly risks of strict liability, for example, for good organizational practice. But, as the economic analysis of law confirms, it is true that risking an accident instead of sustaining a costly organization which tries to anticipate all possible accidents might also pay off (Teubner, 1992). At the other end of the scale, the financial threat of easily claimable compensation might completely block any improvement initiative (Priest, 1991). Therefore, the liability strategy has been accused of causing a certain "stagnation" as well as a lack of improvement incentives[14]. In addition, environmental liability has had to overcome the special problems associated with the identification of damages, their extent, and their cause (Brüggemeier, 1993).

The "classical" regulatory environmental law which addresses environmentally protective requirements to the "operator" of "facilities" requiring public authorization has demonstrated its "bounded rationality". The juxtaposition and rivalry of the various safety and protection concepts of (in Germany) hundreds of laws and directives, and the consequent difficulties of application may paralyse environmental protection policies rather than stimulating them (Schottelius & Küpper-Djindjic, 1993). Equally, public authorities have proved to be unable to implement all these regulations. On the contrary, they increase the environmental risk through the "illusion of correspondence" of the regulative network with real environmental protection (D'Angelo, 1992: 614). In fact, one should never forget the unavoidable residual "ignorance" of all these regulatory strategies expressed in the "insight" of social sciences that the precise regulatory definition of the "contact" between highly interdependent spheres (for example in standards of admissible pollution) necessarily means a disregard of other aspects since it is impossible to regulate all "points of contact" (Perrow, 1987).

Taking advantage of the above description of "regulation by coupling", and bearing in mind law's function of providing for the binding quality of ecological reciprocity by temporarily fixing sustainable quality standards[15], law might find different solutions to increase the ecological responsibility of enterprises without increasing the risk of entrepreneurial indifference, apathy or overload. Which actual messages of the functional (operational) and structural (constitutional) evolution of enterprises may be "constructively misunderstood" by the law if it wants to increase enterprises' environmental responsibility as a part of its ecological responsibility (for all environments)? Perhaps this could be done by ordering procedural requirements (including organizational and behavioural requirements) which further the enterprises' own tendency to proceed in this direction.

The concept of quality might serve the law's regulatory purposes by coupling the normative legal concept of procedural ecological quality with a decisive self-regulatory change of enterprise organization and behaviour which may be described in terms of a company, or group-wide, "quality system" as the case may be. Perhaps the law could use this change to implement TQM in enterprises in the form of differentiated company-wide ecological quality assurance requirements and their improvement (EQM)[16]. The role of environmental officers could then be taken as an element of this total quality organization of enterprise. Its legal "exploitation" could justify the hope that the empirical role-deficiencies of the environmental officer might be overcome within EQM.

"Penser à l'envers"[17]

To meet the state of the art of the new evolutionary (self-)description of enterprise operations a change of scene is necessary. The scene shifts to Japan and the main actor is the Toyota System, "archetype" of the Japanese firm, as it was developed and put into practice by Toyota's chief engineer and director, T Ohno (Corjat, 1991: 17ff.). But this system is more than just an interesting local phenomenon stimulating Western managerial fantasies. The "emergence" of Ohno's "total quality management" corresponds to the "logics" of economic self-organization, leading sooner or later to an intra-firm and inter-firm "network" flexibly "playing" with the allocation principles of market and organization (Williamson, 1975; Imai & Itami, 1984; Favereau, 1989; Teubner, 1990; 1991a; 1992). TQM is the managerial form of the (poly-corporate or contractual) network. It is going to become the new universal "archetype" of enterprise organization, comparable only to the production system of Ford and Taylor that Ohno actually tries to overcome. In fact, this evolution has been described by catchwords such as "post-Fordism", "post-Taylorism", "Toyotism" or "Ohnism" (Dohse et al., 1984; Corjat, 1991). What seems to be just a new kind of managerial "ideology" discloses its instructive information only by confronting our occidental heritage, in Ohno's words, "in an inverse mode" (Corjat, 1991: 13). Legal regulation by coupling must observe this organizational "difference" closely and compare it with the previous model of enterprise organization which determined its legal forms. The following differences may appear.

For Fordism/Taylorism (Wiendieck, 1988: 23ff.) production is based on mechanical technologies with a rigid sequence of operations guided by the principle of "the one and only way to optimal performance". This "way" does not depend on employees' capacities or experiences; it relies entirely on scientific methods which must be implemented by strict working rules. The model implies a strict division of labour into manual and intellectual work. Moreover, it aims at a simplification of labour in order to maintain the independence from personal qualifications and motivations. These are simply substituted by continuous authoritarian external control. Consequently, this plant organization results in the alienation and incapacitation of labour and pushes employees into a position of permanent opposition to management. This coincides with an analogous development of the organization of the "capital side". A complex staff-line system with a rigid distribution of competence tries to optimize both the division of labour and its co-ordination. Simplification of labour is accompanied by complication and bureaucratization of a hierarchical and centralized organization. In the end, the entire system of capital and labour appears to become rigid and inflexible.

As for the "quality" aspect of this "classical" model of enterprise (Galgano, 1990: 1; Migge, 1991: 187ff.; Prätorius, 1991: 237ff.) one finds that cost and delivery, not quality, are the principal priorities. Product performance is defined by competitors and not by customers' needs. High non-quality costs are hidden behind standard costs. Standard accomplishment is important while improvement is of no relevance[18]. Result and inspection-orientation prevail over process. Consequently, end-of-the-line technologies prevail over process technologies for (environmental) product safety (Ladeur, 1988: 312ff.; Zink, 1988: 10). Quality is only a problem of production (at best, of production processes) and not of other enterprise divisions (processes) and there is no direct leadership by top enterprise management with regard to quality. Responsibility is either delegated to the line, or to a quality office, or not considered at all.

However, the increasing competitive gains of Japanese companies, together with radical technological changes in production and the increasing internationalization of the market, have stimulated Western management to reconsider the validity of their own entrepreneurial strategies by (partially) integrating Japanese enterprise organization (Revelli, 1992; Vecchi, 1992; Nagel et al., 1989; Galgano, 1990; Corjat, 1991; Fabbricciani, 1987).

During a first phase in the 1980s, Western enterprises tried to pick and choose between these new organizational strategies, but before long practical experience showed that the simple introduction of additional mechanisms into a management structure which was otherwise unchanged did not pay off (Bockerem, 1991: 243; Galgano, 1990: 3ff.; Wiendieck, 1988: 27ff.; Hill, 1991: 546ff.)[19]:

> The real significance of the Japanese context was not appreciated at that time, however: namely, that quality circles were only one part of the *total* system of quality improvement that we now call total quality management (Hill, 1991: 553).

In spite of strong national differences[20], the integration of the organizational fragments and the (global) universalization of the model of TQM are on the way (Zink, 1988; Zeller, 1988; Galgano, 1990; Wiendieck, 1988; Hill, 1991). For the time being, commitment to quality has found its ("soft") normative expression in the international guidelines on quality management and quality system elements established by the International Standard Organization (Doc. DIN/ISO 9000–9004; i.e. EN 29000–29004), one of those organizations for semi-autonomous economic self-regulation that are increasingly "used" for the delegation and implementation of public regulatory policies[21]. Under these circumstances enterprises are bound for TQM in any case, if their competitive strength is not to be affected[22].

In contrast to the "old" model,

> total quality management assumes that quality is the outcome of all activities that take place within an organization; that all functions and employees have to participate in the improvement process; that organizations need both quality systems and a "quality culture" (Hill, 1991: 554).

Obviously, this "quality culture" or "cultural change" does not imply an attempt to adopt a Japanese mentality. There is nothing mysterious about the performances of Japanese firms and "culturalism" should be forgotten (Corjat, 1991: 9). From the point of view of the "management of uncertainty"[23], Japanese management has simply developed a very successful way of (re-) gaining market shares by learning the inevitable lesson of contingency (uncertainty) and putting it into practice by means of a different concept of innovation (improvement):

> Innovation is a product of the interaction between necessity and chance, order and disorder, continuity and discontinuity. Innovation is the result not only of the planned allocation of resources to meet some predetermined clear objective, but also of some difficulty to predict or duplicate, redundancy, chance, uncertainty, or even chaos (Nonaka, 1990).

In fact, the Japanese definition of "total quality" is "continuous improvement" or the "continuous attempt at excellence" (*kaizen*) (Hill, 1991: 554; Bockerem, 1991). It is based on a redundant overlapping organization, that is, the conscious overlapping of company information (knowledge), business activities and managerial responsibilities (Nonaka, 1990; 1991).

The pillars of Toyotism

These elements and procedures of redundant overlapping quality organization have already been subjected to international attempts at standardization[24], as applied to the production procedures and structures of the individual enterprise. However, there are some general quality elements. Referring to these standards, TQM may be outlined as follows (Deming, 1989; Juran, 1988; Ishikawa, 1985; Mizuno, 1989; Galgano, 1990: 48ff.; Hill, 1991: 554ff.; Nonaka, 1990; 1991).

First, its organizational "hardware" consists in the so-called "just in time" and "zero stock" production (*kan-ban*) where entire production phases are delegated to external suppliers ("minimal factory"; "lean production"). The terms "just in time, just in quality ('zero defects'), just in quantity" refer to a "logistic chain" integrating all steps from customer to supplier to producer.

Second, the main responsibility for TQM and, thus, for innovation is with top management (providing quality policies and "visions", priorities, procedures and resources). Quality improvement takes place within the existing "vertical" structure and within "naturally occurring organizational units" (departments, sections, work teams) and is combined with "horizontal" co-ordination. The entire system is based on the principle of self-control and decentralized responsibility. It includes "cross-functional management" of *ad hoc*, multi-functional or interdepartmental project teams and quality circles together with the idea of the "internal market" ("internal customer", "internal competition": organizational units discuss the quality of their performance with those who receive their outputs; competing groups develop different approaches to the same project; "learning by intrusion"). Members of middle management (team leaders) play the decisive role at these intersections of vertical and horizontal planes. This is where the main responsibility for the quality improvement activities which take place among the rank-and-file employees in the semi-autonomous quality circles lies. The authority for the implementation of improvement is delegated to the employees who actually control the quality of their own work. The system is completed by strategic job-rotation between different technological areas and functions introducing "multiple competences" in the teams that decide what work is to be done and by whom.

As a necessary condition, quality culture training is given to all employees on all levels and the entire system depends on free company-wide information flow.

Third, quality problems (including product safety) are identified and resolved by strict systematic techniques, including statistical methods (for example, statistical process control (SPC)), measures of non-conformity, and cause and effect analysis (for example, failure mode and effect analysis (FMEA)). Detailed quality documentation is necessary to permanently reflect the current state of quality organization.

From TQM to EQM

One can expect these "logics" of TQM observation to discover their "structural affinity" with environmental protection as TQM universalizes entrepreneurial self-commitment to the quality of all organizational levels, all forms of behaviour and all environmental relations. Any TQM limitations would seem purely arbitrary. The following affinities have been described in the literature (Lipp & Riller, 1990; Adams & Löhr, 1991; Lorenz, 1991; Prätorius, 1991: 238ff.; Steger, 1992; Steger & Prätorius, 1992; Dillinger, 1992):

(a) the cross-functional character which includes all dimensions of enterprise;
(b) the need for corresponding top management guidelines, policies and responsibilities;
(c) their organization-wide implementation;
(d) motivational incentives through the creation of quality circles as ecological quality circles and by personal training;
(e) the use and improvement of corresponding technology;
(f) prevention instead of repair ("end-of-the-pipe") oriented management;
(g) and, finally, "environmental sustainability" as a universal quality dimension (that is, improvement of standards aiming at "zero environmental defects" of production, product life span, maintenance, recycling and disposal).

This overview shows that enterprise history is pushed by its own evolution (that is, no doubt, co-evolution with the legal system), to a qualitative self-change in the sense of a revaluation of internal environments (delegation and differentiation) and external environments (customers/consumers, employees, investors and creditors, nature). It is bound to develop a new "corporate culture" (Brennecke, 1992). Yet, as each system experiments with the impact of contingency within its own structured operational logics, the change will necessarily lead again to "asymmetries" of self and other observation. In other words, the enterprise observes managerial and organizational deficiencies as well as the repercussions of serious changes in the natural environment in the form of costs for the loss of competitiveness, market shares and profits and will change its structures to regain more than it lost. In any case, without self-centred asymmetries the system could not fulfil its functions and performances at all.

Nevertheless, it would be too simple to conclude that the "discovery" of (environmental) quality is no more than a strategy. The final evolutionary failure of a certain asymmetric model of self-reproduction leads each system to the momentary rediscovery of "ecological reciprocity" (symmetry) and those other possibilities and chances regarding itself and its environmental relations that had been excluded until that moment. There is the chance of discovering that "more social reciprocity" might pay off, above all in the sense that "profits" are not the only criterion for the survival and prosperity of the company[25]. Finally, the new managerial commitment to cultural self-change (improvement by innovation) as a means of greater competitiveness is inevitably linked to the above-mentioned normative reciprocity content of change as such. This is, in fact, the point of enterprise "sensitivity to contingency" which offers a chance of legal quality regulation without any illusions regarding the (self-centred, "ideological") limits of entrepreneurial quality concepts. It is the role of law to "misunderstand" the competition-oriented total quality attempts of enterprises as enterprise responsibility for (the improvement of) total ecological quality management.

THE RESPONSIBILITY OF ENTERPRISE FOR ECOLOGICAL QUALITY: TAKING TQM SERIOUSLY

But how can this be achieved and become legal practice?

Such an "ecological responsibility" is only the latest in a long series of regulated social expectations regarding the impact of enterprises on their environments[26]: on market structure, dependent companies, minority shareholders, creditors, employees, consumers and now, in fact, on natural environment. In fact, "ecological responsibility" of enterprise, with TQM/EQM as the key concept, refers to the dynamic legal coupling of enterprise with all its environments. Consequently, the various recent examples of flexible legal transformations of enterprise law "principles" in which the law shows its responsiveness to the "*gestalt*-switches" of enterprise appear in a different light. The various forms of employee co-determination (Kübler *et al.*, 1978), the transformation of the law of groups to cope with the changing constructions of groups of enterprises (Schmidt, 1991; Teubner, 1990; 1991a), the tendency to reconstruct the conditions of limited liability (Adams, 1991;

Hansmann & Kraakman, 1991) and the new forms of collective liability attribution through "risk-sharing" ("market-shares", "superfunds", "risk-pools") (Teubner, Chapter 2), all now appear to be concrete contextual realizations of EQM responsibility legally specifying the quality duties of corporate action according to the regulatory context and purpose. EQM can be treated as the appropriate procedural legal criterion for a dynamic enterprise constitution integrating all former stages and their guiding legal definitions (for example, the "enterprise-interest") (Rehbinder, 1989: 354ff.).

However, beyond concrete individual quality aspects of enterprise environmental impact, the EQM procedures must be "exploited" as such in order to cope with the problems created by the impact of enterprise on natural environment and its social effects. We believe them to constitute an answer to the "dilemma" of environmental liability and control and, moreover, to be a realization of our corresponding theoretical "insights" into a sustainable management of uncertainty and ignorance. Consequently, the legal alternative must overlap EQM organizations, procedures and behaviour and exploit the self-created autonomous enterprise "constraints".

A first step in this direction, in the area of liability, has been made by taking the self-regulatory ISO/EN/DIN standards "seriously". In the meantime, these standards have also begun to embrace regulations for environmental protection (BSI Doc. 91/53 252, 53 256, 53 257). These minimum definitions of ecological enterprise organization have been interpreted as specifications of the general tort law rule of organizational conduct, that is to organize the personnel, functional and operational area of responsibility in order to avoid the causation of damages (Brüggemeier, 1991: 59, 1993). They may also define the liability-relevant "ordinary facility operation" of the new German Environmental Liability Act. As a consequence, groups of companies are free to delegate certain environmentally risky operations to a subsidiary (Hommelhoff, 1990; Oehler, 1990; Westermann, 1991) but the top of the group always retains its group-wide responsibility for EQM organization. In Germany, the *BGH*[27] has confirmed this point of view by establishing the general penal responsibility of top management in the field of product and environmental liability (Ebenroth & Willburger, 1991; Schmidt-Salzer, 1992).

This legal reference strengthens the trend within the enterprise to establish a redundant overlapping environmental quality organization. But an organizational and procedural standard based on continuous innovation (improvement) that can go beyond the earlier approaches which merely stressed the assurance of technological standards is only reached if the law is regulating EQM self-regulation of procedures and competence as such! And, in fact, recent regulations have attempted to apply this model. § 52a of the 1990 Amendment to the Federal Emission Protection Act, for example, has introduced the duty to inform the competent authority of which member of the board is responsible for the requirements established by Emission Protection

Law. The communication must include additional information on the steps that have been taken to ensure the fulfilment of legal protective and preventive requirements in the operations of the enterprise. This implies that measures taken to ensure compliance with requirements for environmental protection must be adequately organized with regard to all enterprise operations (Feldhaus, 1991: 928ff.).

But we do not only need compliance with the relevant environmental regulations and standards. In tandem with the ISO 9004 (4, 5, 11, 18) and BSI regulations, the purpose of an EQM regulation would consist in the documentation of an environmental quality constitution of the enterprise with respect to the responsibility of the board, and a clear definition of responsibilities, authorities and delegations, together with interface control and co-ordination measures between different activities. Obviously these regulations do not explicitly require a certain type of organization. Nevertheless, the definition of the required elements implies clear selection impulses for their co-ordination. Furthermore, such a regulation must include internal auditing procedures, equipment control and maintenance and the training of personnel[28]. All in all, one could describe an appropriate EQM regulation as a legally required "environmental quality auditing system", installed to achieve complete awareness and a steady creative transformation of the environmental impact of enterprise business operations, including environmental practices and policies (Van Cleve, 1991; Herz, 1991; Capra, 1992)[29].

The most important step towards the realization of such a model of "social awareness" through EQM regulation, and an alternative to many of the notorious implementation problems of environmental law, has been taken by the EC Commission's Draft of an "Eco-Audit System" (D'Angelo, 1992; Führ, 1992; Knopp & Striegl, 1992: 2014ff.; Scherer, 1993)[30]. In fact, the proposal modifies or even abandons two principles of EC environmental regulation that have reached their implementation limits. The principles of "command and control" (Introduction No. 22) and "*polleur/payeur*" have been replaced by self-regulation in exchange for transparency and co-operation. This regulative focus corresponds to the central requirements for the functioning of EQM and appears to be an explicit application of the concept of "sustainable development" (Introductions Nos. 1 and 4). Moreover, the approach of "classical" environment law is left behind—the ecological awareness of corporate action is of relevance before the occurrence of environmental impact or damage (via air, soil or water). "Trust" in autonomous economic self-regulation is expressed by voluntary participation (Art. 1)[31] and the controlling effect of some kind of "environmental competition". It is hoped that, in the long run, enterprises bearing the "eco-audit label" (Art. 11, Suppl. III) will gain competitive advantages since only audited production would correspond to consumer demand. In order to balance the differential informational knowledge between enterprise and its social environment, transparency is to be created by auditing, documentation and certification by accredited auditors—basically following the TQM/EQM model—and with the publication of the audit-results. Just like EQM this model depends on the free information flow of the "knowledge creating company" (see Nonaka, 1991). Consequently, legal regulations regarding business secrets must also change to allow for the disclosure of environmentally relevant information as a "public good" (Ladeur, Chapter 13).

Let us cast a brief glance at the audit elements:

(a) Registered enterprises must introduce an "environmental protection system" including a global environmental policy, an environmental programme and an environmental management system (Arts. 1, 2, Suppl. IA) whose standards are to be developed by the Committee for European Normation (CEN) (Art. 4).

(b) All ecological problems, effects and performances of enterprise activities in a single plant must be analysed (Art. 2, Suppl. IB).

(c) The value of the performance is established with regard to its compliance with existing regulations, with the 13 requirements of "good environmental management" including supply and distribution, and in view of the "possible clean technologies" (Art. 1, Suppl. IC). Compared with the First Draft's general reference to the "best available technologies" for the reduction of negative environmental impacts, this formulation makes possible the introduction of innovative or alternative technologies at this stage of the audit.

(d) Numbers 2, 10, 11 and 11a of the 13 requirements of "good management practice" designate different forms of necessary "co-operation". The legitimacy of self-organized environmental management depends on the regulatory introduction and confirmation of a permanent discourse co-ordinating the "interests" of management with the "interests" of employees, agencies, customers and "the public" with regard to the environmental impact of a single plant.

(e) The actual audit (Suppl. IE) consisting of the "gathering of relevant evidence", its evaluation and proof-correcting can be carried out by competent personnel assigned by the enterprise ("internal" or "site auditors"—Art. 6, Sect. 1). The audit ends with the yearly (Art. 3, Sect. 1) publication of an "environmental statement" (Art. 5, Suppl. IV) which has to be "validated" by "external auditors" authorized by a national accreditation system (Art. 7, Suppl. II). Associations of industry, consumers and environmentalists as well as the unions are free to participate in the foundation of this system of accreditation. Obviously, it "copies" the already existing "private" quality assurance systems for the fixing and supervising of technical safety standards[32]. Agency permits, prohibitions and conditions for plant operation, as well as organizational liability, will then depend to a great extent on the external auditor's certificate evaluating the enterprise risk management.

To be sure, the proposed EC Eco-Audit System is "incomplete" and has been critically discussed with regard to the elements of voluntary participation, standards for the gathering of the relevant data, the mandatory presentation of alternative technologies, the independence and selection of the internal auditors, deficiencies of audit publication and its public discussion, and the abuse of the eco-audit label (Führ, 1992: 471ff.). But in spite of these criticisms it is, up to now, the most interesting approach which takes EQM and its procedural advantages "seriously". The ecological transparency of enterprise quality organization guarantees autonomous entrepreneurial self-determination and ecological awareness as well as its simultaneous "super-vision" by agencies and the public if the audit elements are closely observed and improved. In any case, it appears to be the most intelligent attempt so far to define standards, competence, procedures and self-reflections/revisions of what we have called "regulation by coupling" of ecological enterprise responsi-bility under the conditions of uncertainty and ignorance of all social actors. In addition, it will probably be easier to cope with the deficiencies in the role of environmental officers if they are embedded into EQM and a corresponding audit system. Then we will finally have gained the criteria for a closer investi-gation of this EQM element.

ENVIRONMENTAL OFFICERS AS PART OF EQM

EQM is where the role of environmental officers comes into its own (Rehbinder, 1988; 1989; 1990a and b; 1991; Feldhaus, 1991: 929ff.; Sander, 1991). The relevant legal regulations for air, noise and water pollution control and waste disposal institutionalize the role of environmental officers as a part of enterprise organization. Therefore, these regulations form a part of the justification of the enterprise's organizational duty as regards EQM.

As the law presently stands, it is up to management to decide where environ-mental officers should be located in the enterprise hierarchy. Empirical studies show that they are usually given a staff position directly below top manage-ment (Rehbinder, 1989: 344ff.; 1990b: 380ff.). Redundant overlapping EQM (and the corresponding EC Eco-Audit System (Arts. 2, 3, Suppl. IC)), with its emphasis on the co-ordination and quality policy of top management, however, demand that the environmental policy and co-ordination function be placed on the board level. Nevertheless, it is questionable whether this func-tion ought to be concentrated in the role of an "environmental director" as has been proposed in Germany by § 94 of the Draft Environmental Code (Rehbinder, 1990a and b). Under current law, the information duties (§ 52 of the Federal Emission Protection Act) seem to provide a sufficient incentive for top management to plan and co-ordinate the environmental quality of all enterprise operations at all levels. It might also be expected that the employ-ment of an environmental director would lead once again to a "separation" of environmental and quality aspects. This argument, however, underestimates the very real pressure to connect both, but overestimates the chances of creating a precise departmental responsibility since EQM is a cross-functional operation.

In any case, it seems advisable to opt for the further introduction of environmental directors in those limited liability companies which are currently obliged to employ environmental officers and to extend the existing audit models in this regard. It would strengthen the environmental dimension of quality, simply because there would be a special board member, subject to liability for this dimension, with the explicit duty and authority to integrate environmentally related aspects of the enterprise at all levels[33]. While the present regulation deals with the environmental problem at top management level, its "personification" in an environmental director would signal a clear shift to environmental quality[34]. Finally, it would contribute to the legitimation and confirmation of environmental quality policy as an element of (German) labour relations (another quality aspect), if the environmental director were appointed by the supervisory board enforcing labour co-determination rather than by an internal managing board, as is required under current law (Rehbinder, 1990a: 233ff.). This type of organization will, in any case, remain the minimum standard for companies with (partial) personal liability. The integrated approach of the EC Eco-Audit System would certainly resolve these problems. This is why it should be preferred to the single introduction of new enterprise functions.

The responsibility for EQM and Eco-Audit organization also implies a re-evaluation and recontextualization of the role of the (non-board member) environmental officer in order to counteract certain failures and "organizational dilemmas" (Rehbinder, 1991) of the old regime. There is certainly a lack of empirical evidence on the role of environmental officers in the current move towards TQM/EQM universalization[35]. In general, it may be said that the trend to couple economic with (various) environmental criteria in terms of "quality" and the corresponding dynamic enterprise constitution justify the expectation of a functional improvement of the role of environmental officers. In fact, from the point of view of EQM, enterprise "management of ignorance" is itself concerned with the continuous creation of knowledge out of "productive irritations". In this sense one can consider "ecological quality of enterprise" a stimulating "metaphor" with the capability to,

> set up a discrepancy or conflict by establishing a connection between two things that seem only distantly related. Often metaphoric images have multiple meanings, appear logically contradictory or even irrational. But far from being a weakness, this is in fact an enormous strength, for it is the very conflict that metaphors embody that jump-starts the creative process (Nonaka, 1991: 100).

Thus, instead of the "reflexive dilemma of the environmental officer" ("inefficiency or over-legalization" (Rehbinder, 1991: 615ff.)) EQM creates "reflexive dilemmas" within the enterprise which, above all, promote innovation by reconciling seemingly incompatible requirements. The "intra-role conflict" of the environmental officer is not the conflict of enterprise. The hopes for a productive convergence of economy and ecology associated with such a conflict underestimate the autonomy of enterprise operation for both the "external" ("public") part of the officers' role as well as its "personal" ("subjective") part (Stone, 1985: 132ff.; Rehbinder, 1991: 616). Regulatory "interest approaches" can generally only be effective if they take care of the "interest", that is of adequate structural coupling. If it is true that the introduction of environmental officers has been a success (Rehbinder, 1991: 612), this is because of its legal integration as "another enterprise organ" (and not an "agency watchdog") and because of the corresponding entrepreneurial "mis-reading" of its functions for the purposes of the differentiated enterprise (Rehbinder, 1989: 350).

Under EQM circumstances it will be more likely that environmental officers will simply be urged to follow their legally expected role (to represent environment) in a qualified way, since it has become apparent that lack of environmental quality is bound to have repercussions on vital enterprise interests, and not only in isolated cases of failure. In fact, demand is the unexpected ally of the environmental officer. We are all consumers and intelligent agencies can take advantage of this linkage by new forms of intervention like the eco-audit system discussed above[36]. From this point of view, the introduction of environmental officers turns out to be an early appeal for the instigation of a policy shift from interest-oriented models of public "participation" (which are merely valid for conflicts regarding the distribution of social welfare) to "more intelligent" models of "instrumentalization of social couplings". Consequently, enterprise is bound to create continuous quality innovations (the core of demand-oriented EQM) and constantly skims off the "dilemma surplus value". And it is because of this that the legal profile of the environmental officer itself demands improvement.

Under present law, organizational responsibility should imply the general obligation to employ the necessary number of environmental officers wherever their presence is required for the fulfilment of their general and special environmental quality functions (surveillance, innovation, communication, representation and certain legal procedural requirements, as in the case of the "emergency officers" of the Federal Emission Protection Act)[37]. This duty is also obviously a prerequisite for the legal task of top enterprise management to promote the functioning of environmental officers[38]. From this perspective, the freedom to organize enterprise operations is qualified by the flexible deployment of environmental officers with regard to the special temporal, local and material requirements of enterprise procedures. This is extremely important for groups of companies and modern "contractual organizations", "poly-corporate networks" or "hybrids beyond contract and organization" (Teubner, 1990; 1991a; 1992). The "synergy of the risks" of decentralized collaboration finds its legal response in the extension of the ecological duty to cover supply and distribution. Companies working in this field are often integrated into the whole by quality assurance systems. That is why the EC Eco-Audit proposal is also extending environmental performance to the entire "logistic chain". The possibility of delegating "risky" activities, combined with the organizational EQM duty of "group" top management, necessarily includes the introduction of environmental officers wherever these are required by the protection purpose of the relevant environmental laws[39]. Consequently, the enterprise and its top managers are liable for the sustainability of their organizational design and it is again this uncertainty of legally sustainable quality in the EQM context that stimulates continuous attempts at improvement. Therefore, the duty to communicate the organizational structure to the agency (§ 52a of the Federal Emission Protection Act) would be counterproductive if it intended a final decision; under present law, major changes must be reported in any case, permitting the agency to move from single control to control of the efficiency of enterprise organization (Feldhaus, 1991: 933ff.).

As central co-ordinators of EQM, charged with the main responsibility for quality improvement, (some) environmental officers should always be located at the intersections of vertical and horizontal planes of enterprise organization (perhaps with the option of delegating a single functional aspect, such as innovation, to one particular officer)[40]. In this position, they should also be a part of the middle management leading the above-mentioned quality teams; the role could also be merged with this management position. A lack of consensus between managers of the same enterprise plane should lead to a transfer of the conflict to a co-ordinating ecological quality committee, functioning as an arbitration board, representing the various quality dimensions of enterprise. If no agreement can be reached, the competence and responsibility for the final decision would lie with top management. Consequently, environmental officers must have decision-making powers and responsibilities of their own. This is not only true in cases of emergency. EQM depends on the strong representation of all quality dimensions, on the principle of self-control and decentralized responsibility. The merely advisory function established by present law does not meet these requirements. Therefore, the legally required design of enterprise organization must include the definition of areas of competence for environmental officers and their relation to each other.

As EQM transforms all dimensions of enterprise the quality of working relations is important. With respect to the environmental officer's task to keep employees informed about environmental matters, to co-operate with them and to give them orders with regard to his/her functions, it would thus be advisable to give the works council a full right of co-determination regarding the employment of environmental officers. This obviously goes beyond the right to be heard by management established under present law (Ig Metall, 1988: 3ff.). However, the "trust" aspect of EQM is not limited only to the relation between environmental officer and management, but necessarily includes the proper relation to the labour force.

It may also be worth considering the question of whether there ought to be a time limit for the functional employment of environmental officers (and directors) thereby aiming at a certain job-rotation in favour of environmental quality improvement.

We have already expressed our doubts as to whether the agency itself is sufficiently skilled to control the changing requirements of ecologically sustainable organization of procedures for enterprise. With regard to the quality benefits to be expected of decentralized public administration, one should consider the improvement of the above-mentioned ecological auditing systems[41] to include an integration of the environmental officer's role. The audit systems oblige the changing organizational EQM devices to be reported to and certified by highly skilled and publicly accredited external auditors. It is usually carried out by internal or site auditors and covers the organizational enterprise structure, the entrepreneurial duties and practices, procedures and resources regarding the definition and realization of the environmental policy. With respect to the similarity or even identity of their cross-functional duties (surveillance, communication, representation, innovation), it seems reasonable to merge the role of the environmental officer with the role of the internal auditor and to consequently develop one system of competences for auditing and the various supervisory functions that are currently divided according to the requirements of the various environmental laws[42]. Then, the success of the audit-system will mainly depend on the professional skills of the environmental officer/internal auditor[43], and on their independence from management with regard to their duty of auditing. Equally, their auditing activity may not interfere with their own competence area as environmental officers.

Independence from the agency must also be maintained. Neither environmental officers nor internal auditors should be employed by and directly responsible to external agencies (Hochgreve, 1988; Führ, 1992: 472). According to our experience such an idea would be likely to become another example of "classical over-regulation" (Rehbinder, 1991: 615ff.) and an obstacle to EQM[44].

Finally, one should at least consider whether a functioning eco-audit system combined with an efficient insurance system might not be more appropriate to resolve the functional problems of environmental protection. The codification of a single environmental code as planned in Germany appears to be highly questionable in this light.

A LAST COMMENT

The role of the environmental officer will gain a lot of relevance under the regime of EQM, profiting from the "synergetic effects" of different overlapping material and organizational EQM devices, some of which have been mentioned here. Enterprise commitment to EQM and the legal regulation of a corresponding responsibility to audit will then give sufficient grounds for the hope that the "ecology quotient" of all "participants" (enterprise and its psychosocial environments) will continue to grow.

NOTES

1 See the "storyline" of sustainable development in Hajer, 1991; Brown et al., 1990.
2 See Die Zeit, no. 7, 7 February 1992: 35.
3 See the International Chamber of Commerce Charter for Sustainable Development, 1992.
4 In my opinion, this change has to be radical, i.e. it has to accept and go beyond Luhmann's "Theory of Self-referential Systems".
5 See for a similar approach, Ladeur (1988), "eigenvalues" are "condensations of identity" resulting from numerous recursive rounds during the life span of the self, see Luhmann, 1991a: 240ff.
6 Following and extending Wiethölter's (1991) approach.
7 See for the following Blecher, 1991: 165ff.
8 In the sense of "Aufhebung", i.e. "preservation by change" without the Hegelian "prejudice" of lifting "higher".
9 A certain historical form of systems and environments is, indeed, bound to emerge, together with a structural definition of their "coupling".
10 While the concept of risk appears to be a (negative) functional reduction of the quality or value aspect inherent in contingency, the reduction of the "ignorance" of "rational" criteria for risk-management, by contrast, can only derive from understanding and incorporating contingency in the above-mentioned sense; see, for a similar approach, Bonß, 1993 and Japp, 1992.
11 The difference to Ladeur's approach consists in the emphasis on the "quality" or "value" aspect of uncertainty and the recognition of its (in unpopular diction) "emancipatory content"; see Hajer's concept of "ecological modernization" (1991: 9ff.).
12 See for this formula Teubner (1991c) and below. However, the normative reference point of the text differs from Teubner's which refers only to the various forms of the self-regulation of different social systems.
13 See for the following, Teubner, 1991b and 1991c. The 1991c paper is, on the one hand, a (self) critique of "reflexive law" and, on the other hand, an implicit reconstruction of the reasons for and remedies of the "regulatory trilemma". See Rehbinder's (1988; 1989; 1991) critique of Teubner's reflexive law approach underestimating, however, Teubner's "autopoietic evolution".
14 See Ladeur's (1988) critique of these "conservative" strategies as well as Schottelius & Küpper-Djindjic, 1993.
15 See the "four steps of procedural law" above. Rehbinder is right when he observes that organizational and procedural interventions "without substantive purposes" are hardly able to regulate the internalization of external interests through enterprise (1989: 309); in any case, one must understand the "substance of certainty" to attempt a description of substantive purposes.
16 Naturally, I refer to the above mentioned meaning of "ecology".
17 See the title and concept of Corjat's book (1991).
18 In fact, even in recent studies the emphasis lies on quality assurance instead of innovation/improvement (including assurance of achieved quality standards), see Arbeitsgruppe, 1988.
19 Hill's empirical study recognizes a "mismatch between structures" (1991: 553).
20 In the UK TQM models have been introduced more than elsewhere (Lascelles &

Dale, 1988; Develyn & Partners, 1989). In Germany essential TQM elements are not taken into account and at the beginning of the 1990s the overall situation "can be characterized as a (rather hesitant) reaction to activities abroad" (Schildknecht, 1993: 19).

21 See EC policy on the harmonization of technical standards which only defines the "basic safety requirements", delegating all details to "private" European and national normation, certification and accreditation. See the Green Book on European Normation, OJ C20/1, 28.1.1991, and the Council's decision on the function of European Normation in European economy, OJ C173/1, 9.7.1992.

22 See Schildknecht's criticism of the German situation (1993: 19). In fact, in 1987, the German ISO representatives dissented from the adoption of the 9000 ff. standards.

23 See above. For the different concepts of "corporate culture" see Brennecke, 1992.

24 See ISO 9004, Quality Management and Quality System Elements—Guidelines 1987.

25 See Hill, 1991: 565: "Finally, there are sound theoretical reasons for believing that top management now has a real interest in making participation work at all levels. These relate to the other changes in organizations over the last few years and give enhanced participation a basis in the material conditions of production rather than just as an ideological prop of management". See on profit incentives for entrepreneurial environmental policies, Ladeur, 1988.

26 See for the "environments" of the "corporate actor", Teubner, 1992.

27 See *BGH, NJW* 1990, 2560—"*Lederspray*".

28 Feldhaus (1991) now wants to interpret § 52a *BlmSchG* in a similar sense.

29 See Hajer (1991) on the "Environmental Performance Review".

30 See EC OJ C76/2, 27.3.1992; and the revised draft Com(93) 97 of 16.3.1993. All notes refer to the revised version of the draft.

31 In any case, Member States are free to introduce mandatory participation (Art. 130t of the EC Treaty).

32 See for the product safety aspect, Joerges *et al.*, 1988: 137.

33 Stone (1985: 127ff) has insisted on the importance of installing "external interests" at all levels of the enterprise.

34 That does not mean that environmental directors have no other management tasks, but, compared with other board members, their observation of enterprise matters is asymmetric, in favour of an environmental priority.

35 Relevant empirical research for Germany is almost ten years old, Rehbinder, 1991: 611ff.

36 But also by other forms such as the negotiable "threat" of "public warnings" (Köck, 1991), and the role of commissions of ethics (van den Daele, 1990) or transparency (Hart, 1991).

37 See Rehbinder, 1989: 327 for officers in line, and 1991: 609ff.

38 See § 55, s. 4 of the Federal Emission Protection Act and Feldhaus, 1991: 930, Sander, 1991: 33 for this assistance.

39 The "environmental officer of the group" is mentioned by Rehbinder, 1989: 328, 359ff.

40 See Rehbinder, 1990b: 390ff. for this enterprise custom.

41 Needless to say, on the other hand, the quality aspect cannot be limited to the social sphere of enterprises or, generally, of economy. (Normative) reference to the quality of the operation of others is, at the same time, self-reference in the sense of self-application, or is of no value at all. Consequently, the "public sector" must also undergo an "ecological quality change" (see Ladeur, Chaper 13).

42 This proposal goes far beyond the Draft of a German Environmental Code

(Rehbinder, 1990b: 389ff.) proposing a public law obligation for the environmental officer to report directly to the agency in emergency cases and to take certain emergency measures; see again the critics of competential rivalry in Schottelius & Küpper-Djindjic, 1993.

43 See Suppl. IE of the EC Eco-Audit Proposal.

44 As a matter of fact, the site auditor must be independent from the externally authorized auditor who does the validation job (Art. 6, No. 2b).

REFERENCES

Adams, Heinz W. (Ed) (1992) *Unternehmerisches Risikomanagement: Bessere Organisation—Mehr Sicherheit*. Köln: Verlag TÜV Rheinland.

Adams, Heinz W. & Löhr, Volker (1991) Bedeutung von Qualitätssicherungssystemen in der entstehenden Haftungsgesellschaft. *Qualität und Zuverlässigkeit*, 1, 24.

Adams, Michael (1991) *Eigentum, Kontrolle und Beschränkte Haftung*. Baden Baden: Nomos.

Beck, Ulrich (1988) *Gegengifte*. Frankfurt/M.: Suhrkamp.

Blecher, Michael (1991) *Zu einer Ethik der Selbstreferenz oder Theorie als Compassion—Möglichkeiten einer Kritischen Theorie der Selbstreferenz von Gesellschaf und Recht*. Berlin: Duncker & Humblot.

Bockerem, J.H.M. (1991) Das Kaizen-Modell-Japanisches-Qualitäts-Management. In Bungard, W. (Ed). *Neunter Deutscher Quality Circle Kongress 1990*. Mannheim: Ehrenhof.

Bonß, Wolfgang (1993) Ungewißheit als soziologisches Problem. Oder: Was heißt 'kritische' Risikoforschung? (typescript).

Brennecke, Volker M. (1991) Technikverantwortung in Staat und Wirtschaft— Unternehmenhskultur als Kommunikation gesellschaftlicher Teilsysteme. In Zimmerli, W. Ch. & Brennecke, V. (Eds). *Technikverantwortung in der Unternehmenskultur*. Stuttgart.

Brown, Lester *et al.* (1990) Profile of a Sustainable Society. In Flavin, C. (Ed.). *State of the World 1990—A World Watch Institute Report on Progress towards a Sustainable Society*. New York: Norton.

Brüggemeier, Gert (1991) Organisationshaftung—Deliktsrechtliche Aspekte innerorganisatorischer Funktionsdifferenzierung. *ACP*, **191**, 35.

Brüggemeier, Gert (1993) Unternehmenshaftung fur 'Umweltschäden' im deutschen Recht und nach EG-Recht. In Martinek, M. & Schmidt, J. (Eds). *Festschrift für G. Jahr*. Tübingen: Mohr.

Capra, Fritjof (1992) Ecologically Conscious Management. *Environmental Law*, **22**, 529.

Corjat, Benjamin (1991) *Penser à l'envers*. Mesnil-sur-l'Estrée: Christian Bourgois Éditeur.

D'Angelo, Sergio (1992) Eco-Audit—vittoria della ragione, o ragioni del più forte? *Riv.giur.ambiente*, 613.

Deming, Edward W. (1989) *L'impresa di Qualità*. Torino: Isedi-Petrini Editore.

Deutsche Gesellschaft für Qualität, Arbeitsgruppe 193 (1988) *Qualität und Recht*. Köln: Beuth Verlag.

Develyn & Partners (Eds) (1989) *The Effectiveness of Quality Improvement Programmes in British Business*. London.

Dillinger, Anton (1992) Qualität und Umwelt—zwei Seiten einer Medaille. *Qualität und Zuverlässigkeit*, **11**, 662.

Dohse, Knut *et al.* (1984) Vom "Fordismus" zum "Toyotismus"? Die Organisation der industriellen Arbeit in der japanischen Automobilindustrie. *Leviathan*, **12**, 488.

Ebenroth, Carsten T. & Willburger, Andreas (1991) Die strafrechtliche Verant-wortung des Vorstandes für Umweltstraftaten und gesellschaftsrechtliche Vermeidungsstrategien. *Betriebsberater*, **28**, 1941.

Fabbricciani, Lucio (1987) Tra Buddha e Confucio. I segreti del miracolo giapponese. *Prima Pagina*, 2.

Favereau, Olivier (1989) Marchés internes, marchés externes. *Revue Economique*, **40**, 273.

Feigenbaum, Armand V. (1983) *Total Quality Control*, New York: McGraw-Hill.

Feldhaus, Gerhard (1991) Umweltschutzsichernde Betriebsorganisation. *NVwZ*, **14**, 927.

Frey, Reiner (1989) *Vom Subjekt zur Selbstreferenz*. Berlin: Duncker & Humblot.

Führ, Martin (1992) Umweltbewußtes Management durch "Öko-Audit"? *EuZW*, **15**, 468.

Galgano, Alberto (1990) *La Qualità Totale—Il Company-Wide Quality Control come nuovo sistema manageriale*. Milano: Il Sole 24 Ore Libri.

Hajer, Maarten A. (1991) Environmental Performance Review as Instrument of Ecological Modernisation. Paper presented at the International Symposium "Environmental Performance Review: A New Tool?" organized by the Fridtjof Nansen Institute, Oslo, 28–31 May 1991.

Hansmann, Henry & Kraakman, Reinier (1991) Toward Unlimited Shareholder Lia-bility for Corporate Torts. *Yale Law Journal*, 1879.

Hart, Dieter (1991) Eigenständige Wirksamkeitsbeurteilung durch die Transparenz-kommission. *AMI Hefte*, 2/1991.

Hart, Dieter (1992) *Towards Risk-Management in Contract Law* (typescript, Bremen).

Herz, Michael (1991) Environmental Auditing and Environmental Management: the Implicit and Explicit Federal Regulatory Mandate. *Cardozo Law Review*, **12**, 1241.

Hill, Stephen (1991) Why Quality Circles Failed but Total Quality Management Might Succeed. *British Journal of Industrial Relations*, **4**, 541.

Hochgreve, Horst (1988) Stärkung des Umweltschutzes durch unabhängige betrieb-liche Anwälte. *Die Mitbestimmung*, **4**, 196.

Hommelhoff, Peter (1990) Produkthaftung im Konzern. *ZIP*, **12**, 761.

Hopt, Klaus & Teubner, Gunther (Eds) (1985) *Corporate Governance and Director's Liability*. Berlin/New York: de Gruyter.

Ig Metall (1988) Umweltschutz im Betrieb. *Positionspapier*.

Imai, Ken-ichi & Itami, Hiroyuki (1984) Interpenetration of Organization and Market. *International Journal of Industrial Organization*, 285.

Ishikawa, Kaoru (1985) *What is Total Quality Control?* New York: Prentice Hall.

Japp, Klaus P. (1992) Selbstverstärkungseffekte riskanter Entscheidlungen—Zur Unterscheidung von Rationalität und Risiko. *Zeitschrift für Soziologie*, **21**, 31.

Joerges, Christian *et al.* (1988) *Die Sicherheit von Konsumgütern und die Entwicklung der Europäischen Gemeinschaft*. Baden Baden: Nomos.

Juran, Joseph M. & Sryna, Frank (Eds) (1990) *Juran's Quality Control Handbook*. New York: McGraw-Hill.

Knopp, Lothar & Striegl, Stefanie (1992) Umweltschutzorientierte Betnebsorganisa-tion zur Risikominimierung. *Betriebsberater*, **47**, 2009.

Köck, Wolfgang (1991) Produtinformationen im Umwelt- und Gesundheitsbereich durch Verbraucherorganisationen. *Verbraucher und Recht*, **6**, 181ff.

Kübler, Friedrich *et al.* (1978) *Mitbestimmung als gesellschaftspolitische Aufgabe*. Nomos: Baden Baden.

Ladeur, Karl-Heinz (1988) Umweltrecht und technologische Innovation. In *Jahrbuch des Umwelt- und Technikrechts*. Düsseldorf: Werner.

Ladeur, Karl-Heinz (1991) Risikowissen und Risikoentscheidung. *KritV*, **74**, 241.

Lascelles, D.M. & Dale, B.G. (1988) A Study of the Quality Management Methods Employed by UK Automotive Suppliers. *Quality and Reliability Engineering International*, **4**, 301.

Lipp, Hans J. & Riller, Peter (1990) Umweltaspekte bei Qualitätssicherungsvereinbarungen. *Qualität und Zuverlässigkeit*, **6**, 323.

Lorenz, Axel D. (1991) Qualität und Umweltschutz. In Bungard, W. (Ed). *Neunter Deutscher Quality Circle Kongress 1990*. Mannheim: Ehrenhof.

Luhmann, Niklas (1984) *Soziale Systeme*. Frankfurt/M.: Suhrkamp.

Luhmann, Niklas (1990) *Die Wissenschaft der Gesellschaft*. Frankfurt: Suhrkamp.

Luhmann, Niklas (1991a) *Soziologie des Risikos*. Berlin/New York: de Gruyter.

Luhmann, Niklas (1991b) *Juristische Argumentation—eine Analyse ihrer Form* (typescript).

Luhmann, Niklas (1992) Ökologie des Nichtwissens. In *Beobachtungen der Moderne*. Opladen: Westdeutscher Verlag.

Migge, Ludger Michael (1991) Praktische Überlegungen der Vorbereitung von Qualitätssicherungsvereinbarungen. *PHI*, 186.

Mizuno, Shigeru (1989) *Company-Wide Quality Control*. Tokyo: Asian Productivity Organization.

Nagel, Bernhard *et al.* (1989) Der faktische Just-in-Time-Konzern. *DB*, **42**, 1505.

Nonaka, Ikujiro (1990) Redundant, Overlapping Organization. A Japanese Approach to Managing the Innovational Process. *California Management Review* (Spring).

Nonaka, Ikujiro (1991) The Knowledge-Creating Company. *Harvard Business Review*, 96.

Oehler, Wolfgang (1990) Produzentenhaftung im Konzern-Deliktsrecht und Haftungsbeschränkung. *ZIP*, **22**, 1445.

Perrow, Charles (1987) *Normale Katastrophen*. Frankfurt/New York.

Pohle (Ed) (1992) *Die Umweltschutzbeauftragen*.

Prandstraller, Gian Paolo (1992) *L'uomo senza certezze e le sue qualitá*. Bari: Laterza.

Prätorius, Gerhard (1991) Qualität und Umweltschutz. In Bungard, W. (Ed). *Neunter Deutscher Quality Circle Kongress 1990*. Mannheim: Ehrenhof.

Priest, George (1991) The Modern Expansion of Tort Liability: Its Sources, Its Effects, and Its Reform. *Journal of Economic Perspectives*, **5**, 31.

Rehbinder, Eckard (1988) Reflexives Recht und Praxis—Der Betriebsbeauftragte für Umweltschutz als Beispiel. *Jahrbuch für Rechtssoziologie und Rechtstheorie*, **13**, 109.

Rehbinder, Eckard (1989) Andere Organe der Unternehmensverfassung. *ZGR*, **18**, 305.

Rehbinder, Eckard (1990a) Ein Umweltschutzdirektor in der Geschäftsführung der Großunternehmen? In Baur, J.F. *et al.* (Eds). *Festschrift für Ernst Steindorff*. Berlin/New York: de Gruyter.

Rehbinder, Eckard (1990b) Umweltdirektor und Umwelt-beauftragter. In Kloepfer *et al.* (Eds). *Umweltgesetzbuch—Allgemeiner Teil*. Berlin: Erich-Schmidt-Verlag.

Rehbinder, Eckard (1991) Reflexive Law and Practice. In Febbrajo, A. & Teubner, G. (Eds). *State, Law, Economy As Autopoietic Systems, Regulation And Autonomy In A New Perspective*. Milano: Giuffrè.

Rehbinder, E. *et al.* (1972) *Ein Betriebsbeauftragter für Umweltschutz?* Berlin: Erich Schmidt.

Repenning, Klaus (1992) Sicherheit durch Beauftragte—Beispiel Umweltschutz. In Adams, H.W. (Ed). *Unternehmerisches Risikomanagement: Bessere Organisation—Mehr Sicherheit*. Köln: Verlag TÜV Rheinland.

Revelli, Marco (1992) La Via Italiana al Post-Fordismo. *Il Manifesto*, 19.2.1992, 11.

Sander, Horst, P. (1991) Betriebliche Umweltbeauftragte als Beitrag zum Umweltschutz. *WVR*, **1**, 31.

Sousa Santos, Boaventura de (1987) Law: A Map of Misreading: Toward a Postmodern Conception of Law. *Journal of Law and Society*, **14**, 279.

Scherer, Joachim (1993) Umwelt-Audits: Instrument zur Durchsetzung des Umweltrecht im europäischen Binnenmarkt? *NVwZ*, **1**, 11.

Schildknecht, Rolf (1993) Total Quality Management: State of the Art. Eine Bestandsaufnahme in bundesdeutschen Unternehmen. *Qualität und Zuverlässigkeit*, **38**, 19.

Schmidt, Karsten (1991) Zum Stand des Konzern-haftungsrechts bei der GmbH. *ZIP*, **20**, 1925.

Schmidt-Salzer, Joachim (1992) Massenproduktion, lean production und Arbeitsteilung—organisationssoziologisch und -rechtlich betrachtet. *Betriebsberater*, **27**, 1866.

Schottelius, Dieter & Küpper-Djindjic (1993) Die Interdependenz zwischen Gesundheits-, Umwelt-, Arbeitsschutz und Anlagensicherheit aus der Sicht der betrieblichen Praxis. *Betriebsberater*, **48**, 445.

Steger, Ulrich (1992) Am Ende der Debatte über Unternehmensethik: Plädoyer für eine Konzeption der leistbaren Verantwortung. In Zimmerli, W. & Brennecke, V.M. (Eds). *Technikverantwortung in der Unternehmenskultur*. Stuttgart.

Steger, Ulrich & Prätorius, Gerhard (Eds) (1992) *Handbuch des Umweltmanagements*. München: Beck.

Stone, Christopher (1985) Public Interest Representation, Economic and Social Policy Inside the Enterprise. In Hopt, K. & Teubner, G. (Eds). *Corporate Governance and Director's Liability*. Berlin/New York: de Gruyter.

Teubner, Gunther (1984) Verrechtlichung—Begriffe, Merkmale, Grenzen, Auswege. In Zacher *et al.* (Eds). *Verrechtlichung von Wirtschaft, Arbeit und sozialer Solidarität*. Baden Baden: Nomos.

Teubner, Gunther (1990) "Verbund", "Verband" oder "Verkehr"? *ZHR*, **154**, 295.

Teubner, Gunther (1991a) Unitas Mulitplex—Das Konzernrecht in der neuen Dezentralität der Unternehmensgruppen. *ZGR*, **20**, 189.

Teubner, Gunther (1991b) Die Fremdproduktion von Recht oder: Wie die Wirtschaft das Recht zur Ko-evolution überredet. *ZfRSoz*, 161.

Teubner, Gunther (1991c) Steuerung durch plurales Recht. Oder: Wie die Politik den normativen Mehrwert der Geldzirkulation abschöpft. In Zapf, W. (Ed). *Die Modernisierung moderner Gesellschaften—Verhandlungen des 25. Deutschen Soziologentages in Frankfurt am Main 1990*. Frankfurt/New York: Campus.

Teubner, Gunther (1992) Piercing the Contractual Veil: Social Responsibility of Contractual Networks? Presented at the international conference on Critical Contract Law, Tuusala, Finland.

Van Cleve, George (1991) The Changing Intersection of Environmental Auditing, Environmental Law and Enforcement Policy. *Cardozo Law Review*, **12**, 1215.

van den Daele, Wolfgang (1990) *Die Kontrolle der Forschung am Menschen durch Ethikkommissionen*. Stuttgart: Enke.

Vecchi, Benedetto (1992) Una Fabbrica al Minimo. *Il Manifesto*, 19.2.1992, 12.

Westermann, Harm Peter (1991) Umwelthaftung im Konzern. *ZHR*, **155**, 223.

Wiendieck, Gerd (1988) Quality Circles and Corporate Identity als Instrument eines strategischen Personalmanagements. In Bungard, W. (Ed). *Sechster Deutscher Quality Circle Kongress 1987*. Mannheim: Ehrenhof.

Wiethölter, Rudolf (1989) Materialization and Proceduralization in Modern Law. In Teubner, G. (Ed). *Dilemmas of Law in the Welfare State*. Berlin: de Gruyter.

Wiethölter, Rudolf (1991) Zur Argumentation im Recht: Entscheidungsfolgen als Rechtsgründe? (typescript, Frankfurt).

Williamson, Olivier E. (1975) *Markets and Hierarchies: Analysis and Antitrust Implications*. New York: Free Press.

Zeller, Hermann (1988) Organisation des Qualitätssicherungssystems im Unternehmen. In Masing, Walter (Ed). *Handbuch der Qualitätssicherung*. München, Wien: Carl Hanser Verlag.

Zimmerli, W. Ch. (1991) Verantwortung für Technik. Perspektiven einer Unternehmenskultur von morgen. In Verein Deutscher Ingenieure (Ed). *Deutscher Ingenieurtag 1991 "Der Ingenieur in der Verantwortung"*. Düsseldorf: VDI.

Zimmerli, W. Ch. & Brennecke, V.M. (1992) *Technikverantwortung in der Unternehmenskultur. Von theoretischen Konzepten zur praktischen Umsetzung*. Stuttgart.

Zink, Klaus J. (1988) Total Quality Management, Versuch einer Begriffsklärung. In Bungard, W. (Ed). *Sechster Deutscher Quality Circle Kongress 1987*. Mannheim: Ehrenhof.

CHAPTER 12

Environmental Officers: A Viable Concept for Ecological Management?

Leonie Breunung and Joachim Nocke[1]

Hannover

INTRODUCTION

The talk about the ecological responsibility of the firm is of such moral weight that one is almost compelled to accept it without further ado as the target formula for reform strategies in the sphere of economic organization. Looking closer, however, one very soon comes up against a well-known problem: can a system operating under the conditions of competition accept goals other than that of profit maximization—or be compelled "from outside" to concern itself not only with its capacity to pay, but also with interests of society that follow a logic other than that of profitability? As we know, the question is a very old one.

In conditions of omnipresent competition, how can the values associated with the term "social welfare" still have any chance? After various failed attempts to give those concerned a moral duty, liberal theoreticians, it will be recalled, came up with a superficially rather amoral answer. Adam Smith thought he had never seen "those affecting to engage in trade for the common weal ever bringing anything good" (1976: 14)[2]. He had reasons for not wanting to see the common weal as depending on the individual's altruism, and made self-interest into an economic virtue. There is, as Mandeville noted nearly 300 years ago, clearly no place in an economy in which even "private luxury and extravagance" bring, by roundabout ways, public benefit, for deliberately responsible actions by a business firm in relation to society. This has remained the position in the liberal tradition to date. If we read, for instance, that "the social responsibility of business is to increase its profits"

Environmental Law and Ecological Responsibility: The Concept and Practice of Ecological Self-Organization
Edited by G. Teubner, L. Farmer and D. Murphy
© 1994 John Wiley & Sons Ltd

(Milton Friedman, cited by Teubner, 1985b: 174), then it is not the responsibility of economic actors that is appealed to, but the very needlessness of any special motive for responsibility that is expressed.

Yet, however the action of the invisible hand which is alleged to reconcile self-interest with common good is explained, it has not always been entirely successful, as is confirmed by the state of our environment. There is accordingly a need, according to normative political theories, for a reconciliation of these aims through a specialized body. The state, from which the necessary corrections are expected through its legal restriction of individual freedom in order to assert public interests that have come off worst in free competition is, from this point of view, an institutionalization of social responsibility. Its sole competence for legislation and, if need be, coercive defence of the common weal seems, accordingly, to assign it a monopoly of responsibility. The impression is created of a body acting, as it were, outside society, taking on responsibility for the whole, equidistant from every particular interest.

But even this construction of a state regulating individual freedom in planned fashion proves to be fragile. How the state is to know what is in the properly understood interests of society is something that not only liberals are asking. And even when it thinks it knows, it would seem that legal commands alone are doing less and less to secure this interest. The public too, is noting increasingly frequently that (if anything is brought about at all) something quite different happens from what was really wanted. This raises directly the issue addressed by this chapter. If the free play of economic forces is just as unreliable as a guarantor of a "socio-economic optimum" as is state intervention, then at what point in society can the differing interests of ecology, industry, the unemployed or pensioners be reconciled, if at all? The answer increasingly given today is that the state must share responsibility for the whole with those social entities responsible for "social problems", in the sense of potential or actual causation. Thus, we hear of the social responsibility of science, the church, education and other social instances; and in our case, we hear about the social responsibility of firms.

The following discussion will not enter into the ethical debate. Academic interest in what has been for some years cultivated in the managerial echelons of the economy as "business ethics", has diminished more and more in the great years of ideological critique—perhaps also because its effects, as we see, remain within certain limits. Nor is the concept of the social responsibility of firms intended for use as a moral formula in the debate we shall be concerned with below. The expression here lays claim to an analytical, if not empirical, content.

The viability of the concept will be tested in the following ways. First, the legal instruments, seen from a post-interventionist viewpoint as an application of social or ecological responsibility of firms will be presented. This has to do with legal changes in corporate governance, and the newly introduced environment protection officer will be taken as the model figure (the common good and the market). The attempt to justify the effectiveness of this sort of "institutionalization of public interests" within firms leads, in the legal micro-economic debate, to the assumption of a social responsibility of firms which is in turn supported by a theory of pluralist company objectives (economic and ecological responsibility). The theory of pluralist company objectives is then similarly subjected to criticism from systems theory premises (company objectives and systems theory), as is the associated concept of reliability (social responsibility and social systems). There follows a discussion of reflexive law, which seeks to build the concept of responsibility into a systems-theory conception of the economy and of the firm (bringing social responsibility back). The question is then generalized to consider how far not only the economy, but also all other social subsystems in a functionally differentiated society can assume any responsibility towards society as a whole (organizational and economic functions). The concluding assessment will look at the costs arising for theory as such when, with practical intent, it seeks a corresponding extension of functions.

THE COMMON GOOD AND THE MARKET

The liberal error that the "common weal" is created of itself by free competition can be no longer simply corrected by the concept of a state intervening in planned fashion. Law as an instrument of policy comes up against limits in crucial areas. These have been exhaustively discussed and empirically confirmed, specifically in relation to environmental law. Regulatory environmental law treats economic enterprises on the stimulus/response patterns of a machine. A mechanical model of this nature buys its legal clarity only at the cost of extreme simplifications, not only of organizational reality but also of the environmental issue itself. It must necessarily keep all possible marginal conditions constant, and work with extremely selective, that is to say ultimately arbitrary, attributions of causality. This concentration on partially isolatable circumstances may admittedly lead to notable "hits", but the fact that these are in principle incalculable calls the usefulness of the model as a general solution into question. Its linear causal principle is dependent on the allocation of individual responsibilities, although environmental problems typically result from the combination and cumulative effect of activities whose effects are in themselves negligible (think of acid rain, produced by an unknown multiplicity of nameless actors). Since, however, neither connections nor all individual activities can be forbidden, one has to have recourse to necessarily arbitrarily threshold values[3], or look for other support. These may, if a firm takes on this cost factor in compliance with the law, have definite and valuable effects. Moreover, the claim of regulatory law to bind its addressees to predefined situations increasingly fails in the ecological area because of the non-availability of information that is needed by the legislator in order to solve the problem (that is, to specify an exact legal consequence). In other words, politics acts, but does not solve the problems. Similar informational asymmetries arise from the imposition and verification of unilateral governmental mandates. Consequently, in practice the state has to stick with what, if anything, seems likeliest to solve the problem. From being a sovereign authority, it becomes a negotiating partner. Consensus, co-operation and exchange-based negotiations often replace unilateral legal commands. This is so whether the economy offers information to specify what really should be made the standard (for instance, what is technically and economically "feasible" in the area of environmental protection), or whether in applying and verifying a given solution it maintains its interests through particular offers of collaboration.

The traditional legitimation of state and law, however, clearly comes into difficulties here (Ritter, 1990). The application of democratically legitimated law, neutral as to interests, can be questioned if, in those negotiating processes and other forms of situational solution of problems, all that happens is that social relations of forces are reproduced (Maus, 1986: 400ff.). For negotiations are carried on only with those that have something to negotiate. Politics can no longer maintain its monopoly on the definition of the public interest in these conditions. It is at the very least hard to see why general interests should be upheld only because the seal of the law is retroactively placed on the outcome of co-operative state action (Nocke, 1988). Nor is there any spontaneous reason for the assumption that industry, citizens' movements and other participants in negotiating processes will act as representatives of general social interests and not in their own.

There is a consistent demand for solutions that, on the one hand, avoid the drawbacks of the regulatory command model and, on the other, bring the public interest to bear. In the economic sphere, hopes turn towards a reshaping of corporate governance, with the objective of "involving representatives of diffuse (public) interests from the sphere affected by the firm in its decisional process" (Rehbinder, 1989: 306). A typical figure in such experiments, which the discussion below will take as an example, is the company environment protection officer (see paras. 53ff. of the German Pollution Control Act). He is appointed by the firm and tied in to its decisional organization, but at the same time legally obliged to bring the interests of environmental protection to bear in the planning, execution and verification of environmentally relevant objectives of the firm, though without having the traditional punitive power of regulatory law.

For our discussion, it is the environment protection officer's contribution to the development of environmentally favourable solutions to problems that is of greatest interest. This is because this "innovative function" (Rehbinder, 1989: 347) comes closest to meeting the claim of post-interventionist law to make possible a solution to the problem through those who have the necessary competence and information. Accordingly, this constitutes a more demanding test for the post-interventionist model of law than simple compliance with control objectives. The classical control function can of course similarly be interpreted in post-interventionist terms, as long as the state refrains from forcing direct access to critical information. But this is done in the expectation that this reticence will be honoured by the firms, through effective self-control. The classical policing machinery, with the possibility of monitoring compliance with statutory instructions by coercion where necessary, accordingly formally remains in existence. The firm must respect the self-control imposed on it through the environment control officer in such a way as not to provoke the use of governmental measures. It will make its conduct depend on whether the avoidance costs involved in environmental investments are calculable and in the end lower than the consequential costs of governmental instructions, or than the costs of non-compliance (Terhart, 1986). Accordingly, at a control level there is still a potential regulatory punishment in the background that could hinder measurement of the success of "pure" post-interventionist law. In respect of the innovative function of the environment protection officer (the core of the whole philosophy) one must, however, ask how the image of a "counter-force" to the egotistical and accordingly socially blind economic interests can be maintained even where corresponding punitive powers are lacking. Mere legal formality will scarcely be able to guarantee that company environment protection officers act as "bridgeheads of the public interest in private enterprises" (Ritter, 1990: 77), and not, say, as defenders of company interests against a public pressing for more environmental protection. Empirical proof of a separate contribution to the solving of the problem, in the sense of "ecological innovation", is at any rate still absent. The available studies are based solely on estimates by the company people themselves (Rehbinder, 1989: 346ff.).

In this sort of situation, the question that suggests itself is how far this sort of success can be expected at all, from this approach. What is supposed to cause a firm to keep itself within the limits of the socially acceptable? The law can no longer mark the point at which the illicit begins. By what criteria is one to establish that a firm is not making the necessary ecological efforts? If it is not obeying its environment officer, this could be treated as an omission, contrary to law, only if the environment officer could be assumed to be an ideal ecologist, a sort of *homo oecologicus*, with full knowledge of the right solutions to problems. The law simply cannot work with this sort of fiction. But even the legal precept to seek the most practical solution and uphold it with penalties, would be hard to reconcile with the overall logic. The sole remaining sanction lies in the latent threat to reactivate the old instruments of coercive governmental law.

ECONOMIC AND ECOLOGICAL RESPONSIBILITY

Variants of "post-interventionism" have accordingly sought a different basis for the effectiveness of non-directive law. The concept of an external compelling force was to be replaced by an internal one. Something or other has to be supposed within the firm, compelling it to take social interests into account itself. It is at this point that the theory brings in the concept of the social responsibility of firms. This is not shorthand for business ethics. Firms are not supposed to take on responsibility for society; they already have it.

This accordingly questions the exclusivity of profit as the sole motive for entrepreneurial action. If the social effects of this action can no longer be controlled by political and legal means alone, but only by the introduction of the responsibility concept, profit necessarily becomes one reference point among many for entrepreneurial decisions. This is the assertion of the microeconomic theory of "pluralization of company objectives" (Raiser, 1980; Brinkmann, 1983a & 1983b; Hopt & Teubner, 1985).

The argument in brief is that a firm's object is no longer confined to profit maximization because of the fact that as they increase in size they act "with effects for society". This leads to the conclusion that they have an expanded social function. Firms grow into a quasi-social institution (Ulrich, 1977), which is said to be expressed in an expansion, or pluralization, of company objectives. In relation to our subject, the formulation would accordingly have to be that large companies act with consequences for our ecological environment that have repercussions for the company itself. This connection must lead to an expansion of the list of company objectives towards taking ecological interests into account.

The debate would end at this point, if the argument said no more than that firms today, in their properly understood own (economic) interest had also to concern themselves with "social interests": that the businessman had, in the interests of his budget, to deal gently with politicians; that in certain circumstances art patronage pays off; and that they would be poorly advised to continue using sulphur-treated raisins when the competition has long since shifted to fruit with no pollutants. But this adoption of non-economic interests for the purposes of exploiting them economically does not raise any theoretical problems from our point of view. Nor does practice require any further help here. As far as the exploitation of market opportunities is concerned, one can rely on the entrepreneur's flair. And it has the additional virtue of publicity value for the entrepreneurial self-image.

The responsibility concept can only take shape on other conditions, namely when an additional criterion is introduced into a company's decisional organization alongside economic calculus. If the solution to the environmental problem is sought from the outset in a framework of economic rationality, then "ecologically responsible conduct" cannot become visible as such, for the decisional criterion in these cases remains the prospect of profit, not protection of the environment. There would be no contrast with what constitutes the core of all economic activity. Ecological responsibility would have passed its test only if, in our example, those extremely profitable eco-raisins were no longer sold because the refrigerators needed to conserve these more perishable fruits were a burden on the environment.

Ecological responsibility accordingly presupposes the distinction between ecology and economy, with the relationship between the two values including the possibility of competition, so that a sound environment can be had only at the expense of profit, and *vice versa*. Ecological company responsibility can, furthermore, only come into play where this conflict is not decided externally, say by politics or law. If, as often happens, moral or responsible conduct is still attributed to those who keep to rules where breaking them would be more profitable (see Tietzel, 1988), then the responsibility concept evaporates into thin air. Fiduciary action in favour of the environment is thus necessarily associated with industry's self-restraint. It can be realized only by developing internal barrier rules whereby the company sets itself limits on the exploitation of opportunities of profit, in the interest of the environment. Taking up ecological responsibility presupposes above all that a firm does not orient its responsibility exclusively to economic criteria. But can there be anything like a co-existence of two or more decisional criteria within an organization?

This question splits the legal micro-economic debate into the so-called monist and pluralist factions. The latter believe that they can see competing decisional viewpoints in current business reality, and thus reject the exclusivity of the profit criterion:

> The profit motive too ... has lost its unambiguousness. In the conditions of modern economic and enterprise structures, it can *a priori* no longer mean maximal, but only optimal, profit, guaranteed in the long term. Inclusion of this time dimension also implies including those other factors in the business decision process, making the profit motive relative. Long-term profit maximization in fact means partial renunciation of profit from regard for the firm's "social environment". The profit approach is limited by principles of controlled growth, maintenance of market share and stability of the company (Ott, 1977: 167).

This observation, conceived as a rejection of the monist position, clearly shows why this dispute can scarcely be resolved empirically, so that false fronts have been set up.

The pluralist position is often supported by the argument that big firms are subject to less pressure of competition and, therefore, not dependent on securing "maximum profit" (see Hopt & Teubner, 1985). It is hard to see whether this assumption derives from a description of reality or from a model working with such hypothetical quantities as "maximum profit" (as distinct only from "optimal" or "appropriate" profit), "perfect competition" and the like. If, however, a firm can, after achieving a given size, afford not to make possible profit in the short term, in the light of advantage expected in the longer term, this need not necessarily imply "relativization" of the profit motive freeing, as it were, the firm's hands to pursue ecological or other non-profitable social values. It would be much more plausible to assume that large firms are forced to change their strategies by the present social conditions. They are required to show more respect for the time dimension, and to take into account factors in their political, social and ecological environment that can no longer be ignored as "economically irrelevant" to the company's calculations. This can be the result of changed social awareness but is also because of the increased mutual dependence of industry, politics and other social functional systems. "Maximum profit" then becomes what is possible under the given circumstances. And there are no obvious reasons why (and in view of the aggressive fight for worldwide market shares one must add: alleged) the reduced pressure of competition could keep large firms away from fully utilizing the market opportunities available, in the light of competing (and therefore not really profit-making) values. The fact that company concentrations are as a rule associated with massive restrictions on free competition is of course not disputed (Hilferding, 1968: 246ff.). What is queried is the logic that associates the idea of imperfect competition with the profit motive in such a way as to arrive at consequences in the direction of a "pluralization" of company objectives.

The fact that the environmental problem presents itself differently in large companies than in medium-sized ones can also be explained by other circumstances. There are, for instance, arguments in favour of the view that big firms are more likely to be able to finance environmental investments, and also to have an economic motive for doing so. In oligopolistic markets, they can transfer the associated costs through "administered prices", particularly through agreements with other firms. The explanation can do without ecological responsibility, and for that very reason has perhaps the advantage of greater closeness to reality (Zohlnhöfer, 1981).

It is argued that, under present social conditions, firms are forced to alter their strategies. Their decisions ought to take account of developments in their political, ecological and social environment, and this for economic considerations alone. "Marketing", for instance, for big firms currently includes the need to comprehensively influence the climate of opinion; and the need for legitimation has shifted from "profit", which has recovered from the attacks it suffered in the 1970s, to "environment" (Salazar-Volkmann, 1991: 718). In this connection it has been recognized that a short-term renunciation of profit will pay in the long term "if the moral argument is rewarded by the public" (Mark-Siemons, 1992). Means and ends can be separated so much, in time and materially, that the connection is no longer recognizable to the observer. Environmental activities are then perceived as selfless ecological patronage. It is on this illusion, among others, that business ethics thrives. It harmonizes with the "gospel of responsibility" (Ott, 1977: 169f., with reference to Earl F. Cheit), in which enterprise hymns its responsibility for the whole, in scientific harmonies (Mark-Siemons, 1992).

Not that the pluralists wish to set ethics up as the ultimately decisive viewpoint instead of profit, when it comes to deciding as to the multiplicity of the firm's interests. For no other criterion can be seen that would be as unambiguous as profit. Too many competing interests appeal to "social responsibility" for this formula to be able to give any, even partially verifiable, decisional viewpoint. Thus, the ideas for the institutionalization of diffuse social interests within the firm consistently point to the organization as a "clearing house" of interests. The legal constitution of this has to ensure primarily that those differing social interests have equal opportunities of access and involvement, given appropriate opinion formation and decision making processes.

In our view, this tends to obscure the practical and theoretical problems. The latter disappear behind a harmonizing formula of responsibility, supported by concepts of doubtful analytical value, such as the term "social interests". Its use assumes the existence of social *loci* from which the problems of society can still be perceived directly and dealt with in their "totality". The pluralism of social interests is not, runs the argument, reduced by industry or by individual firms to purely economic criteria, but preserves its original multiplicity. The firm as a quasi-public institution is trusted to see problems of politics, science, ecology and so on not merely as questions of cost. The social problems remain in their specificity and thus require decisional criteria that accord with this specificity.

Given this assumed multiplicity of decisional rationalities, questions of principle arise. How, under the conditions of a functionally differentiated society, can the area of competence of a social subsystem (in our case the economy) be defined for solving a problem such as the preservation of the environment? By what means can a corresponding assignment of responsibility be verified theoretically? Is it possible to extend the specific function of a social subsystem in the direction of solving further social problems? Connected with this is the question of whether the claim that the concept of ecological responsibility which includes the need to include non-economic viewpoints in the entrepreneurial decision making process, can be met given market-economy premises.

COMPANY OBJECTIVES AND SYSTEMS THEORY

Dealing with these questions presupposes a theory that can separate itself from the internal perspective of individual functional systems and make possible a point of observation that can maintain equal distance in relation to law, the economy and politics. The theory of self-referential systems which, in its sociological variant, is particularly associated with the name of Niklas Luhmann (1984), is a suitable observational instrument in this sense. For our argument, what initially matters is what observing and describing mean in the context of this theory. These terms do not aim to present a pre-existing reality, independent of the observer. Instead, the reality exists only for an observer—a system that makes its own image of the outside world for itself. In the process of observing, the outside world is not reflected or represented, but constructed: and in a way that makes the system's continued existence possible. The descriptions that come about in this way are not due to a process of transformation, in the sense that the system processes information coming from outside into images of its own (Roth, 1987). The system does not receive any information at all from the environment. All that it receives is signals to which it does or does not assign a meaning according to its own decisions. In this way it produces all its information itself (Roth, 1987). The allocation of meaning can take place only in the context of the particular system's inherent possibilities[4].

For us, it immediately becomes clear that the treatment of a firm's goals, objectives and interests must be set-off from their traditional meanings. On this view, these terms can no longer be assumed as the basis on which a firm orients its internal structures and processes. This sort of organization, fixed through its objectives, would suffice only for a machine controlled from outside, as in traditional theories the organization is nothing but a tool in the hands of a "master", who pursues particular objectives through the organization. The sociological approach to organizational research (Cyert & March, 1963: 26ff.) and the concepts of evolutionary (Weick, 1985: 27ff., 130ff., 178ff.), and self-organizing systems (Bauer & Matis, 1989) that developed in its wake had already gradually taken the ground from under the feet of that theory. From the viewpoint of a theory of self-referential systems, these teleological formulae have their remaining "objectivity" removed. This theory treats them as internal and external descriptions of the social systems involved. Purposes help the firm to demarcate itself as a unit in relation to the environment, and in relation, for instance, to other organizations that each have their own set purposes (Luhmann, 1968: 39ff.). At the same time, organizations thereby create internal orientations intended to guarantee their operationality. The system restricts the choice of means by fixing on one or a few definite ends; that is, it restricts the number of its alternatives for action. It is only through this restriction, this selection from the multiplicity of the possible, that an organization acquires the capacity to decide. In its self-descriptions, in other words, the system simplifies its image of itself and its environment in such a way that it can exist in a complex environment.

The environment also includes the demands made of an organization. Here also it is exclusively a matter for the firm whether it notes or adopts these demands. It can only react to the environment by demands on itself. By opting for the objective of producing umbrellas, the firm is not incorporating the social need for umbrellas in its purpose. A firm adopts this purpose because it hopes thereby to bring about something in its environment that will serve the maintenance of its own capacity to pay. It must find purchasers for the umbrellas. Thus, in everything it does the firm is always satisfying needs of its own, although this does not rule out that it thereby also meets the needs of its social environment.

The problem for us now lies in the fact that an organization may be caught in the embarrassing position of defining contradictory demands. It is expected, say, of a modern criminal enforcement organization that it should lock up its inmates and at the same time prepare them for a life of freedom. This discrepancy cannot be shown in the formal purpose of an organization[5]. For an organization cannot define itself through a manifest contradiction. The conflict can be balanced out only beneath a higher formulation of purpose, kept sufficiently general, using additional ("secondary") decisional criteria that allow a separation in time of the contradictory demands (in our example, therapy and incarceration that continually creates new occasions for therapy). In this sense contradictory demands are not objective data which the system can only accept or reject. The system itself decides what signals from its environment it treats as demands and whether it may give these demands a form that sets off an internal conflict. In any case the organization takes a risk. If it ignores particular demands, it is threatened with danger from outside; if it cannot deal with the contradictions that it admits, it is threatened from within[6].

With the intensification of environmental problems, businesses are increasingly exposed to contradictory demands. On the one hand, the main responsibility for threats to the environment is attributed to manufacturing industry, and it is required to stop its environmentally harmful activity. On the other hand, almost all economic action, at least in the manufacturing sphere, is associated with some environmental impact, so that to fully meet that demand would amount to stopping production. No compromise can remove this contradiction, for firms are forced by the conditions of competition to keep their costs as low as possible, and environmental protection measures are usually costly (Sprenger, 1981). Here too it is up to the individual firm whether it sees these demands as a problem for itself or not. It cannot be deprived of the decision of whether it will bear the costs of possible environmental investments or else run the risk of simply ignoring ecological expectations. In both cases it decides according to the criterion of its continued existence, which ultimately amounts to guaranteeing adequate prospects of profit. Is it possible to bear these costs, to shift them onto others, to convert them into sales-promoting strategies ("ecological products"), or whatever?

The internal philosophies of big firms are increasingly committed to the upholding of social interests, including protection of the environment (Salazar-Volkmann, 1991). Whether this has consequences that go beyond the handing out of pamphlets by PR divisions can simply not be determined at this point, since it is impossible to see whether the officially proclaimed objectives of firms are simply a sham or symbols with no consequences, or whether they are thereby letting themselves in for a more or less costly conflict of objectives[7]. The consequences for an organization when it incorporates contradictory environmental demands in its own construction of reality are borne out only at the level below the formal goals. But the actual references of contradictory decisions can no longer be displayed as official subsidiary purposes. This, as stated above, would call into question the unity-creating function of the organization's formal purpose.

The economic and legal debate produces more questions than answers with its pluralization formula. Is the extension of the list of company objectives by the formula "ecological responsibility" meant to constitute an empirical finding? According to what has been said, that could only mean that firms today generally perceive what society feels to be an environmental problem as a demand on themselves, and take care to do more for the environment—quite independently of legal coercion. This limitation has to be made because the treatment of the concept of responsibility has shown that the term loses its capacity to make distinctions if formal coercion lies behind it for the assertion of the ecological responsibility of firms was supposed to establish the possibility of compensating for the poor functioning of regulatory law through the firm's "voluntary" action. We have already seen that the formula would dissolve into nothing if it also covered ecological action that was anything but a contemporary variant of profitable business. This exploitation of ecology for purposes of profit is readily visible. Scarcely a single advertisement now misses this opportunity. But it is not yet proven whether firms generally do more than their duty, do more for the environment than is legally required or economically opportune; and there are few plausible arguments to support such an assumption. Why should a CFC factory shut up shop as long as sales still pay and the legislator has not yet banned the manufacturing of the substance[8]? Responsibility that includes self-sacrifice in favour of the environment can be justified only in ethical terms, and even the defenders of the pluralist position do not wish, with good reason, to move on to this level.

What remains is the conclusion that the responsibility formula itself constitutes a simplification, namely of the system of science. Science simplifies the complex ecological inter-relations by ascribing the capacity to solve problems to a body it regards as both responsible and competent. Its construction of both entrepreneurial and ecological reality follows the simplifying logic of the ends/means schema and the causality principle by which firms also operate. It is assumed that the cause of the undesired effects lies in the entrepreneurial sphere, the removal of which is a matter for the one causing it. Coping with this task then appears as a company purpose, that is, something that cannot be neglected without consequences for the firm.

This summary is, of course, in itself a great simplification of the state of debate, but by exaggeration it does allow the underlying theoretical problem to be made somewhat clearer. For a closer look shows that the academic, legal and economic disciplines involved take the self-descriptions of their object, that is, the practice of company law and of company reality, into the premises of their own theoretical work. They define enterprises by goals and interests, or conceive of organizations as a community of rational individuals whose actions are co-ordinated by a common organizational purpose. In short, they do not observe their thematic area from outside, but develop their reflexive potential from the internal viewpoint of the thematic area concerned. Niklas

Luhmann (1990: 469ff.) introduced the concept of "reflexive theories" into the debate for these functionally specific subdisciplines, describing their close links with "practice" in the following words: "The functional systems describe themselves using reflexive theories, for which a more or less scientific cover is sought" (Luhmann, 1990: 475)[9]. This sounds rather bold, for a mayor or a manufacturer would hardly spontaneously recognize themselves and their actions in contemporary political theory or in the incomprehensible formulae of economic theory. But from the distanced view of systems theory in which each actor is abstracted only into the environment of a separate communicational context, politics or the economy, the statement takes on a certain plausibility. Legal doctrine (which in turn describes itself at a "higher" level in legal theory) and economic science (Luhmann, 1990: 479f.) do indeed observe legal and economic practice from the viewpoint of "correct" law and "correct" business practice. They are accordingly oriented towards increasing the capacity of these functional systems for self-description, since in a world that is getting more complicated they are presumably dependent on such accomplishments. These efforts are "subject to the code of the system they describe, and to that extent operate *within* the system" (Luhmann, 1990: 485, our emphasis). In other words, they lack the distance that is the precondition for external, system-independent observation.

The theoretical status of the debate on enterprise purposes now becomes clearer. As long as the theory of the enterprise seeks to define the identity of its thematic area using the concept of purpose, it uses the simplification through which the enterprise itself guarantees its decisional capacity. Accordingly, from this point of view, the function of a company purpose cannot come into consideration as a mechanism for self-simplification. A functional analysis that reveals that the purpose does not, at least from the theoretical viewpoint, provide what practice must believe in, endangers the orientation effect of the purpose. It cannot, therefore, be valid as a direct guide for practice. The discovery of a "multiplicity of enterprise purposes" (Luhmann, 1968: 39f.), and *a fortiori* the emphasis on the associated possibility of contradictory purposes, inevitably brings the traditional theory into difficulties. On the one hand, the economic and legal debate, if it is not to become unrealistic, has to take account of the fact that firms, like other social systems, must set themselves various, increasingly incompatible, demands. But they can only record this fact if they free themselves of their action-theory fixations. For "no-one can act two ways at once", and consequently in "action theory ... freedom from contradiction ... is a binding requirement" (Luhmann, 1968: 157). A move to systems theory, which can get along well with the presence of contradictions, is hardly possible without abandoning the claim to remain faithful to economic and legal practice. In other words, one cannot simultaneously stay with the self-simplifications of firms, law and the economy in order not to lose relevance for practice, and emphasize the simplifying nature of, say, the concept of purpose, with its theoretical costs. It is hard to defend a company

interest in contradiction. If both are wanted, the argument necessarily starts
to lurch:

> In order to avoid misunderstandings, it should be explicitly pointed out that
> actions and firms' room for manoeuvre in action are *primarily* oriented economi-
> cally, and accordingly *also* towards the market. But the decisive point is that con-
> duct is not controlled for firms either *solely* by the "invisible hand of the
> market", nor can it be understood and explained *solely* by economic criteria
> (Ott, 1983: 25; our emphasis).

This quotation brings out the dilemma plainly. On the one hand, the distinction between economic and non-economic decisional criteria acknowledges the
existence of differing demands; on the other, their relationship to each other
is left hanging. Are incompatible things competing with each other here? If so,
then the question arises of the criteria on which this conflict of objectives is
ultimately to be decided. Or, is it possible to act simultaneously economically
and politically? If so, one might come up with the idea that the one thing can
be reduced to the other so that the distinction would collapse as a genuine
difference—as, in fact, has been stated elsewhere, "economics is politics" (Ott,
1972: 387).

SOCIAL RESPONSIBILITY AND SOCIAL SYSTEMS

The systems-theory arguments reported so far leave little room for the conception of the "ecological responsibility of firms". This would be true even if this
formula were not used as an analytical category, which may, in the view of
those who brought it into the debate, perhaps also be seen as overstraining a
single concept. But even if it is treated only as a normative project aiming for
practical consequences, it can have no sociological foundation within systems
theory.

From this perspective, the question of what happens with society's problems
is not an optimization task, say about where a corresponding responsibility
should best be located. Systems theory, with its concept of functional differentiation, claims to supply a description of how society in fact copes with its
problems. Society must, and this is the starting point, rely on the functioning
of its subsystems, namely politics, the economy, law, science, education,
health and so on. Each subsystem is focused on the exclusive carrying out of
its specific function. Only in law can programmes be developed that bring in
a binding decision as to (legally) right and wrong; only the health system
decides, on self-generated rules, whether one is sick or well, and what help to
give; and the same applies to all the other functional systems. There is no
super-ordinate body with the capacity of entrusting other bodies with social
responsibility. In functionally differentiated societies it is no longer possible to
assume the existence of a centre that first of all determines whether or not there
is a problem, and if so, whether it is a political, legal or economic one, in order
then to refer it to a subordinate body to be dealt with. Politics may well think

that ailing forests are a case for the law. But the decision to make a law is itself a political decision that is a reaction to shrinking electoral support, that is, to a threat to one's own existence. What the law does with this political request lies outside the influence of politics. Perhaps it will arrive at a dogmatics of environmental law that blames the poisoning of the forests on business, or else calls politics wrong for not acting. As far as the economy is concerned, it includes law in its calculations only as a cost factor, since it can only think in terms of prices. Perhaps production will be partly suspended. Perhaps because of the new Forest Act the usual donations to government parties will be cut so much as to create a further problem for politics, and so on. In other words, the social subsystems do their various jobs by operating within their own "code" (Luhmann, 1986: 75ff.), thereby guaranteeing their own continued existence. And, society itself continues to exist only through the continued existence of its subsystems.

All of this has to have consequences for the concept of responsibility, which always includes an aspect of representation, that is, action for others. An authority acting in trust for the whole can no longer be assumed. The state, which in normative political theories bears overall social responsibility in this way, is deposed from its throne by systems theory, and regarded only as a legal organization of the political system alongside the other functional systems of society. The overall interest claimed "by the state" is, from the point of view of systems theory, a production of the political system itself, which does not have an overall social need in view, but only a political one.

Saying farewell to responsibility in this way does not solve the problem. It cannot be assumed that the functioning of the subsystems can guarantee the solution of all social problems. A system of science engaged primarily with problems that interest no-one outside science certainly functions, to the extent that it devotes itself to the production of new knowledge following the distinction between true and false. But this cannot be regarded as a contribution to solving social problems as long as the system is deaf to demands for knowledge from other systems. The same applies to what the legal system does for the economy. This is largely dependent on the production of reliable decisions—something that can no longer be assumed as a matter of course, particularly in environmental law. Admittedly, in the end the practice of environmental law is still formally oriented to the distinction between (legal) right and wrong. But the question of right and wrong can, in individual cases, look increasingly arbitrary because, given the pressure of circumstances, adequate knowledge on which the rules can be based can no longer be established (Nocke, 1988). And equally, it cannot be assumed that the economy is still generating so much surplus that health, science and policy tasks can still be funded. This makes clear, first, that the autonomy of the individual social subsystems is associated with a high degree of mutual dependence, and second that the abstract functioning of one subsystem does not necessarily also guarantee adequate output for the other systems.

Finally, there is much to support the view that the carrying out of a function and the providing of a service may be mutually limiting (Luhmann, 1977: 54ff.). The more science comes under pressure to meet specific demands for exploitable knowledge, and accordingly becomes dependent in relation to content (and perhaps also to funds), the harder it is for it to defend its autonomy. It is only as autonomous science that it can do its job of giving its own answers to questions it sets itself. Self-referential systems cannot be forced to take up problems from outside, and in order to become possible its boundaries would have to be taken away—with the consequence that "practice" would be answering its own questions, under the appearance of being science[10]. In the same way the law is under growing pressure from politics to take up problems for which it does not yet have any rules. It therefore increasingly faces the embarrassing situation of no longer being able to present a conflict as a legal problem (that is, a question of legal right or wrong), and thus takes a decision that, clearly considered, could equally well be taken by a parish priest: what the law does specifically as law can no longer be made visible[11].

BRINGING SOCIAL RESPONSIBILITY BACK

At this point the concept of social responsibility, already almost buried by systems theory, unexpectedly makes its presence felt, knocking loud enough for even systems theoreticians to hear. In particular Gunther Teubner (1983 and 1985), Helmut Willke (Teubner & Willke, 1984) and (with somewhat different theoretical premises, but similar conclusions) Karl-Heinz Ladeur (1984 and most recently 1991: 186ff.) have asked how functional systems that can observe the world only in relation to themselves and from the viewpoint of their own functions can sharpen their eye to social problems. In what follows we shall make particular reference to Teubner's work, because his articles relate directly to the functional system of the economy and to the role of business firms. His work is related to legal practice, that is, it asks about the possibility of an economic and environmental policy using the means of law. It thereby necessarily comes up against the problem of how the plan-based utilization of law can be established on the basis of a theory of self-referential systems.

The problems arise as a consequence of the autonomy of systems. The economy and business cannot be treated as simple machines that can calculably be induced to follow particular courses of conduct through particular inputs and transfer mechanisms. As self-referential systems, they can only control themselves (Luhmann, 1991). Then the most massive attempt at intervention cannot bring a system to act in a way that does not accord with its own possibilities, but with those of its environment. Even the highwayman, often cited in this connection, cannot take away from the prospective victim the decision between his (or her) money or life (Vanberg, 1982: 56). Robber and victim can each solve only their own problems, hoping to profit from the other's solution.

This is of course not to deny that there may be manifold types of influencing relationships between social systems and between social and psychic systems. Even system theorists will drive on the left in Britain and on the right in Germany, as the respective laws require. But these unmistakable effects, which may be highly correlated with particular legal commands, cannot be explained by the strong or weak controlling power of law. In the given example, the mere existence of suicide drivers going the wrong way up motorways confirms that the law does not control, but has to rely on internal mechanisms of its addressees that the control concept cannot reach. The conditions on which a system reacts to stimuli from its environment must be sought in the internal circumstances of the system itself, although the state of the environment is of some significance to that decision.

The social environment of a money economy looks different from the environment of a pure natural economy. But this connection, based on past observation, cannot be formulated as a sort of law of nature, making relationships calculable for the future (and making political planning possible in this connection). Systems are termed "autonomous" because they guide themselves by their own criteria and not those of the environment. They respond to changes in their environment, but do so in a way that in principle cannot be calculated and, therefore, cannot be used for planned external control. At most one can expect social systems to adjust to changes in their environment. Politics can seek to set some other system moving, but can in principle only act as a stimulus to the system's changing itself.

Reflexive law and related conceptions now plainly see the possibility of setting the boundaries for a "corridor" of directed self-change, thus making the system "evolve" in a particular direction (Ladeur, 1989: 179, 186), in order to get beyond blind experimentation in the hope of an occasional hit. The claim is reduced from the controlling of particular targets to that of bringing under control mechanisms "that select structural patterns from blind variation, make them last and thus control the development of systems" (Teubner, 1989: 63). By a particular arrangement in the legal environment of the system, business (or the individual firm) is to be made sensitive to the upholding of social interests. The difference from the hard and soft control variants of regulatory law would then lie in the fact that the law is no longer seen as a transport mechanism conveying legal content, however defined, from one system to the other. Nothing leads "from the law's conceptions of the economy into the reality of the economic system" (Teubner, 1989: 99). In these circumstances "contextual control" (Teubner & Willke, 1984) can only mean change in the environment. The legislator's attempt to bring "diffuse social interests" to bear within the decisional organization of firms, by changing corporate governance, then appears as the law's expectation of the individual firm to take up its "social reference" (Teubner, 1985a: 472).

Even in relation to this concept, one can of course ask why the firm should be interested in society, just because the law wants it that way. What leads out of the circle of self-reference in order to open up the argument to the concept of social responsibility?

At this point Teubner (1985b: 160ff.) uses Luhmann's suggested distinction between three types of system relationship of a social subsystem. At a functional level the system enters into relationship with the whole system of society; the performance level is concerned with relationships to other subsystems; and the reflexive level with a feedback relationship of the system to its own identity (Luhmann, 1981a). All three types of system relationship have already been discussed, so that all that remains to be filled in is how the situation looks specifically for the economy. Its function is seen by Luhmann as the "guaranteeing of supply, stable for the future, associated with various distributions in the present" (1988: 64). Under the conditions of a money economy, the securing of supply need no longer take the form of hoarding goods for later use. It is enough to have money that is a guarantee for the future without having to decide already on one's future requirements. It is only through the money mechanism that economic activity can be emancipated from all social ties. One does not ask money where it came from, a fact that already fascinated the philosophers of antiquity, nor is it fixed where it is going to. Spending it is of course, as a rule, associated with the satisfaction of particular needs. But it can no longer be said that the modern economy is there in order to meet "the needs" of society, because it largely determines what needs seek satisfaction, or what needs are "possible" at all. Thus the, at first sight rather odd, notion that the code of the economy, the sign by which one recognizes economic action, is the distinction between payment and non-payment gains significantly in plausibility. The condition for the functioning of the whole is that payment remain possible, that there be grounds to trust the value of money. This requires a number of precautionary measures inside and outside the economy. For instance, it must be ensured that the enormous expenditure we have to meet in order to repair damage to the environment is not paid for simply by printing money. But does one have to bring a special responsibility of firms to bear? This is obviously what happens. For instance, an invariable component of annual wage negotiations is the appeal to the sides of industry to avoid wage agreements that drive inflation. But that does not lead to the theoretical conclusion that the function of the economy is extending in the direction of social responsibility.

But this is just what has to be assumed in the concept of "corporate responsibility". This is the only way the divergence of function and performance that is assumed can be presented. The economy's function, providing for the future, is extended in the direction of satisfying "collective interests". The concept starts with the idea that the *performances* of the economy (servicing the demands that other subsystems place on the economy) do not happen in such a way that relevant social interests (like environmental protection) are sufficiently taken into account. It is claimed that private profit-seeking (performance) and economic provision for the generality (function) pull in opposite directions (Teubner, 1983: 52, following Raiser, 1980: 225). Reflexive law sees this as a problem of balance, no longer to be restored by the regulatory machinery of governmental economic policy, but through "regulated autonomy". This is the process of "harmonization between doing justice to the firm's function and doing justice to its performance" (Teubner, 1983: 53; 1985b: 160ff.), which cannot be set in motion by appeals to the businessman's conscience or other moral feelings. The change in corporate governance is supposed to stimulate reflexive activity within the functional system itself (here, the individual firm). "Corporate responsibility" would then stand for a legal policy programme aimed at harmonizing function and performance through enhanced reflection within the system.

We can scarcely disagree with this goal. But in our view it can only be justified by weakening the theory at central points, though the very advantages of this theory lie in its oft-criticized rigidity. Meeting the need to protect the environment can be accommodated at the level of neither function nor performance. "Providing for the future" as a function of the economy does not mean responsibility for everything that society's fate depends on. Such a broad version would no longer allow demarcation from other functional systems. Education, science, health protection and other subsystems have also in their various ways to provide for the future, but do so according to different rules from the economy, even if money is needed for everything. Ecological issues admittedly have the special feature that no differentiated functional system has developed for dealing with them (Luhmann, 1986). They are dependent on being handled by the existing functional systems, each in their own way. But this cannot be done by the subsystems adding, as it were, an ecological appendix to their various functions. This would have little prospect of success, since neither the individual subsystem nor any other body can decide to extend or restrict its social functions or dispose of them in any other way. All that is left is the decision whether, given the conditions of its specific function, a system takes up the ecological issue or not. Certainly, there are multiple demands for this. Science is asked for explanations of the origins of the hole in the ozone layer, medicine for effective protection against its effects, politics and the law for energetic action against those causing the damage, education for a sharpening of ecological awareness, and finally the economy for environment-friendly manufacturing and products. Irrespective of the question of whether meeting these demands would still be performance in the sense of the theory, if the subsystems took up the demands they would not be handling social responsibility but their own problems. Ecology only appears in the systems' various internal representations as a legal, political or medical object. In the case of the business firm this is a question of cost—are we to pay for the environment, or preferably not (Teubner, 1989: 98)? What other alternatives could the harmonization processes of reflexive law in the area of the economy refer to? Supervisory boards could well ask themselves whether "looking after the Bonn landscape" (a phrase from the scandal over contributions to political parties) should not take priority over looking after the environment, or whether agitating shareholders or some other interest ought to take priority. But however it is looked at, it is not costs versus environment or environment versus art patronage that is up for decision, but always cost versus cost, and the decision is taken exclusively in the firm's interest. What decision will cause the firm the least costs in the short, medium or long term, including costs arising from omitted environmental measures? Enhanced self-reflection will not be able to induce the firm to break out of this logic and conclude a voluntary self-restraint agreement with itself. It may be that the firm will miscalculate and its decision will not only ruin the environment, but also itself. But to be able to assess (self-)endangerments of this nature better, what is needed is a self-reflexive entity, as developed in economic theories, which may very well take on an entirely new form, such as environmental officers or other experts that perhaps understand just as much of ecology as they do of economics. A reflexive theory is subject to the code of the functional system and is consequently not a suitable instrument for drawing a balance between any kinds of social interests. Theory (like the company) cannot treat various social interests as different situations. It is not the case that ecological interests look green to it, health needs white and political interests red. It sees everything in the uniform grey of cost. Consequently, we cannot expect the harmonization of social interests. This perspective can only attempt to bring order to the complicated structure of prices and costs. One cannot, therefore, object when the modern economic analysis of ecology pitilessly translates everything that we hold sacred about nature into money terms (Horlitz, 1989: 137).

This rather sober approach cannot stop at the environment officer and other attempts (Stone, 1975) to bring ecological interests to bear in the firm's decisional process. What was said about interests also applies, of course, to their various representatives. The expression "representative within the firm" draws attention away from the fact that, being prescribed by law, the officer is primarily nothing but a demand by law on the individual firm. It is exclusively for the firm to decide how to respond to this legal signal. Where reflexive law claims this as a form of "indirect control", there is no difference on this point between reflexive law and regulatory law (Rehbinder, 1991: 604; 1989). The firm records this cost factor and will act accordingly. The reflectors will not be left standing on the doorstep. Since people are easier to trace than the consequences of using banned materials, the firm will have to include a high likelihood of discovery in its risk calculation. What would then be decisive would be the likelihood of penalty and its severity (Terhart, 1986: 44ff.), presuming that sanctions were provided by the law. How the outcome of this balance would look in an individual case is a straightforward question for (empirical) implementation research, which we need not go into at this point. The same applies to the question of how successful an environment officer can be in influencing the firm's decisional process. Here too everything depends on how much influence the firm allows him, that is, how the firm treats the demands made on it and with what "job description" it responds (as distinct from the one that may be laid down by law[12]). If it sees no need for action it will perhaps, in the extreme case, let this foreign body sweep the yard, or at least employ him in such a way that he is not a useless burden on the firm. In present circumstances that is not very likely. Since ecological pressure can no longer be ignored and might gradually take on existential importance for individual industries, firms often no longer need the legal stimulus to equip themselves with environmental knowledge, or to increase their self-control in relation to ecologically critical situations. The first cases of "environmental executives" at the head of firms are being reported from Germany[13]. But for the company this does not alter the fact that it can think only in terms of cost and revenue, that is, it will only sell eco-cars at prices that drivers accept. Even a firm with an "environmental executive" must ask in each case whether it can afford to take account of the interests of the environment. The "modern eco-manager" (Stitzel, 1988) is not reaching the higher ranks of the firm because he is regarded as an honest broker between social and company interests[14]. Instead, he is likely to advance his career because the increasing need for information produces a further division of labour in the decisional organization of the firm. Environmental themes call for specialized competence. However, the firm's self-reflection which is possibly also advanced, does not serve to harmonize its performance and its function (which is, in any case, not possible at company level as can immediately be shown). It takes place in order to ensure the conditions for the firm's existence even under enhanced environmental demands.

We may now generalize this criticism to cover all attempts to make firms sensitive to ecological or other social interests. All proposals that seek to distinguish themselves from regulatory law by setting only framework conditions and not the result itself start, as far as the issue of control goes, from an illusory alternative. The "external stimuli" involved, whether the case is one of the strictest possible threshold value or a procedural provision neutral as far as the result is concerned, are distinguished exclusively by the intensity with which they are required to be considered (not complied with) by the firm. The penalty potential of a threatened factory closure will presumably be treated differently from an appeal to corporate social responsibility. The criteria, namely the balancing of costs and benefits, nonetheless remain the same. One can therefore look at the interventionist and post-interventionist machinery together and assess the *de facto* effects from this viewpoint. As long as this has not been done there is no reason to be more optimistic about the effectiveness of procedural provisions than of substantive norms (Maus, 1986: 400ff.). For the firm, they are both outside data. The distinction between formal and substantive does not enter into its decisions. There is only its own forecast of how its operations will affect its environment. From this viewpoint procedural norms have no greater likelihood of being followed than regulatory law. The plunging into the shadows of law, customarily regarded as a weakness of regulatory law, is usually associated with avoidance of prescribed procedures, just as the "informal rule of law" (Bohne, 1981) can be described as deviation from *procedural* norms.

Summarizing, it is impossible to distinguish the claims that law makes on a firm according to their control intensity, control direction, mediacy or immediacy. In every case the firm as a self-referential system controls itself. The artificial distinction between direct and indirect, the metaphor of frameworks, and the assumption of "meta-rules of self-change", overlook the consistency of the theory appealed to in order to keep open legal policy options that cannot be justified by that theory. The programme of reflexive law is a variant of the "third way" that sees the limits to governmental economic and environmental policy, but does not wish to plump for the alternative of the ethical control of the economy ("planned economy"). The firm is to play out the competing social interests within itself. This is a corporatist revival of ideas from the first third of this century[15] (Teubner, 1987: 82ff.). The programme cannot be carried out. On the one hand, one cannot slip social interests into the organization, either through the portal of regulatory law or the postern of contextual control, without their becoming interests of the organization. On the other, neither is the recourse to a state commissar within the company (with necessarily political decisional powers) a desirable option.

ORGANIZATIONAL AND ECONOMIC FUNCTIONS

A fundamental aspect of this, that can only be addressed briefly here, but which deserves more attention concerns the relationship of individual economic organizations to the functional system of the economy. Despite all the close connections, these are two levels that must fundamentally be distinguished. It follows that what has to be said in relation to function, performance and reflexiveness of the economy does not necessarily also apply to the individual firm (though Teubner consistently says so, e.g. 1985b: 163).

This concerns, first, the function of the economy. The "guaranteeing of future satisfaction of needs" (Luhmann, 1981b: 394) can no longer, under prevailing social conditions, be handled at the level of simple interaction, on the model of a vegetable market. The keeping up of payments calls for manifold precautionary measures which can henceforth be coped with only in organized form, and therefore by systematic decision. For instance, the overcoming of an archaic storage economy (accumulating supplies to last through the winter) presupposes not only the money economy but also the possibility of making production independent of the time of consumption. This takes organization, just as there has to be organization to see that "customers can arrive at unforeknowable times with unforeknowable desires and nonetheless be brought to find what they thought they wanted" (Luhmann,

1981b: 399). The scarcity of goods and of money, moreover, makes cost-conscious action necessary, increasing the need for organization. "It is only the two demands *together* that classify an organizational system under the functional sphere of the *economy*" (Luhmann, 1981b: 399). "Classifying", however, does not mean "make identical". Organizational and functional systems remain at different levels and are not linked even by a common function. One can only speak of economic organization if "organizational systems are employed to secure at present, over and above immediate activity and its point, an economic contribution to guaranteeing an as yet undefined future" (Luhmann, 1981b: 401). This sort of definition is given from the viewpoint of the economy as a functional system, not that of the organization. That is, the economy, which cannot itself act, is in a position to do its job only if there exist organizations that produce that contribution through economic activity. That does not mean that a firm acts from some sense of responsibility for society and in that sense carries out a social function. It acts for itself and not for society. Certainly, to solve the above-mentioned timing problem, typically associated with the money economy, it has to form capital, but only in relation to securing its own future. One could talk of securing society's future as a "function" of firms at most if the social reference arose of itself from the action of the individual economic organization. But this is just what is not the case. One cannot presume that firms make Art. 14(2) of the Basic Law ("Property imposes duties. Its use should also serve the public weal") into the maxim for their own action. Providing society with money (profits, wages, taxes) is instead required of the firm by the means of law, backed by coercion (Luhmann, 1981b: 402). To speak of social functions in this connection would mean that the firms, which must be concerned with keeping taxes, wages and profits (insofar as they do not remain within the firm) as low as possible, wished, and were able, not to be concerned with their social function. It is closer to reality to assign firms, if you will, the "function" of "serving the function of the economy"[16]. Here one would admittedly have to bear in mind that the organizational function and the economic function are not at the same level, that is, the organization as an economic organization does not itself become a functional system and thus identical or partly identical with the economy[17]—just as the widow carrying her mite to the bank (this time not from compulsion, but from expectation of interest) and thus making provision for her own future, while she serves the function of the economy, does not herself become a bearer of the economic function.

A clear separation of the plane of the economy from that of its organizations takes the ground away from the idea that it would be within the sphere of possibilities of a firm to calibrate the performance and function of the economy, in the sense of a social optimum. An organization from which the "common weal", that is the diversion of surpluses to the social environment, has to be coerced is, quite apart from the theoretical objections mentioned, hard to see as a broker of that common weal. No entity can be conceived of that could be seen as providing any sort of balancing or equilibrium here, for no-one can move the function of a social subsystem or its performance towards any imaginary point of equilibrium. To religion, politics, science, the health system or art, this is immediately obvious. The ideal system situations that were continually aimed at in history, like Aryan art, socialist realism, German science or the Lysenkoism of a marred Marxism, each presuppose a state-dominated society and, therefore, an extreme de-differentiation. They are all more or less of unhappy memory. This is not meant to cast the efforts of current economic and financial policy in a questionable light. But in the circumstances of a differentiated society, formulae such as the "overall economic equilibrium" that the Grand Coalition (1966–69), with its Keynesian minister for the economy, Karl Schiller, had inscribed in the German Constitution in 1967 (Art. 109(2) of the *GG*) can have the function only of political declarations of intent. The contradiction-ridden quadrangle of objectives through which the Stability Act gives effect to that equilibrium is rightly called "magic". Between individual firms, between firms and the authorities, between business associations and the trade unions, there are agreements, "concerted actions" and all sorts of other contacts. But what is up for harmonization in each case is not social interests, but the interests of firms, associations and authorities. No organization, no association, no individual, can negotiate with effects for a social subsystem, or act for it in any other way. Such activities may have the outcome of effects on wages and prices, taxes and profits, the prestige of an association or electoral favour for a party—just as a failed harvest or a political event like reunification can have the same effects. This is not to say anything against concerted actions, economic policy, the pronouncements of experts wishing in the name of science to engage in the affairs of politics, or the reflexive efforts of economic theory. All that is to be contradicted is the notion that at the level of individual firms anything other than the firm's own interest can be balanced out. This is as true of internal harmonization processes as it is of the "external activities" of the firm. The securing of social interests is an illusory alternative, since any system can act only within the limits of its possibilities and, therefore, not in its environment. One way or the other, the only thing that can keep a successful firm in equilibrium is its balance sheet.

CONCLUSIONS

What for some is pluralization of purposes is for others pluralization of functions. It comes about in intentions of legal practice which seek to extend the sphere of responsibility of the economy towards "diffuse social interests", with theoretical underpinning. Similarly, pluralized law is supposed to make an image within itself of the broad spectrum of social interests, and thereby take on responsibility for the drawing of a balance for which it has no rules at all (Ladeur, 1984). In our view this makes theory a blunter instrument for observing society. It can no longer employ the concept of function as a means of comparison if it "pluralizes" the reference point to which it refers, or adapts it to the various movements of the observed object, and thus no longer keeps the conditions of observation stable. It no longer grasps the functional losses of law if not only law but theory along with it shows tendencies towards softening. The change in classical criminal law in relation to the "treatment concept", as a result of which therapists and lawyers can only be distinguished by the colour of their working clothes (often leading to a *de facto* loss of legal possibilities of defence), has had consequences that have long been overlooked by science because, under the slogan of treatment, it brought therapy into the function of law as a purpose of punishment. The changes arising for religion with the problem in bringing its message to the people, leading to moves sideways into politics or non-denominational social work, can scarcely be grasped by a theory that is itself interested in increased political commitment by the Church and, therefore, starts by defining the Church in terms of a political mandate. Finally, the micro-economic legal debate on our topic has led to empirical and normative airy-fairiness not least because the talk of social functions of firms *inter alia* blurs the boundary between politics/law and the economy to such an extent that ultimately, if the blend is looked at closely, it makes either the concept of the economy or that of politics superfluous.

In society, there is no system that has specialized in ecological communication. The ecological problem, which is of course felt outside the existing functional systems, is omnipresent, but is nowhere at home (Luhmann, 1986). Formally, it imposes itself on existing subsystems. Politics has no other option but to take up this theme. The attempt to foist it off on law can succeed only imperfectly, because the law can only accommodate the problem more or less provisionally. Legal practice often hands it back uncoped-with to society. In this situation, it takes us no further if science attempts to give ecology a place to stand by seeking to make the law, the economy or itself into an ecological subsystem, by pluralizing the respective function. On the one hand, "practice" will not be impressed, and on the other, theory would also be ill-served by this proposal. The system-specific solutions to the ecological problem or its migration from one system to another can no longer be recorded if the reference point for problem solutions, that is the function, migrates along with it, or is expansively adapted to the changes in reality. The theory is then seeing double, or no longer seeing at all. These are the costs of an attempt to employ theory too directly for a legal policy programme. The proposal would accordingly be to do without the concept of ecological or social responsibility as an element in theoretical work. By including the term in its own conceptual apparatus, theory makes it harder for itself to define the very problem it wishes to grasp, the relevance of which is beyond all doubt.

NOTES

1 Translated from the German by Iain Fraser.
2 "But man has almost constant occasion for the help of his brethren, and it is in vain

for him to expect it from their benevolence only. ... It is not from the benevolence of the butcher, or the baker, that we expect our dinner, but from their regard to their own interest. We address ourselves, not to their humanity or their self-love, and never talk to them of our own necessities but of their advantages" (Smith, 1976: 14).

3 The "state of the art" is scarcely valid as an objective defining characteristic, because the corresponding power to define it largely lies with industry itself (Wolf, 1986).

4 "The system introduces its *own* distinctions and with the help of this distinction grasps situations and events which then appear for the system itself as *information*. Information is accordingly a quality that is purely internal to the system" (Luhmann, 1986: 45, emphasis in original).

5 Though para. 3(3) of the Criminal Enforcement Act does come very close to a contradictory formulation of objectives: "Enforcement shall be oriented towards helping the prisoner to integrate into life in freedom".

6 In our example, thus, the prison organization must ensure, even in formulating and applying secondary criteria, that it does not appear as a sanatorium, but not as a jail either. In one case the threat is withdrawal of public support, in the other a revolt on the prison roof.

7 Even with the least binding declarations of commitment to the environment, there will always be a residue of "textification" within the system that has consequences for the system's own operations. The company will no longer be able to afford "obvious" sins against the environment. An example is the case where workers refuse to carry out particular tasks, in their view risky to the environment, appealing not only to their consciences (Art. 4 of the Basic Law), but also to company objectives. See the Federal Labour Tribunal's decision in *Betriebsberater* 1985: 1853.

8 While these lines were being written, the news came over the radio that (a) the hole in the ozone layer was in the process of extending to the northern hemisphere, and (b) the association of CFC manufacturers did not at present see itself as being capable of stopping the CFC supposed to be chiefly responsible early (i.e. before the manufacturing ban came into force), because the main customers, refrigerator manufacturers, had not yet found any suitable alternative material.

9 The same is true of, say, education or theology, but also of the theory of science (Luhmann, 1990: 479f.).

10 This situation can be seen in its purest form in the case of scientific legal reports in connection with trials, for which usually not only the question but also the answer are given in advance.

11 In our view this is the main objection to Ladeur's (1984) conception of the drawing of a balance; for a criticism see Nocke, 1990.

12 See e.g. paras. 54 and 58b of the Third Act amending the German Pollution Control Act of 11 May 1990 (*BGBl* I 870), and on this Feldhaus, 1991.

13 Most recently Volkswagen announced the appointment of an "environmental executive" (Ulrich Steger). A little later the group advertised a Golf under the slogan "Your Environmental Officer".

14 Environmental executive Ulrich Steger had this to say on this conflict: "It becomes critical when environmental protection costs money, because we are doing it voluntarily—not if the competitors also have to do it".

15 Rathenau, 1917. His conception of corporate social responsibility is here plainly under the influence of the "sticking together" then advocated by the war economy.

16 By coping with their own problems of scarcity (of resources and customers), and

continuing to do so (since otherwise they could not survive on the market), organizations at the same time serve the economy's function of coping with the problem of provision for the future for society by controlling scarcity. Firms thus primarily solve scarcity problems for themselves alone, without necessarily making generalized provision (Luhmann, 1981b: 406).

17 The individual subsystems of society are distinguished from each other (and from the overall system of society) by the fact that they specify the capacity to link up with communication (in the economic system, in accordance with the code of payment or non-payment). Organizations produce their identity (in relation to their environment) through a quite different selective mechanism, namely that of membership (associated with particular conditions).

REFERENCES

Bauer, Leonhard & Matis, Herbert (1989) *Evolution—Organisation—Management*. Berlin: Duncker und Humblot.

Böhm, Hans (1976) Idee und System gesellschaftlich verantwortlicher Unternehmensführung. *Management International Review*, 35.

Bohne, Eberhard (1981) *Der informale Rechtsstaat—Eine empirische und rechtliche Untersuchung zum Gesetzesvollzug unter besonderer Berücksichtigung des Immissionsschutzes*. Berlin: Duncker und Humblot.

Brinkmann, Tomas (1983a) Aufgaben des normativen Unternehmensinteresses in der puralistischen Unternehmensverfassung. In Kießler, O. *et al.* (Eds). *Unternehmensverfassung, Recht und Betriebswirtschaftslehre*. Köln/Berlin/Bonn/München: Heymanns, p. 67.

Brinkmann, Tomas (1983b) *Unternehmensinteresse und Unternehmensrechtsstruktür*. Frankfurt/M./Bern: Peter Lang.

Cyert, Richard M. & March, James G. (1963) *The Behavior Theory of the Firm*. Englewood Cliffs, New Jersey: Prentice-Hall Inc.

European Management Forum (Ed) (1973) *Davoser Manifest*. Davos.

Feldhaus, Gerhard (1991) Umweltschutzsichernde Betriebsorganisation. *Neue Zeitschrift für Verwaltungsrecht*, 927.

Hilferding, Rudolf (1968) *Das Finanzkapital. Vol. II*, 1st edn. 1908. Frankfurt/M.: Europäische Verlagsanstalt.

Hoffmann-Riem, Wolfgang (1982) Selbstbindung der Verwaltung. *Veröffentlichungen der Vereinigung der deutschen Staatsrechtslehrer*, 40, 187.

Hopt, Klaus J. & Teubner, Gunther (Eds) (1985) *Corporate Governance and Directors Liabilities*. Berlin/New York: de Gruyter.

Horlitz, Thomas (1989) Monetäre Bewertung von Umweltschäden—Ein geeignetes Instrument zur Erfassung ökologischer Folgekosten? In Donner, H. *et al.* (Eds). *Umweltschutz zwischen Staat und Markt*. Baden-Baden: Nomos, 125.

Hucke, Jochen & Ullmann, Arieh A. (1980) Konfliktregelung zwischen Industriebetrieben bei der Durchsetzung regulativer Politik. In Mayntz, R. (Ed). *Implementation politischer Programme*. Königstein im Taunus: Athenäum, 105.

Ladeur, Karl-Heinz (1984) *"Abwägung"—ein neues Paradigma des Verwaltungsrechts*. Frankfurt/M./New York: Campus.

Ladeur, Karl-Heinz (1989) Zu einer Grundrechtstheorie der Selbstorganisaton. In *Festschrift für Helmut Ridder*. Neuwied, Frankfurt/M.: Luchterhand, p. 179.

Ladeur, Karl-Heinz (1991) Die Neuordnung der Telekommunikation—Zur Funktion eines öffentlichen Unternehmens in hochkomplexer Umwelt. *Kritische Vierteljahresschrift für Gesetzgebung und Rechtswissenschaft*, 177.

Luhmann, Niklas (1968) *Zweckbegriff und Systemrationalität*. Tübingen: J.C.B. Mohr (Paul Siebeck).

Luhmann, Niklas (1977) *Funktion der Religion*. Frankfurt/M.: Suhrkamp.

Luhmann, Niklas (1981a) Theoretische und praktische Probleme der anwendungsbezogenen Sozialwissenschaft. In Luhmann, N. *Soziologische Aufklärung 3*. Opladen: Westdeutscher Verlag, p. 321.

Luhmann, Niklas (1981b) Organisation im Wirtschaftssystem. In Luhmann, N. *Soziologische Aufklärung 3*. Opladen: Westdeutscher Verlag.

Luhmann, Niklas (1984) *Soziale Systeme*. Frankfurt/M.: Suhrkamp.

Luhmann, Niklas (1986) *Ökologische Kommunikation*. Opladen: Westdeutscher Verlag.

Luhmann, Niklas (1988) *Die Wirtschaft der Gesellschaft*. Frankfurt/M.: Suhrkamp.

Luhmann, Niklas (1990) *Die Wissenschaft der Gesellschaft*. Frankfurt/M.: Suhrkamp.

Luhmann, Niklas (1991) Steuerung durch Recht? Einige klarstellende Bemerkungen. *Zeitschrift für Rechtssoziologie*, **12**, 142.

Mark-Siemons (1992) Das nette Produkt. *Frankfurter Allgemeine Zeitung*, 29.1.1992, M3.

Maus, Ingeborg (1986) Perspektiven "reflexiven Rechts" und Deregulierungstendenzen. *Kritische Justiz*, **19**, 390.

Nocke, Joachim (1988) Rechtsproduktion der Juristen im Umweltrecht—Eine Profession stößt an ihre Grenzen. In Bryde, B.-O. & Hoffmann-Riem, W. (Eds). *Rechtsproduktion und Rechtsbewußtsein*. Baden-Baden: Nomos, p. 181.

Nocke, Joachim (1990) "Alles fließt"—Zur Kritik des strategischen Rechts. *Jahresschrift für Rechtspolitologie*, **4**, 123.

Ott, Klaus (1972) Die soziale Effektivität des Rechts bei der politischen, Kontrolle der Wirtschaft. *Jahrbuch für Rechtssoziologie und Rechtstheorie*, **3**, 345.

Ott, Klaus (1977) *Recht und Realität der Unternehmenskorporation*. Tübingen: J.C.B. Mohr (Paul Siebeck).

Ott, Klaus (1983) Die politische Dimension des Unternehmens als Problem des Unternehmensrechts. In Kießler, O. *et al.* (Eds). *Unternehmensverfassung, Recht und Betriebswirtschaftslehre*. Köln/Berlin/Bonn/München: Carl Heymanns, p. 123.

Raisch, Peter (1976) Zum Begriff und zur Bedeutung des Unternehmensinteresses als Verhaltensmaxime von Vorstands- und Aufsichtsratsmitgliedern. In Fischer, R. *et al.* (Eds). *Festschrift für Wolfgang Hefermehl*. München: C.H. Beck, p. 1347.

Raiser, Thomas (1980) Unternehmensziele und Unternehmensbegriff. *Zeitschrift für Handelsrecht*, 206.

Rathenau, Walther (1917) *Vom Aktienwesen*. Berlin: S. Fischer.

Rehbinder, Eckard (1989) Andere Organe der Unternehmensverfassung—Die institutionalisierte Vertretung diffuser Interessen im Unternehmen unterhalb der Gesellschaftsorgane. *Zeitschrift für Unternehmens- und Gesellschaftsrecht*, 305.

Rehbinder, Eckard (1991) Reflexive Law and Practice. In Febbrajo, A. & Teubner, G. (Eds). *State, Law, Economy as Autopoietic Systems: Regulation and Autonomy in a New Perspective*. Milano: Guiffrè, p. 604.

Ritter, Ernst-Hasso (1990) Das Recht als Steuerungsmedium im kooperativen Staat. In Grimm, D. (Ed). *Wachsende Staatsaufgaben—sinkende Steuerungsfähigkeit des Rechts*. Baden-Baden: Nomos, p. 169.

Roth, Gerhard (1987) Erkenntnis und Realität. Das reale Gehirn und seine Wirklichkeit. In Schmidt, S.J. (Ed). *Der Diskurs des radikalen Konstruktivismus*. Frankfurt/M.: Suhrkamp, p. 229.

Salazar-Volkmann, Christian (1991) Weltbilder, Unternehmensbilder, Menschenbilder. Die Unternehmens-Philosophie transnationaler Konzerne. *Das Argument*, **189**, 711.

Smith, Adam (1976) *The Wealth of Nations*. Oxford: Clarendon Press.

Sprenger, Rolf-Ulrich (1981) Kostenbelastung der Sektoren durch Umweltschutz und ihre wettbewerblichen Auswirkungen. In Gutzler, H. (Ed). *Umweltpolitik und Wettbewerb*. Baden-Baden: Nomos, p. 165.

Steger, Ulrich (1992) Interview in *Die Zeit* (Nr. 13) 20.3.1992, 33.

Stitzel, Michael (1988) Ökologisches Management. In Simonis, U.E. (Ed). *Lernen von der Umwelt—Lernen für die Umwelt*. Berlin: Edition Sigma, p. 1287.

Stone, Christopher D. (1975) *Where the Law Ends. The Social Control of Corporate Behavior*. New York: Harper & Row.

Terhart, Klaus (1986) *Die Befolgung von Umweltschutzauflagen als betriebswirtschaftliches Entscheidungsproblem*. Berlin: Duncker und Humblot.

Teubner, Gunther (1983) Corporate Responsibility als Problem der Unternehmensverfassung. *Zeitschrift für Unternehmens- und Gesellschaftsrecht*, 34.

Teubner, Gunther (1985a) Unternehmensinteresse—das gesellschaftliche Interesse des Unternehmens "an sich". *Zeitschrift für das Gesamte Handelsrecht und Wirtschaftsrecht*, 470.

Teubner, Gunther (1985b) Corporate Fiduciary Duties and Their Beneficiaries. In Hopt, K.J. & Teubner, G. (1985) *Corporate Governance and Directors' Liabilities*, Berlin/New York: de Gruyter, p. 149.

Teubner, Gunther (1987) Unternehmenskorporatismus. *Kritische Vierteljahresschrift für Gesetzgebung und Rechtswissenschaft*, 61.

Teubner, Gunther (1989) *Recht als autopoietisches System*. Frankfurt/M.: Suhrkamp.

Teubner, Gunther & Willke, Helmut (1984) Kontext und Autonomie: Gesellschaftliche Selbststeuerung durch reflexives Recht. *Zeitschrift für Rechtssoziologie*, 4.

Tietzel, Manfred (1988) Ethische und theoretische Probleme interventionistischer Wirtschaftspolitik. In Cassel, D. *et al.* (Eds). *Ordnungspolitik*. München: Vahlen, p. 77.

Treutner, Erhard (1986) Verwaltung zwischen Recht, Bürger und Unternehmung. In Simon, J. (Ed). *Regulierungsprobleme im Wirtschaftsrecht*. Neuwied, Darmstadt: Luchterhand, p. 107.

Ulrich, Peter (1977) *Die Großunternehmung als quasi-öffentliche Institution. Eine politische Theorie der Unternehmung*. Stuttgart: Poeschel.

Vanberg, Victor (1982) *Markt und Organisation*. Tübingen: J.C.B. Mohr (Paul Siebeck).

Weick, Karl E. (1985) *Der Prozeß des Organisierens*. Frankfurt/M.: Suhrkamp.

Wolf, Rainer (1986) *Der Stand der Technik*. Opladen: Westdeutscher Verlag.

Wolf, Rainer (1987) Zur Antiquiertheit des Rechts in der Risikogesellschaft. *Leviathan*, 356.

Zohlnhöfer, Werner (1981) Umweltschutz und Wettbewerb. In Gutzler, H. (Ed). *Umweltpolitik*. Baden-Baden: Nomos, p. 15.

SECTION IV

TOWARDS A THEORY OF ECOLOGICAL SELF-ORGANIZATION

CHAPTER 13

Coping with Uncertainty: Ecological Risks and the Proceduralization of Environmental Law

Karl-Heinz Ladeur[1]
Bremen

INTRODUCTION: EXPERIENCE AND DANGER

Danger and trust in normality

The environmental crisis in its various manifestations requires no further proof. It is less readily accepted that environmental law is overtaxed and, as a result, is coming up against its limits. But in this way, the environmental crisis is also a crisis in environmental law. This can be demonstrated by one of its key terms, the concept of "danger" which has been adopted from police law. "Danger" connotes a situation whose further development would probably result in damage, that is to say, a change in the normal continuing existence of an object of legal protection. This requires police "intervention" in the rights of the perpetrator of the danger. The concept refers to a central idea of "normality" against which the danger appears as a deviation, and the defensive intervention as its re-establishment.

A closer investigation of the category of "normality", however, shows that the concept of danger goes beyond the apparently formal recourse to the size and probability of the damage and is based on social conventions, the foundations of which have been shaken by economic, scientific and technical development. These conventions are aimed at both the classification of the object of legal protection (and of its "normal" state) and the identification of the perpetrator and the endangering action. The normality of the object of

Environmental Law and Ecological Responsibility: The Concept and Practice of Ecological Self-Organization
Edited by G. Teubner, L. Farmer and D. Murphy
© 1994 John Wiley & Sons Ltd

legal protection is more accurately characterized as that, such as property, health and so on, which can be privately and also publicly classified. It must be capable of being isolated by a legal rule as a defined object of legal protection. The extent of this classification is also limited by fluctuations in the possibilities for action and availability (Müller, 1930; Ladeur, 1986; Preussisches Oberverwaltungsgericht, Preussisches Verwaltungsblatt, 1884/5: 380). This means that extraordinary sensitivities or claims to self-definition are not recognized as normal, just as it shows that police decisions are also determined by the question of whether or not to intercept a causal chain of events. Finally, the endangering action is not regarded as having contributed to a collective or general effect, just as it cannot be the result of the interlocking of a number of causes.

This conventionalism remains to a large extent unexpressed in public law because protection in police law, even where private objects are concerned, serves public security above all else. As a result a certain discretion is conceded to the police. For a long time the individual has not possessed a right to intervene against technical dangers, and so the police have been able to exercise a monopoly of definition behind the veil of official secrecy. This dependency on convention is, thus, perhaps more explicit in the private law protection against nuisance, which is based on the recognition of local custom (Winter, 1983: § 906, no. 55ff.). The obscurities are clear in the definition of the endangering action. The threshold of intervention apparently refers to the probability and size of the damage. It also contains, however, an element of the social discounting of the future in the interests of technical progress. This was especially the case in the nineteenth century with the growth of ever more virulent technical dangers such as, for example, the use of pressurized boilers. The assumed benefits of technology became the yardstick for the stipulation of the threshold of intervention, both for normal operations (dampness, smells, noise and so on) and for disturbances such as fires and explosions (see Wolf, 1987). There is an implicit assumption that the danger of certain technologies can be assessed by means of knowledge acquired from accidents, and this has gained the paradoxical nature of a self-fulfilling prophecy.

This social convention about the benefits of technology had to be explicitly formulated in private law. Lawsuits between neighbours required a legal decision on the competing claims of neighbours who were, in principle, of equal status. However, the English common law, for example, distinguished between public and private involvement in order to make the substantive option for progress disappear behind the formal legal division of spheres of action. In many cases the courts did not relate the fear of future environmental damage to the individual property, but defined it as public nuisance (öffentliche Belästigung) in order to withdraw it from the legal disposal of private property. This question was further related to the definition of local custom (ortsüblich), which affected the fixing of a danger limit in the normal operation of a plant. Thus, for example, the problem of whether smoke was a nuisance did not arise but became the question of whether it was "a nuisance to a person living at Shields" (Silver, 1986). Judgment on the dangers of accidents, which could also affect particular individuals, operated by allocating the burden of proof, and on the basis of the delimitation and anticipation of the future. Uncertain future incidents could only then be countered by preventative defence claims if their occurrence was "free from doubt" (Silver, 1986: 81). In general, neighbours could only make claims for compensation in the case that damage had actually occurred.

This preference for dynamic over static property is often expressed (as in the German Industrial Code, *Bundesimmissionsschutzgesetz (Bisgz)*) in the exclusion of civil law injunctions against certain types of activity. This exclusion is also associated with the state authorization of dangerous facilities (§ 14 *Bisgz*) and public law authorizations thus form the private law. It is, however, quite rare to find explicit reasons given for the limiting of preventative private law claims against the dangers originating from technical facilities. An example can be found in an English decision from 1900. The plaintiff was referred to the fact that, in general, it is necessary to wait "until the dread is justified by the event" because,

> experience has shown that a meddlesome interfering policy represses the spontaneous energy and many-sided activity which arises naturally from self-interest, differences of taste and inclinations among men and constitutes the true springs of progress (*Holke* v. *Herman* cited in Silver, 1986: 81).

This reveals the option of the law to deal with risks by balancing the public benefit against possible damage on the basis of experience or social convention. This goes beyond the simple delimitation of property through local customs although, as we have said, in many cases it conceals itself behind rules governing the burden of proof. It should not, however, be reduced to the search for a Pareto-optimum by means of a welfare economic perspective, as the economic analysis of law seeks to argue. In the judgment quoted above, there is also the expression of trust in the reliability of experience and the possibility of future progress (see also Willard, 1991).

The limits to danger and the concept of danger

The concept of danger in this form contains a dynamic element which looks beyond the legal framework. This element gives the possibility of confidence in the viability of new "moves", to be aggregated situatively and co-operatively, through the creation of new connections and constraints (Bratman, 1992: 1). Experience is thus guaranteed as a form of knowledge which can be distinguished from those legal rules which are counter-factually stabilized expectations. This knowledge remains implicit in the rules and is nowhere available in a concentrated form, for it is bound to co-operative networks of action and can only be very incompletely freed from its context. It is oriented to continual change and extension through the practice of co-operation. Knowledge by orientation, by contrast, is formed by public institutions and expresses itself in legal rules which create expectations and are codified according to a homogeneous standard (Willard, 1991: 101; Bunge, 1974: 35). It is a sphere of "general" deliberation and disposition about heterogeneous practices, abstracted from context, treating its subject matter under formal procedures.

Nonetheless, these different variants are dependent on each other in the creation of confidence. Both operate on the basis of experience, because even the normative modelling of possibilities remains bound to the network of prior "moves" and has to continually make sure of this relationship to its context by interpretation (Ladeur, 1991). The only alternative to (a largely implicit) knowledge by experience is ignorance (Polanyi, 1962). Mistrust must remain bound to the limits of conceptually disposable knowledge which are, of necessity, narrow.

Confidence in a certain stock of knowledge which is not fully subject to reflection is apparent in the example of the regulation of school curricula. It is a paradox that a state premised on the freedom of the citizen should educate for that freedom by prescribing certain elements of knowledge and proscribing others. And the possibilities for the opening up of alternatives within the teaching syllabus, and thus breaking the expectations of the economic system, the culture and so on are slight. The overstepping of these narrow limits can have catastrophic consequences. A similar problem arises in the broadcasting system, which is governed by public law, and must take socially relevant interests into consideration in the organization and production of programmes. Formal civil liberties, in the traditional sense, presuppose an equality in cognitive capabilities and opportunities (as normative assumptions), in a similar way to the market in the traditional equilibrium model. However, such normative presuppositions are not possible in education, public broadcasting or similar fields of cultural reproduction organized by the state. Here, the legal system depends, both in fact and law, on complex, organized networks of knowledge, values and interests. The abstract universalism of freedom and equality must not only presuppose the existing history of restraints and possibilities, but must also partially continue it, thereby simultaneously excluding alternatives. This exclusion, which is also a legal exclusion, is only tolerable insofar as other institutions make possible a greater wealth of alternatives, for example, freedoms of individual opinion and of the press, and religious and aesthetic freedom. These find their limits, if at all, in "general laws".

We can now come back to the concept of danger, which also refers to an implicit contextual knowledge, bound to practice, and developing on a case to case basis. Through the connection with experience, the concept of danger becomes the link between the normative safeguarding of expectations and cognitive learning. It thus maintains the interplay of the inductive and deductive variants of knowledge.

It must be added that police intervention to ward off danger is also bound to a model of a linear equilibrium which must be maintained by damping down individual fluctuations around a resting point. Police law does not permit more complex strategies which do not aim at the restoration of normality. This conception, described here in a simplified form, is thus based on a balance between institutional forms of learning and paradoxical forms of "non-learning" (Luhmann, 1988: 335). It is guaranteed by legal norms and aims to produce and maintain a complex, confidence-generating network of social co-operation. It can admit possibilities for new (inter)actions, based on the division of labour, which presuppose and use that knowledge and distribute it through the practical context of actions.

This police law concept of danger has been further developed in environmental law, in various respects. The danger is reactively and systematically examined by means of permit procedure, but the "average wealth of experience" (Müller, 1930) which the police formerly believed themselves to have at their disposal, has had to be supplemented by expert knowledge. The immediate and free availability of information can no longer be assumed. The object of legal protection is globalized. It is "nature's household" (ecosystems) that is endangered and must be protected. The threshold of reaction is no longer determined by danger but by the principle of prevention (*Vorsorge*), and it is thereby disconnected from suppositions of probability (Kloepfer, 1989: § 7, no. 52; Hoppe & Beckmann, 1989: § 5 no. 11ff.; di Fabio, 1991). Technical experience is, for its part, increasingly modelled secondarily through public, private or mixed forms of organization and is made explicit in norms (standards) whose classification causes difficulties for the legal system (see Salzwedel, 1991; Breuer, 1989: 43). The unity of experience and its capacity to be universalized (for the formation of either conventions or legislation) has, thus, been pluralized. It is ranked from the "generally recognized rules of technology" (*allgemein anerkannte Regeln der Technik*), to the "state of science" (*stand der Wissenschaft*), which is the furthest removed from general experience.

The instruments of intervention have also become highly differentiated. To a certain extent they have also been temporally stretched since, according to many environmental laws, follow-up measures can also be required. Decision-making can, thus, be self-revising. This has also entered the legislation, in the constitutional form of a duty to make subsequent rectification of defects, which also acts under conditions of uncertainty (*BVerfGE* 50: 290, 330ff.).

The reverse side of this process, however, has become increasingly visible. The various institutions for the generating and conventionalizing of knowledge, which formerly shaped police law, can no longer be harmonized. The traditional structural principles of environmental law, oriented towards experience, have reached the limits of complexity. According to figures from 1984, for example, there is almost no available knowledge about the toxicity of 80% of the 48 000 chemicals used commercially in the USA (a fraction of the known total) (Lyndon, 1989; Applegate, 1991). In addition, quite apart from the problems of the combined effects of several materials, there is no practical method available for assessing the effects of the better-known pesticides and chemicals on the complex content of ecosystems. This limit on quantity must be seen as signalling problems in defining the normality of the object of legal protection: how can "nature's household" be delimited? How can the normality of a network of relations, itself subject to dynamic change, be determined? Another aspect of this is the increasing cost of the generation of information. A model for the regulation of social risks which has been separated from the continuity of experience and confidence in technological progress must gain a large part of the information necessary for the evaluation of risk through experiments and modelling. It thus gives rise to costs which increase with the complexity of the context. And, given the uncertainty in the test procedure and over whether an adequate basis of evaluation can be found, all this occurs without the reliable expectation of an adequate gain in security. It is not even clear which substances are worth investigating and which are not. Equally, it cannot be assumed that the degree of available information matches the extent of the risk. Accordingly, the development of a public awareness of chemicals in the environment, itself dependent on chance and coincidence, comes to be important.

This uncertainty finds its legal expression in the American law on chemicals and plant protection where the possibilities for the prohibition of chemicals are limited by the required burden of proof. However, if there is a remote probability (of the second order) that there is a danger to the environment (thus, of the first order), this is sufficient to found a demand for additional tests (Applegate, 1991: 32). In this way the examination of chemicals can be very expensive for their manufacturers. At the same time, old substances which are not subject to the registration procedure and may still be used, are possibly much riskier—which demonstrates that linear thinking in terms of temporal development processes can lead to a new kind of perverse results. These can themselves be traced back to the deficient co-ordination of the institutions for the generation and control of social knowledge. Old substances are not merely a transitional problem. They are of great significance because the intensive control of new materials means that they continue to be widely used as the producer is deterred from innovating by the costs of inquiry proceedings. Thus, the increasing differentiation and dynamism of technological and material development represents an additional problem because the firms which generate knowledge have an interest in suppressing or neglecting the search for information about risks (see Stewart, 1981; 1985; 1988; Ladeur, 1987; 1988). The problem of the accessibility of the available knowledge is increasingly overlaid by the limits on the generation of technological innovation. And so risk-avoidance technologies become increasingly unavailable since the authorities can only order their precautionary application after the "state of the art" has been established (Salzwedel, 1991: 45). Equally, information about the "side effects" of dangerous substances can no longer be collected by means of experience.

The introduction of liability for the endangerment of the environment does not necessarily lead to the problem of information being shifted onto the firms. This form of liability is based on private rights, the traditional indicators of environmental damage, but firms are collecting a mass of information in order to avoid liability claims. This information is only available to the authorities to a very limited extent. However, given the number of possible risk situations, it is completely impossible to transfer sole responsibility to the authorities for the collection of that information about risks which is necessary for environmental decision-making. At the same time the practical development of technological knowledge by firms has lost its orientation because of the disintegration of the traditional institutions for the generation, control and evaluation of knowledge. This is also true of those which produce environmental technologies as such and which are not merely assessing the damaging side-effects of technology. The market for environmental goods is difficult to assess, partly because the degree of integration of environmental interests into general market preferences is uncertain, but also because the legal support of such technologies, and thus the creation of the market, depends on the state.

The reverse problem arises for the state if it wishes to carry out a policy of "technology forcing" (Stewart, 1988), because it has to reckon with firms reacting strategically to state decisions. It cannot know whether and to what extent a particular environmental technology can be realized. The uncertainty of the information on which decisions can be based arises from the difficulty of assessing the economic consequences, in different sectors, of state intervention in technological development, given that these decisions are no longer co-ordinated with the innovation policy of the firm. The technical or economic consequences of imposing an obligation on firms to invest in environmental technology cannot be predicted. Conversely, for the firms, the decision-making procedure of the state, assessing risk from case to case and separated from experience-based general rules, is also difficult to calculate. Decisions on the avoidance of "unreasonable risk" (Applegate, 1991; see also Fischmann, 1992) can no longer presuppose the existence of common concerns (see Luhmann, 1991: 54ff.; Bechmann, 1991). The perpetrator has an interest in claiming a greater level of security than has actually been ensured. Equally, a third party may dramatize the risks.

The difference in perceptions of risk by those affected is further accentuated by the fact that a shared confidence in technical progress has long since disappeared. Risks are traced back to visible social decisions, instead of being looked on as natural or cultural. A great number of alternative strategies are argued for, distancing themselves from technology. Yet technological knowledge, bound as it is to networks of interpretation and its particular applications, cannot admit all these strategies which seek to subsume it within a different type of knowledge, based on context-free judgments and reproduced in public institutions. This is also a symptom of the disintegration of connections of support and reference between various types of knowledge, which were formerly maintained by implicit rules of harmonization.

There have been some pragmatic attempts to make different types of knowledge and social values compatible by the explicit institutionalization of agreements between them. There is, for example, the "consensus-workshop", or the procedure of "mediation" (Di Mento, 1986), in which co-operation is organized across the boundaries of different types of knowledge. It is far from clear, though, whether these attempts have met with any success.

The reaction of the state to these information problems is paradoxical. It uses the powers of intervention that have accrued to it under the new environmental law, and in particular the instrument of follow-up orders, as a means of bargaining with firms through a process of informal consultations (see Kloepfer, 1989: § 4 no. 100ff.; Hoppe-Beckmann, 1989: § 25 no. 95ff.). Although conceived with the aim of galvanizing technical knowledge, this instrument becomes a bargaining chip whose value is not fixed until after the consultations have been made. This response is connected to the problems of knowledge insofar as the actual effects of pollution and, thus, the possibility of formal legal resistance by the firms involved, are difficult for the authorities to assess, while the relinquishing of the use of the decree may be worth something in return. Although the arrangements thus achieved may be constitutionally questionable, they recognize, through their informal character, the necessity for the formation of inter-organizational, practice-bound processes for the harmonization of different types of knowledge (Püttner, 1991). The reduction of the formal, legal protection of confidence has changed nothing here, since it has been compensated by situative, inter-organizational, cooperation. However, as the individual elements of environmental law, which have traditionally been oriented towards the generation and application of social knowledge, become increasingly differentiated there has been produced a tendency to over-estimate the degree of self-co-ordination between various types of knowledge and the institutions associated with them (Florida, 1991: 569). Thus, while the increasing refinement of the machinery of environmental law has been successful to a certain extent, the accumulation of individual reforms has led to a series of perverse effects which produce the danger of self- and cross-blocking. This raises the question of whether a new and more complex, order-forming, structural principle is thinkable, which could satisfy the need for co-ordination and harmonization in an environmental law based on the principles of provision and precaution. And this raises the issue of whether the tensions described require a more precise conceptual description, going beyond the pragmatic straightening of boundaries, and allowing the transition to the complex model for the co-ordination and compatibilization of different types of knowledge that is necessary for making decisions under conditions of uncertainty.

One alternative to the existing legal order is the recently developed market model (Stewart, 1981; 1985; 1988; Roe, 1988). This aims to internalize the hitherto externalized costs of "environmental consumption". We must take this approach seriously because not only does it aim at the superficial relief of the state but it also stresses the problems of generating and processing knowledge. The limitation of this model, however, is that environmental goods, like public goods, must be evaluated by the state and so the adjustment of environmental law to the strategy of economization can simply run up against new limits of knowledge. Alternatively, a strategic intervention which reduced the possibility of differentiating between single environment-endangering substances, for example, would have to be accepted which would be to partially bargain away some of the flexibility gained by the economization strategy (Ladeur, 1987: 22). Without underestimating the value of a market model, in what follows the cognitive approach is more sharply emphasized. We will look, in particular, at possibilities for making firms more open to the perception of the problems in their natural and social environment, as well as means of making state activity more flexible under conditions of uncertainty, thus leading to a perspective-changing combination of self- and cross-observation.

In order to do this it is necessary to find, under conditions of greater complexity, a functional equivalent to the continuity given by the role of experience and for the implicit co-ordination rules of the traditional types of knowledge that it generated.

THE SEARCH FOR KNOWLEDGE OF RISKS AND ITS STOP RULES

Technical knowledge and the closure of technical systems

In this section we will attempt to analyse more precisely the causal assumptions which underlie the concepts of danger and experience. This will principally seek to demonstrate their dependence on, and connections with, social conventions. These act as "stop rules" which lend the knowledge a pragmatic character bound to forms of application (Rasmussen, 1991).

The knowledge processed in the technical system is based on the isolation and inclusion of the behaviour of physical systems by means of the "application" of natural laws (Rasmussen, 1991: 47). The classical idea of causality, based on the establishment of relationships between events, has been superseded in more recent technological and scientific thought by the measuring of variables and the methodical construction of relationships between them. Events are no longer described by means of the attribution of characteristics, but as relationships which have been isolated from a context. This is true for the experiment, but also for the way that technical cause-effect relationships are "included" in a world of their own, defined by technology itself, whose contact with the diffuse stream of natural events is treated as "frictional losses". An example of this would be the management and control of energy development in a steam engine. This is characterized by precisely defined relationships between the technologically related internal processes and undefined, diffuse, interactions with the non-technical environment.

This corresponds to the thinking of the law, described above, on the respective interests of neighbours where the inner technical world is left at the disposal of the owner and a claim of defence is conceded to the neighbour only when there are indisputable interferences in his rights. Factually, then, this amounts to a parallel between the spheres delimited by rights and the inclusions and exclusions of technical thought (Luhmann, 1991: 97ff.). The side-effects of normal operation can only be determined within the framework of the "normal" (as this has been determined by technology). The causal description of a chain of events can only be resorted to retrospectively in the case that there is a breakdown and the technical system collapses. In this situation the technical system has broken down and become simply an element of the context in which inside and outside can no longer be differentiated. The description of events can be made in terms of a different form of knowledge, although this remains bound to the context of application of that theory. The theory of "adequate causation" (*adäquate Kausalität*), for example, is relevant for the purpose of a meaningful limitation of the attribution of causal chains to responsibilities.

However, this reversal of the two forms for the description of causal relationships (variables/events) is still bound to maintenance of the technological mode of thought, because the actual risk-event may then be generalized as a variable and technically mastered which, thus, makes future inclusion more certain. One could, of course, substitute hypothetical single events for all variables and thereby come close to the complexity of real life (Rasmussen, 1991: 254). However, the number of possible border-crossings is theoretically unlimited, so the recourse to modelling relationships between variables is equivalent to a stop rule. The mode of functioning of the stop rule remains bound to the practical availability of alternatives (Rasmussen, 1991: 255; see also Breuer, 1976). Thus, intuitive cost-benefit analyses are included in the development of generally recognized standards of technology. They are normatively controllable but are, however, also attuned to requirements articulated through "disciplines" ("reasonable diligence") (Willard, 1991; see also Polanyi, 1962: 215ff.). Additionally such stop rules are indispensable in the search for information. This is because without confidence in a knowledge-aggregating network, which cannot be made fully accessible either to deductive examination by the standard of universal laws or to the complete inductive description of the attributes of single events, decision-making under conditions of uncertainty and temporal constraints would not otherwise be possible. Thus, those strategies which aim at the virtual institutionalizing of mistrust and wish to annul pragmatic stop rules must, of necessity, shift the problem of the delimiting of knowledge to a meta-level where trust in ideologies or the reasonableness of decision-making processes is required. Equally, it must be stressed that the artificial, constructed character of the knowledge by which societies reproduce and modify themselves should not foster the illusion of the possibility of a central perspective arrived at by means of inter-subjective agreement (see Gauchet, 1979). This assumption has been undermined by the increasing plurality and heterogeneity of the various types of knowledge, which no longer admits the possibility of any unified rationality.

This insight allows a focusing of the question of the function of stop rules which delimit, and must delimit, the search for "safety". These rules do not presume security but, paradoxically, allow the taking of risks for the purpose of learning from accidents. Beyond the apparently objective calculation of danger as a product of the extent of damage and the likelihood of occurrence, lies the assumption that the maintenance of the limits of the autonomous technological system must depend on practical testing. This is implicitly made possible by the diffusion of risks: for technical reasons, medium and small-size pressurized containers are built instead of gigantic ones so that one breakdown can contribute to the avoidance of others. This spontaneously generated strategy for the avoidance of risks is also transferred back to nature: risks which cannot be measured against single objects of legal protection are regarded as sufficiently widely distributed. These are, so to speak, the manifestations of an unavoidable increase of entropy in the world. This is the strategy which functions as the stop rule in the harmonization of different types of knowledge and is built into their institutional co-ordination. Yet this is contradicted by the fact that the accumulation of single, minimal emissions in the environment can damage individual objects or even so alter the ecosystem that whole classes of objects can be negatively influenced (changes in climate, soil erosion and so on) (see *Bundesgerichtshof*, NJW, 1988: 478 on damage to forests). In these cases, for example where there was a major accident in a large nuclear power plant, the shift from the prospective construction and modelling of relations and variables to the retrospective description of chains of events is neither possible nor meaningful. The traditional system of environmental protection could be developed here, to some extent, by lightening the burden of proof or moving towards collective systems of attribution (fund model, super fund etc.) (Gaines, 1990). This soon meets its limits, however, in relation to cumulative effects, because these call into question the whole way that separation and attribution are described (Rasmussen, 1991).

The same problem arises with technologies which can be included in the models but which, if they "cross the border", damage their environment so severely that the expectation of benefit to the technical system is disappointed. Learning by breakdown is, thus, too expensive. This problem has arisen in the use of nuclear technology. Here, there have been attempts to make thinking in models and thinking in chains of events compatible by the simulation of "fault or event trees", which run both forward from the cause to the damage or, in reverse, from the damage to the possible causes (Renn & Kats,

1990: 65). However, even here, the co-ordination of different types of knowledge is based on the assumption that in nuclear power stations many minor accidents will occur allowing the accumulation of the knowledge necessary for avoidance of major breakdowns. Thus, accidents such as that on Three Mile Island have contributed to improvements in the safety of the technology (Rasmussen, 1991: 255).

Epistemic risks in the modelling of complex technologies

Nuclear technology has brought a further change in the process of the generation and use of technical knowledge. It is so specialized and, in comparison to earlier technologies, develops so quickly that there is not merely a smaller temporal gap between scientific research and technical development, but it also excludes the communal formation of an experience accessible to a wider circle of participants. This points to a general problem, beyond the field of nuclear technology: the blurring of the boundaries between science and technology changes the way that they interact. In the evaluation of risks by experts it can be seen that the balance has shifted in favour of theoretical model construction to the detriment of the causal description of events. Beyond the superficial problem of the relationship between qualitative and quantitative evaluations of risk, the recourse to subjective assessments of likelihood by experts, as well as the methodical-formal operation with large amounts of data, turn out to be unavoidable, because of a lack of real experience of rare incidents. It becomes increasingly evident in complex technologies, which cannot necessarily be broken down into individual elements and well-structured relations, that the multiplicity of possible relations within the system reaches the limits of "modelability" of a micro-world separated from its environment. As a consequence a new type of epistemic uncertainty appears as the uncertainty of real life returns to the model from which it was to have been driven out (Rasmussen, 1991: 255; Latin, 1982; 1983: 339, 357; Flournoy, 1991: 336). Faced with highly complex technologies, the assessment of the probability of known possibilities of damage and the determination of the possibilities of damage are confronted with great methodical problems. Modelling tries to take this into account in that it even quantifies the risk of the occurrence of unknown risks reserving variables with unknown, but possibly at a later date determinable, value. Such scenarios require large amounts of data which can only be processed by computer and are either not understood or misunderstood by the non-scientific public. Science contributes to this by not disclosing the element of uncertainty and the role of intuition in the operation of risk evaluations.

Science also remains bound to a co-operative, a-centric and heterogeneous network of ramifications and connections in which each only understands one part and that develops further through confidence in the reliability of the whole (Polanyi, 1962: 215). This makes partial control possible through the overlapping of sub-units, just as the separation from other social systems allows a relative autonomy in the formation and reinforcement of standards and the institutionalizing of criticism. However, the "truth" which the scientific system generates, to which recourse is increasingly had in the evaluation of risks, is bound to its own rationality rather than experience. It cannot guarantee any higher objectivity to which the *eigenvalue* of other partial systems must be subordinated—a feature which is often misunderstood (Lewis, 1980).

The social organization of technical knowledge is, to a considerable extent, determined by pragmatic stop rules which keep the search for knowledge of risks compatible with the need to create confidence in the durability of the networks for the generation and evaluation of knowledge. New knowledge is only possible on the basis of this confidence (see Wildavsky, 1988). The processes of harmonization of earlier types of knowledge could largely remain implicit as a result of the dependency on experience. However, the development of new, more complete forms of knowledge puts the traditional forms to the test, forcing reflection on, and explication of, the function of these stop rules. If one takes this into account, problems in the harmonization of new types of knowledge, and the social values that are thereby mobilized, may be better described, particularly since it has been shown that these stop rules make an indispensable contribution to the reduction of complexity. This may be demonstrated. In the face of the limits to the mastering of uncertainty described above, an insistence on a precise description of all possible chains of events connected with the use of complex technologies is synonymous with a renunciation, and thus, with a blocking, of social knowledge systems through social values (see Böhler, 1991: 999). It is hardly questionable that this cannot be viable without the social knowledge systems as a whole being damaged. On the other hand, the recourse to a normative, consensus orientation is hardly more promising in the face of epistemic risks which are connected with the differentiation of knowledge systems. It is highly improbable that the dissolution of the experience of the one reality can be compensated for through more common values. This is underlined by the fact that the new moral universalism has to make use of very small protest groups which are separated from the technical and economic systems.

It is conceivable, however, that a stop rule for the harmonization of different types of knowledge could be reformulated, making the (formerly implicit) co-ordination function explicit by means of proceduralization. Behind the stop rule described above is concealed a need to guarantee the continuity of technological learning by maintaining the diversity and flexibility of a self-organizing and steering network of practical knowledge (see generally Allen, 1988; Probst, 1987: 13). In practice this was bound to experience, which assumed a linear model of equilibrium dampening down the fluctuations around a resting point (see Allen, 1988; Day, 1987; Csanyi & Kampis, 1987). Technological forms of knowledge, by contrast, can only be described by a non-linear disequilibrium model (see Arthur, 1990 for economic evaluation). It cannot be assumed either that there is a continuous technical experience developing across a multiplicity of participants, or that damaging consequences of technology will eventually be offset by the benefits. A stop rule which adapts itself to the conditions of technological knowledge could more usefully aim at the facilitation of the expectation of the unexpected through the conscious maintenance of flexibility within and between types of knowledge. And it could presuppose the maintenance of diversity and variety in the modelling of technical micro-worlds and in the observation of the consequences of boundary-crossing.

In a model of disequilibrium it is conceivable to increase the flexibility of technological knowledge in such a way that the separation of the internal, technical world and its environmental effects need not lead to a loss of differentiation. This can be taken into consideration in more complex secondary modellings. A strategy for state control can possibly also be developed in such a way that an evaluation of risk is not made on a case by case basis (or from substance to substance). Rather, in each case the possibility of revision and reversibility and the extent of the increase in the expenditure of diversity and variety in the environment would be taken into account as a criterion. (This would speak strongly against the existing form of the use of nuclear technology.) In bio-technology or in plant protection, the "consumption" of biological diversity and variety could be compensated by the pressure to introduce more variety to help the self-stabilizing ability of the environment in the face of unknown dangers (Luhmann, 1991a: 101; Wildavsky, 1988: 30ff., 185ff.; Probst, 1987: 61). This could be made possible through a change in agricultural policy, though the details will not be dealt with here. Instead we shall test the proposed reformulation of the stop rule by looking at whether, and how far, the internalizing of environmental protection interests by (technology oriented) firms can be extended with the help of a more procedurally determined legal regulation which increases flexibility and diversity. This would be a strategy which both voids the over-burdening of the state and makes the innovation strategies of firms sufficiently open to environmental protection interests that the self-steering of economy and technology is not hindered (for a legal theory of the firm, see Teubner, 1987). In the following, therefore, the conditions of technical innovation will be examined, with a particular focus on the production of pesticides.

Considerations of the dynamics of social innovation

From economic and legal perspectives the availability and progress of technical knowledge can no longer be presupposed as a condition of production (Amendola & Gaffard, 1986: 473). Accordingly, contrary to the assumptions of neo-classical theory, entrepreneurial options cannot proceed on the basis of complete prior knowledge of possible results. Following Schumpeter (1934: 8),

who insisted on the necessarily uncertain character of entrepreneurial innovation, other writers have also seen innovative behaviour as structured by a
sequence of "technological paradigms" (Dosi, 1982), and have questioned the
dependence of decision-formation on the mechanical principles of the maximizing of successes on the bases of given options (Nelson & Winter, 1977).
Recently the conditions of technological innovation have been characterized
by a transition from the product industry to the "function industry" (Amendola & Gaffard, 1988: 21). This opens another perspective on the transformation process outlined above, from which the separate phases of production
(and, within the firm, the different departments) become increasingly permeable. This is connected to the "dematerialization" of production and the
increasing sharing of knowledge (Herman *et al.*, 1990). In some analyses, this
has already led to the interpretation of material resources themselves as "information" (Ayres, 1987; 1988). The neo-classical approach to the definition of
technological innovation has always presupposed the availability of a certain
state of technology which is regarded as capital. This has to be built into a
given economic structure and used for a specific production capacity
(Amendola & Gaffard, 1986: 473). Technology, however, cannot be separated from the network of general knowledge, its development having a
procedural character which remains bound to the network as its "environment". Expectations which determine the acceptance and development of
technology develop on this basis. And although, in a world of disequilibrium,
these do not make permanent stabilization possible, they can trigger imitative
processes which make future developments difficult. The model of disequilibrium describes the transition to a faster self-modification of technological development as the "environment" of the production system. However,
this leads to the production system being increasingly dependent on a general
cultural pool of possibilities, as a result of the greater knowledge content of
production (Fehl, 1983; Kunz, 1985). This can only be saved from self-
blocking through flexibilization and an increase in unspecific learning ability
(Amendola & Gaffard, 1986: 478; Ayres, 1987).

Recent technological development has clearly revealed the significance of the
knowledge-generating network as a source of technological innovation. The
rapid self-transformation and differentiation of this network implies that technical change is increasingly leading to options for flexibility as such, that is to
say the creation of more possibilities for the future. In the face of uncertainty,
the procedural creation of "option-domains" replaces a substantive reality
which presupposes the availability of options (Favereau, 1989a; 1989b: 121).
The maintenance of adaptability, and thus of the ability to learn, becomes central to the long-term strategies of enterprise (Amendola & Gaffard, 1988: 26;
Kirat, 1991: 42).

This means that processes of innovation cannot be reduced to the effect of
profit stimuli alone (Nelson & Winter, 1977: 48; Boynton & Victor, 1991;
Day, 1987; Foray, 1991). An innovation is generated through a co-operative,

procedurally self-transforming network of possibilities (Silverberg, 1991: 69; Florida, 1991). This means, conversely, that the strategic power of big business to influence innovation tends to favour a certain passivity. Confidence is placed in the possibilities of strengthening innovations developed by others, as well as a developing tendency to block innovation. This, incidentally, is encouraged by the fact that there is only a slight private interest in the maintenance of knowledge, which is in part a public good, since the ability to appropriate it is limited (Mackaay, 1986; Imai & Itami, 1984). This status of knowledge as public property contrasts with the ever-increasing need to generate knowledge in private forms. The need to harmonize the relationship between publicly available and privately appropriable knowledge leads to new self-reflexive forms of organization which provoke a secondary modelling of the institutions of the business organization and the exchange contract (Amendola & Gaffard, 1986: 474; Imai & Itami, 1984).

A COGNITIVE VERSION OF PROPERTY

It is of decisive importance for us that the development of technological knowledge is apparently reproduced through an a-centric and discontinuous network of constraints and possibilities. Although this cannot be *steered* by the state, the question arises of whether, to promote environmental knowledge, it can be sensitized to "irritations" which reduce the danger of the processes of technological search and innovation blocking themselves. Given the complexity of technological innovations, environmental law can no longer treat the business organization as a black box. At the same time, the capacity of technological development for self-modification leads to the consideration of whether, and to what extent, legal interventions which do justice to that complexity could affect the reproduction of the network so that its immanent flexibility can be mobilized for the internalization of environmental protection interests. To this end, we will attempt to reformulate the right of property, taking into account the increasing significance of knowledge, and emphasizing its function in the generation of knowledge and the formation of expectations.

From a functional perspective property appears as the guarantee of a decentralized hierarchical social form of decision (Coral *et al.*, 1986: 462; Schmidtz, 1991: 143). The maintenance of inequality, perpetuated by the law of inheritance, continually brings ownership into conflict with formal universal freedom. However, it is precisely the separation of ownership from material resources, and the way that this guarantees the functional decentralization of laws of decision, which connects with the need to maintain the historically generated "population of ideas". These are the behavioural and thought patterns, removed from an individual availability, with which the exercise of freedom must be co-ordinated.

The formation of common expectations is only possible on the assumption of a common, historically accumulated and legally institutionalized, network of knowledge. Ownership makes possible the use of common knowledge which is distributed over the network (Hayek, 1973). However, it is nowhere mutually available and, thus, admits no external standpoint from which the justice of the distribution of property may be assessed. Ideas of justice are generated through institutions themselves and so are bound to historical practices and the network of the population of ideas (Day, 1987: 254; see also Hayek, 1973). Thus, paradoxically, justice can only assert itself as a self-fulfilling prophecy, processed by its own institutionalization (Schmidtz, 1991: 163). The legitimation of the "beginning" in the presupposition of a state of nature, revoked through a social contract of the founders, which repeatedly causes problems for the theory of ownership, naturally cannot plausibly explain the fact that succeeding generations are also bound. It is the historically emergent institutions which generate the possibility of self-determining individuals as decision-makers.

Ownership always allows for new knowledge and new possibilities (even for non-owners) in that it allows the appropriation of new possibilities, connected to new ideas. It reduces the need for consensus and makes separation (of ownership and non-ownership) into a pre-condition for co-operation (contract). The emergence of patents, as secondary modelling of the ownership of things, is revealing of the manner of functioning of ownership as a whole (Childs, 1990: 590): through the inequality of decision-making possibilities it supplies diversity and variety to the social knowledge system, thereby making self-organization possible. Ownership is institutionalized as a form of the "memory" of constraints and the aggregation of possibilities, whereas contractual freedom only produces discontinuity (Luhmann, 1991b). The full use of the knowledge distributed through the social body is only made possible by the productive interplay of continuity and discontinuity, order and disorder. This was exemplified by the case of collective ownership in Yugoslavia. This did not allow public interests to be considered as part of a wider perspective on decision making, but rather tried to absorb the benefits of innovations as quickly as possible in the form of higher wages. The collective would even postpone the payment of present costs if this was possible (Pejovich, 1987). Conversely, of course, the management's willingness to innovate was under-developed because, in a crisis, appeal could always be made to the state.

The adjustment of the theory of ownership to the conditions of the information society demonstrates the problems that are involved in harmonizing the conception of knowledge as a public good with the necessity of maintaining private ownership, that is to say the capacity for innovation. Given the increasingly complex nature of social development, the harmonization of the production and acquisition capabilities of knowledge seems more and more problematic because of the destabilization of the relationship between different types of knowledge and their modes of application. As a result the uncertainty impedes the formation of expectations and leads to the efficiency of the legal system being called into question.

This is similar to the problem of the harmonization of different types of knowledge (modelling versus causal chains of events) (Rasmussen, 1991; see generally Stachowiak, 1983; Allen, 1988). As we have already seen, it is not possible to break down the separation between technical "micro-worlds" and their environment. These forms of differentiation are essential. They are not indicators of the indifference of the separated to each other but require a *harmonization* of the different models of order. The problems cannot be solved by the dissolution of the separation between knowledge as a public good and as an object of private acquisition. Rather, it consists in the mastery of the need for a secondary modelling of the harmonization of the separate forms. This produces the situation where elements of the contract, which are typical of market spontaneity, are reproduced within the business organization based on ownership. Conversely, there are new flexible types of organization such as joint venture within the contractual system (see Teubner, 1990a; 1990b; Imai & Itami, 1984). Together these give potential for the production of greater possibilities than are aggregated in the market.

The relevance of these considerations for environmental law and policy lies in the fact that the possibilities of a more complex legal model of regulation can be tested for their ability to reformulate and adjust to the conditions of the production and acquisition of technical knowledge. The function of ownership has to be adjusted to the necessity for the maintenance and generation of diversity and variety in a world of disequilibrium and uncertainty (Fehl, 1983: 80; Kunz, 1985).

Against the historical background of equilibrium and continuity, property could be regarded as a right of domination over specific material objects whose separation from the public sphere and non-owners was attenuated by implicit co-ordination rules (tort) and structures for co-operation (contracts). The new paradigm should be constructed following the, formerly exceptional, concept of the "patent", where property is valued for its dematerialized function of generating knowledge. The frame of reference is thus the model of disequilibrium which is more openly dependent on the redesign of co-ordination rules between different types of knowledge and the forms of their appropriation. The legal regulation of property can thus endeavour to broaden the public sphere and to impose limits on the dominion over objects. It can, thus, tune its functions to the conditions of a knowledge-*generating* society and relate more closely to the complexities of the present situation.

LINKING THE INTERNAL AND EXTERNAL ENVIRONMENTS OF THE ENTERPRISE

A cognitivist perspective can also give shape to company law as it responds to the transformation of enterprise organization (see Teubner, 1987; Ladeur, 1989). The knowledge-based enterprise that must turn away from a fixed orientation to products and services (Boynton & Victor, 1991: 58) to the self-transcending goal of generating knowledge in an open and rapidly changing field can only be conceived of by the model of disequilibrium. The traditional enterprise organization which aimed at economies of scale and the fragmentation of property functions between management, shareholders and capital market is no longer adequate as the phases of innovation and application become increasingly interchangeable (Florida, 1991: 567). This transformation reveals the "proceduralization" of enterprises (Vardaro, 1991: 226) which transcends the limits of the single enterprise and embraces groups as "poly-corporate networks" (Teubner, 1990b). New forms of "dynamic networks" are created within and between enterprises which break down traditional hierarchical forms of organization and, at least partially, substitute them by variable heterarchical interrelationships (Miles & Snow, 1986: 62). These new networks are characterized by processes that replace the systematic divisions of work, based on experience and rules, with flexible forms of self-construction within a poly-contextual world. The separation of the enterprise from its environment becomes permeable insofar as its heterogeneity must be remodelled, by decentralization, within the organizational forms of the enterprise (Nonaka, 1988a; 1988b). The organization transforms itself as it reproduces (and transforms) within itself the flexibility of the logic of the market and that of using a type of knowledge which is distributed over a number of participants and which can no longer be aggregated by division of labour alone.

It should be stressed once again that this is not a process of de-differentiation, but one of interference (see Teubner, 1989: 110, 125). The organization opens itself up to disorder and diversity by forming temporary, overlapping and decentralized groups which are to generate information by redundancy and competition. The cognitivist perspective stresses knowledge generation because the mechanistic model of "information processing" is not compatible with the idea of self-generating and modifying enterprises (Nonaka, 1990; 1991). However, this does not mean that the enterprise should be able to "plan" itself, or that the open network of information that is the market has been wholly introduced into the enterprise. It is still dependent on specific organizational conditions beyond the parameters of the market, requiring more complex forms of co-ordination to sustain itself as a collective actor. An increasingly informed demand for products, for example, requires a reintegration of information on customer use into the production process. Production embraces handling, repair, and ultimately recycling—in legal terms substituting property by leasing systems (Imai & Itami, 1984; Rosegger, 1991; Mitroff, 1990). Products are increasingly transformed into a kind of service because of their dependency on knowledge. In a process designed to slow down the bureaucratization of enterprises, groups work like enterprises within the enterprise (Nonaka, 1988a) and generate a creative tension which enables the use of the "implicit knowledge" of the network of cooperation and communication by "overlapping" mutual irritation (Nonaka,

1988b: 58; Silverberg, 1991). The organization is punctuated by disorder and so is fit for flexible responses to turbulence in its environment. Diversity and variety is introduced into the organization, raising its learning and innovative capacity by augmenting internal fluctuations. And, the tension is kept within limits by the acceptance of a shared overall "vision" by the participants (Senge, 1990: 7). In the words of Le Moigne (1987) one could say that the enterprise is "self-producing".

It should be emphasized that the main causes of this transformation are, on the one hand, the rising dependency of enterprises on knowledge and, on the other, the problem of the sustainability of an effective market for information (Imai & Itami, 1984: 299). Innovations are the product of a collective process which can only partially be steered. The developing form of joint venture operates with the paradox that production is more and more dependent on knowledge which is itself only partially subject to appropriation. Through common research and development organizations try to combine the effects of market and organization by introducing elements of the organizational generation of knowledge into the "market place of ideas" that comes with processes of self-organization (Imai & Itami, 1984: 299; Miles & Snow, 1986: 64; Foray, 1991). The transformation is managed by remodelling contractual co-operation and internal processes of organization. A contractual "exchange" takes place, but its object will be accessible to a plurality of participants and for purposes that may lie outwith the joint venture itself. While traditional forms of contractual co-operation presupposed the long term building of confidence on the basis of experience, the new forms of joint venture guarantee the accessibility of a pool of knowledge greater than the immediate needs of the individual participants and which is not capable of being generated (at the same cost) by the individual organization (Miles & Snow, 1986). This procedure allows organizations to produce greater internal diversity than they themselves would be capable of generating internally. This demonstrates the rising dependency of the enterprise on its turbulent environment and also, as a consequence, the permeability between the internal and the external environment (Teubner, 1990b).

All this points to the possibility of taking an "institutional ecology" into account (Weise & Brandes, 1990). In the face of an unmanageable complexity, individual institutions seeking to preserve their internal functions and efficiency are constrained to contribute explicitly to the conservation of the "viability" of the whole population of institutions which make up their common context. This is the pool of unspecified and multiple possibilities upon which a generative process is based (Probst, 1987: 62). This process demands the remodelling of the *eigenvalues* of the institutions. The wider time horizon of decision-making and the greater flexibility will allow for the construction of new domains of options rather than restricting choice to a fixed set of possibilities.

The current debate on business ethics is a symptom of a new level of, hitherto unmanaged, complexity. If this were to lead to de-differentiation, rather than a revision of the differentiated but structurally-coupled subsystem *eigenvalues*, then this could create a risk. The combination of economic rationality and "ethically conscientious forms of organization" by means of "argumentative discourse processes" (Ulrich, 1991; 1989; for a critique see Priddat, 1989), does not seem to be very promising in the light of the rising level of uncertainty which has accompanied the dynamization of social knowledge and the shattering of values. The proposed reordering of the disorganized relationships of the economic system and the natural and social environment by means of a change from "exclusive" property to differentiated rights of participation and disposition (Ulrich, 1989: 140, 147) does not seem adequate to the complexity of the situation because of its assumed consensus on values.

According to our argument, the redefinition of the profit interest as a search and learning mechanism, and thus in the long run also creating more possibilities for non-owners, could be more significant. Rather than being suppressed, it could in this way be "punctuated" with more intelligence (see also Coffee, 1991: 1279; Buxbaum, 1990; Fitzroy & Kay, 1991; Albert, 1991). The point is not to subject the enterprise to a "higher purpose" but to exploit its inherent potential for flexibilization through legal intervention. By the generation of more diversity and variety this can strengthen the capacity for self-maintenance of the environmental field composed of different subsystems and their environment.

To draw a provisional conclusion, then, the accent is to be put on the possibility of describing and co-ordinating processes for the dynamization of knowledge, the differentiation of property, and the flexibility of enterprise organization. This is essential before the possibility of reintegrating the hitherto externalized environmental problems can be examined. From the perspective of environmental law this requires a harder look at the limits of interventionist models of regulation. This has been neglected as a result of the more general failure to look at the internal structure of the enterprise. And this, in turn, must be attributed to classical theory and its description of the market as the primary institution for the generation of knowledge (Williamson, 1991a: 269; see also Williamson, 1991b; Jarillo, 1988; Thévenot, 1989: 180). The development of these new co-operative networks for the generation of knowledge has reinforced the momentum of the process of the self-production of collective knowledge. This has given rise to the possibility of an enterprise law adapted to ecological reflection and, as a corollary, a model of environmental law adapted to the infrastructure of the enterprise. The condition for this is the continuing development of those heterarchical forms of relationship, between market and organization, that transcend the boundaries between institutions by new forms of proceduralization and which are apparently open and sensitive to environmental interests. The "ecological field", within which social knowledge and practical constraints for decision-making are generated, also offers possibilities for external intervention, by the state, for example. It could raise the potential for self-observation and modification by adjusting the context, interrupting or creating of constraints, or by introducing diversity from a societal pool of possibilities, always being wary of the risk of blocking the flexibility of enterprises by the imposition of abstract rules or single interventions and, thus, producing a solution by the creation of a greater problem (see Godard, 1990: 228; Orléan, 1985: 150; Favereau, 1989a: 310, 327; Bienaymé, 1988). Especially important among these is the process of adjusting relationships and networks because the monitoring function of enterprises, which seems to be under internal pressure from the fragmentation of property functions (see Masten, 1988), can and should be restructured with respect to environmental criteria, in order to make the enterprise more permeable to complexity.

REFORMS AND THE IMPROVEMENT OF ENVIRONMENTAL PROTECTION

As we noted earlier, there has been a tendency to replace the traditional form of external legal regulation with informal arrangements between public administration and private enterprise. These can be regarded as a form of interference (see Teubner, 1989: 110, 125) between state and market which give enterprises a new flexibility and planning potential (Harter, 1982: 23). Of course, it must also be borne in mind that by following this path the state runs the risk of the ultimate surrendering of its *eigenlogic* if it fails also to develop a reflected strategy that structures these quasi-contractual arrangements according to certain goals and procedures. Without this the benefits to be gained from such a complex administrative strategy will be of only minor importance because enterprises will tend to keep information secret.

Another strategy, the introduction of environmental officers at the level of the plant or the higher management of the firm, could give rise to an opposite sort of problem. This appears to be an attempt to remodel a version of external regulation within the logic of the enterprise and to profit from a kind of (quasi-administrative) decentralization (Rehbinder, 1989; 1990). But if the environmental interest is a distinct office within the institution the work of environmental officers is limited in its information and influence. And, the introduction of a high level environmental manager is not necessarily equivalent to a strategy of integrating environmental interests into corporate management for they might still present themselves as a kind of (quasi-public) external interest. A parallel can be drawn with the German concept of the "labour manager" (*Arbeitsdirektor*) in co-determined enterprises, where it may also be the case that external social purposes are to be integrated into the goals of the enterprise (see Dierkes, 1985). Problems begin with the definition and description of the post: *Arbeitsdirektor* is not a representative of the workers; he has to support the plural interests of the enterprise; and he remains dependent on the enterprise management. The problems all derive from the fact that the enterprise is a corporate actor which cannot be dissolved into a coalition of groups (Teubner, 1987; Ladeur, 1989).

While we cannot undertake a thorough analysis here, it would seem that the possibility of introducing environmental criteria into enterprise decisions by the institutionalization of environmental officers must be viewed sceptically, as must all attempts to install environmental interests within the enterprise on the basis of an analogy with public administration. And the opposite also holds true, where the design of environmental policy follows economic models, such as the transfer of marketable pollution rights (see Boucquey, Chapter 3), or the formation of a "central environmental bank" following the example of the national currency bank (Staehelin-Witt & Blöchliger, 1990). The analogy remains superficial because the independence of the latter is complemented by a complex network of interrelationships with the private bank system, which is without a parallel in environmental law (Baecker, 1991: 173ff.).

The problems in contemporary environmental law stem, first, from the conflict between its strong orientation to the traditional doctrine of "negative" liberal rights and accumulated elements of a public "protective duty" (Klein, 1989) and, second, from the decline of confidence in technological progress. Its recent evolution has made it more diffuse and complicated but not more complex (Schroeder, 1986; O'Leary, 1991). The balancing model strengthens the orientation towards discretionary decision-making, operated on the basis of the traditional alternative (safe/unsafe) and provoking a politicization of information and research. The liberal right to health or life is reinforced by the implicit presumption that inaction is always equivalent to a gain in safety (Flournoy, 1991). But, considering the many unknown risks, the necessarily uncertain knowledge of the effects of toxic substances, and the privileging of traditional toxic substances which are not subject to regulation, this may be an illusion. Given the unavoidability of risks in an artificial society based on industrial evolution we must accept the need for a change in our conception of risk-management that goes beyond the "safety limits" of traditional experience-oriented models of danger. For this reason we must concentrate on the potential for self-observation and modelling of the "ecological field" as a whole (Weise & Brandeis, 1990). This could further the willingness to make information on risk publicly available. Appreciation of the character of this as a public good could even be stimulated by legal and financial support for joint ventures pursuing risk analysis, risk assessment and risk management (see Huber, 1985; for possible conflicts with competition policy see Pernice, 1992). Such a strategy could be favourable to enterprises in their willingness to keep organizational limits permeable because of their dependence on information and knowledge.

In this way the pharmacological and epidemiological analyses done by the chemical industry could be used more efficiently in a co-operative manner which enables the design of prospective and flexible models of risk paths. As we have seen the assessment of technology has traditionally been based on a strong methodological separation of the general model assumptions and the individual causal interrelation of events. This cannot be sustained, particularly for ecological risk assessment, because closed technical micro-worlds whose borders are transgressed by accidents or "normal" pollution cannot be taken as the conceptual point of departure for the assessment of complex "eco-systems". Complete knowledge of the side-effects of toxic substances on the environment cannot be attained, and we can only operate with incomplete models which remain open to self-revision (Sharples, 1987; Latin, 1983). On the other hand, traditional technical systems had an "intersection" which could also function as a kind of sensor for spontaneously generated knowledge about risks. The knowledge generated in joint ventures could possibly be more accessible to administration and the public at large if environmental policy had itself developed a greater reliability, calculability and continuity in the process of formulating its purposes and strategies (Flournoy, 1991; Latin, 1982; 1983; Shapiro, 1990).

ENVIRONMENTAL MANAGEMENT SYSTEMS AND THE GENERATION OF KNOWLEDGE AS A PUBLIC GOOD

Environmental regulation and second order proceduralization

One potentially fruitful approach to the generating of greater sensitivity to environmental issues is to attempt to profit from tendencies inherent in technology and production processes that lead to a more flexible "proceduralized" firm. This could allow the integration of new forms of self-observation and self-control of product development with respect to environmental risks.

One of the main targets of this strategy of flexibilization could be the adoption of an internal environmental management system (Aalders, 1991; Steger et al., 1991). This should be imposed on firms by legal norms that themselves allowed for flexibility. The main difference of this from traditional forms of safety control would be their integration into the complex process of product development, while end-of-the-pipe technology is normally modelled retrospectively. In its initial phases, this type of model could develop forms of environmental auditing or controlling (Aalders, 1991), staff instruction, recruitment of professionals and so on (see proposed EC Regulation on Environmental Audits, OJ C7612, 27.3.92; also Klein, 1991; van Kleve, 1991). It should also enable firms to integrate environmental concerns systematically into monitoring functions and post-market control—a process which would be adapted to the provisional character of risk assessment.

The development of a cyclical conception of products as services ("leasing instead of selling"), on the other hand, could be encouraged by financial incentives and public information on environmental products. By placing the emphasis on the informational rather than the material resources, this could help, for example, to combine the use of chemicals with greater technical assistance up to and including sophisticated expert systems. Additionally, it could enable firms to collect more specific information on soils and the effects of pesticides, for example. The organizational integration of production and use, and of the roles of the producer and consumer, could lead to the generation of new knowledge not obtainable either from the market or the organization on its own.

The main principle of these environmental management systems should be the generation of more risk *information*. Yet it is questionable whether this strategy is compatible with the type of administrative control strategy that uses regulations to strengthen emission standards and exposure levels. This is because the control strategy lacks information. It is always confronted with this self-destructive paradox: it tends to suppress that risk information within firms that it is unable to generate itself (see Ashford *et al.*, 1985). Equally, it would involuntarily promote the survival of inflexible old technologies which would exert political pressure to ensure their preservation. As a result, standards would only be imposed on the new flexible industries (see Stewart, 1985: 8; Huber, 1985: 320; Trauberman, 1983; Ladeur, 1987; 1988). Given these obstacles, legal incentives to deepen flexibilization and the dematerialization of production and, in the longer term, move towards a cyclical conception of intelligent services, could consist in the provision for new organizational and contractual forms, designed to facilitate the generation of more complex forms of knowledge within and between organizations.

Technical innovation strategies that aim to reintegrate externalized environmental problems into the organization could be supplanted by private forms of control that remodel traditional administrative forms. This might take the form of environmental consulting firms, for example, whose use would be mandatory for the chemical industry (and supported by fiscal means), and so on. A new market for environmental control services along the lines of consulting and controlling firms could be created by the state (see the proposed EC Regulation on Environmental Auditing OJ C7612, 27.3.92).

In adapting environmental law to the end of creating environmental knowledge as a public good, use could be made of procedural mechanisms to reorganize enterprises. This would allow for a new approach to property and its legal form. The rise of complex new forms of finance capital ("paper capitalism") has blocked the "memory" of the economy and its function of self-observation that was made possible by traditional forms of property. This has created opportunities for short-term speculative profits (management buyouts and so on) that pose a risk to the long-term innovation capacity of firms (Albert, 1991; Coffee, 1991). Environmental policy is also challenged if the profit orientation is redesigned by company law to protect the autonomy of the corporate actor from the short term interests and motives of shareholders. Rights of shareholders could vary according to their time horizon. We can also reconsider the recent interest in business ethics. The inherent tendency of capitalism to generate an abstract monetary "wealth" turns itself against its true foundations: the ethic of work and achievement. As the market economy becomes increasingly artificial, it is increasingly confronted by the need to reintegrate the implicit presuppositions of its reproduction into the explicit targets of its institutions. The secondary modelling of a market for the control of corporate actors can, thus, be likened to the problem of the overtaxing of the resilience of nature, and the legal design of the market for the organizational control of environmental problems.

Administrative regulation and flexibilization

The internal administrative perspective must take into account the fundamental change in the nature of risk information now that it is no longer governed by experience, continuity and familiarity. This cannot be compensated for by ever larger numbers of standards based on ever greater scientific knowledge or by a case-by-case approach to new risks (Huber, 1985; Trauberman, 1983). This would only work on the basis of the rather problematic assumption that a decision against risky technology is always a decision in favour of safety. This is less than certain because of the degree of interpenetration of private and public technological decisions and old and new risks, new and unknown. The risk of avoiding risks must always also be taken into account because decisions binding resources reduce the capability to take further decisions should greater risks subsequently arise—not to mention the perverse effects on old technologies.

This new, and confidence shattering, situation places a heavy burden on administrative risk assessment. This is especially important in the case of the law relating to toxic substances and pesticide registration because those "stop rules" relating to the borderline, technical risks give no help in making decisions on hazards that stress eco-systems. Yet, trying to compensate for the absence of practical experience by referring to scientific knowledge in the formulation of "exposure standards" must not ignore the fundamental differences between scientific and administrative perspectives. Scientific experiments are only as productive as the range of possibilities that are put to the test. Regulatory models aim at a practical set of tests based on *present* knowledge (Zervos, 1989: 653). They are not designed to promote new knowledge. In this way they are related to the older concept of experience, although the "closure" of the technical system and the resilience of nature to small-scale transborder pollution are no longer self-evident. Although the presuppositions of regulation cannot be taken for granted, it must be reliable in the long term. In fact, the regulatory perspective must be oriented to a reduction of data (Zervos, 1989: 655). This is inevitable because new practical knowledge can only be generated on the basis of stop rules, limiting the search for new causal information.

Any new attempts to co-ordinate the internal perspective of the firm and its practical knowledge with the external administrative perspective geared towards the controlling of the side effects of chemicals must also take the potential openness of the enterprise to environmental problems into account, and integrate this into its own perspective. It must also be open to the alteration of its regulatory instruments. All this is to say that the regulatory perspective of the state must be adapted to a second-order proceduralization. This is a version of proceduralization that both presupposes and assists the processes of search and learning at the level of the enterprise. The first step towards the proceduralization of risk management can already be seen in the imposition of a set of predetermined legal standards. This is not satisfactory because it can create perverse effects, through the costs it imposes on the product development process, without effectively creating more safety (see generally Gaines, 1990: 289; Shapiro, 1990: 39; Magat, 1982). Thus, for example, the development of less risky (and less profitable) pesticides may be suppressed by the costs of registration. The uncertainty inherent in the process of registration of chemicals and pesticides under the American Toxic Substances Control Act and Federal Insecticide and Fungicide Act is highlighted by the fact that the Environmental Protection Agency (EPA) possesses a wide competence to impose expensive new tests on firms without knowing whether these are the essential ones, or whether negative results are equivalent to a higher safety level. At the same time, the effect of this uncertainty on the innovation processes of firms is incalculable. The possibility of having the requirement to undertake new tests imposed on them does not give firms an incentive to make public any risk information that they might possess. New information may simply give rise to the risk of potentially unending burdens of proof, and more and more expensive tests. And the number of chemicals is so high that systematic testing, even of new chemicals, is simply impossible (for cost-benefit analysis see Sater, 1990). A change should be considered. Would it not be more useful to operate beyond the traditional forms of intervention on a second order level, with presumptions for (or against) enterprises according to their procedural level of environmental management (as a kind of self-definition and self-observation of dynamic enterprises)? Procedure could be the basis for the confidence that risk information is being generated in a reasonable way, and thus turn administrative control away from the imposition of tests and towards the evaluation of the risk management of the enterprise itself.

This move towards the control of environmental management systems will favour innovation if the burden of standard tests is reduced (Mazmanian & Morell, 1991). In any case, any new strategy must accept that product development is too diverse and too complex to allow for general control once practice can no longer be guided by experience. Monitoring practices should systematically allow for the observation and control of presumed and pre-modelled cause-effect paths of chemicals released into the environment after registration. Additionally, the capacity of information technology to operate with large amounts of data should be used for the construction of a complex model of the networks and paths of those substances (Borries, 1990). This could also be supplemented by computer-modelling of sophisticated new methods integrating quantitative analysis and the description of chains of events. These new methods allow for more complex forms of co-ordination between the types of knowledge on which economic and administrative decision-making is based and, in this way, redesign the traditional forms of co-ordination that were based on experience shared by discipline and practice. This is particularly important because it could mean that the information bases of firms could be broadened by access to public sources of knowledge. In return for the access to publicly held risk knowledge, enterprise should accept the lowering of the protection of commercial secrets.

Citizens and environmental groups should also be granted access to information, although this need not be unlimited. The balance between the freedom of information and the protection of commercial secrets (see Stenzel, 1987; Winter, 1990) could be made by putting the process under the control of a kind of "mediator" (see generally Susskind & Madigan, 1984), to whom the disclosure of secrets (under restrictions on public use) should be mandatory. A procedure should be institutionalized under which environmental groups could challenge the use of a substance if they establish a "reasonable doubt" as to its safety.

Setting priorities in environmental policy

In the long run, a more complex model of information could provide a better foundation for the formulation of strategies which aim at the reduction of certain risks. Problems could be more reliably identified, and the sort of alternatives formulated that are normally neglected if substances are eliminated on a case-by-case basis, at high cost, and where other substances do not receive any attention at all. Priorities could be set to differentiate between substances of primary and secondary concern (Shapiro, 1990: 48; see also Latin, 1982; 1983), or to specify guidelines for the testing of certain groups of chemicals. Ultimately, this could lead to allowing the use of certain risky substances for a limited set of purposes, where the risk of provoking unintended consequences is low (e.g. pesticides in agriculture). This would be the case where the complexity of involved purposes and values is not too high.

Without going into details, I think that a conception of second order proceduralization advocating the *generation of diversity* and flexibilization of product development would also allow for a co-ordination of, and interference between, the proceduralization of the internal structure of the enterprise and administrative risk management. A more complex regulatory perspective should try to redesign the "ecological field" (Probst, 1987) that lies between private and public institutions and the natural environment. A new regulatory approach cannot just rely on the pragmatic reform of existing control strategy but is bound to remodel the whole network of information and the distribution of knowledge within which technology is developed and reproduced. If we can no longer draw stable boundaries because we must take into account the dynamic of repercussions between regulation and its "object", the regulatory model must be adjusted accordingly. It must be made sensitive to the purpose of generating a fruitful process of interrelated self-modifications and reciprocal irritation between enterprises, technology, natural environments and administrative regulation.

In this process, small firms should be encouraged as generators of innovation, diversity and dynamism and they should be regarded as counterweights to large enterprises which are better able to generate the political pressure that eventually leads to the blocking of the dynamic of technological innovation. This tendency of small firms to organizational flexibility and innovation could be encouraged by a programme that set lower levels of proof for the registration of promising innovations and production processes (an "innovation waiver") (see Ashford *et al.*, 1985).

The capacity of existing control strategies is largely exhausted. They cannot be made any more flexible. The integration of cost-benefit analysis into risk assessment cannot alter this because neither cost nor benefit can be easily evaluated. This is also true for strategies, such as that proposed by the American EPA (see Baker, 1988), to integrate local specificities into the risk assessment of chemical substances (soil qualities and so on). This would lead to an increasing expenditure by the administration but without the guarantee of compensation. And the administrative agencies would still not have solved the problem of how to prevent the blocking of innovation because of their lack of sufficient information.

A strategy which was adapted to the self-organizational potential of the "ecological field" could, for example, aim at a productive linking of the different types and phases of decision-making in order to introduce more diversity and a greater resilience to unexpected developments. Environmental pollution could also be assessed with reference to the target of creating or preserving a natural "pool of diversity" (see generally Allen, 1988; Taylor, 1987; Ulrich & Probst, 1988), acting as a buffer against the spread of unknown chains of cause and effect (Müller, 1989). Such a procedural approach could also be used to relate the policy of natural preservation to the purpose of regenerating the self-stabilizing potential of nature. The separation of different domains of administrative decision-making, or the case-by-case approach of agencies, must be made open to self-reflection on the inter-relationships between decisions, and to the necessity of reconsidering their incomplete informational base.

Agricultural policy can serve as an example here (see *Bundesamt für Landwirtschaft*, 1990; Bauer, 1990). Government policy in this area is, on the one hand, directed towards the creation of an incentive for the intensification

of agriculture while, on the other hand, it tries to set limits to the use of pesticides. This latter aspect is not easily managed by traditional instruments,
because registration and restrictions on use will not be sufficient. An alternative
approach could be the political setting of priorities for risks. In an industrial
society one could consider whether the taking of risks by (the chemical) industry
is more acceptable than the taking of risks in agriculture. But perhaps both
cannot be allowed because the consumption of natural diversity and the options
contained in the "pool of possibilities" may simply be too high.

To give another example, it can be argued that in the field of genetic engineering, the criterion of generating and producing (or reducing) genetic
diversity could be integrated into decision-making. Formerly, it was natural
boundaries that limited the process of reduction of species by pests and so on.
But now the artificial reduction of crops and the elimination of hedges (Müller,
1989) have made a reduced set of crop varieties more sensitive to pests. This
could be considered to create incentives for the use of a wider range of grains,
and more frequent changes between them, because the possible reduction of
profits could be compensated for by a gain in ecological stability.

This sketch should make it clear that it is possible to conceive of a strategy
for the production of diversity and a reciprocal irritation between administrative and corporate economic actors. This new conception of management
could be part of a second order proceduralization based on the co-ordination
of perspectives between different social institutions, subsystems and organizations. A complex design of multiple worlds of options and possibilities could
compensate for the loss of the continuity of a reality that is identical to
itself.

Science and regulation

These considerations should have made it clear why the problem of environmental law cannot simply be solved by substituting a reference to science for
the loss of practical experience. In fact, to do so is misleading, because the
openness of science suggests an endless search for safety, unless this is specifically prevented by pragmatic, administrative rationality. This unresolved tension between scientific and administrative rationality has been hidden by
unsatisfactory compromises, such as the so-called administrative "preroga-
tive" of standardization in the case of conflicting opinions between scientific
experts, formulated by German administrative jurisdiction over issues of
nuclear safety (see *Bundesverwaltungsgericht* 72, 300, 318; di Fabio, 1991;
Breuer, 1989; for historical analysis, Wolf, 1987). In these cases a range of
possible scientific conjectures is taken as the basis for administrative discretion. It would be more convincing to acknowledge the political character of
decisions on risk, rather than to hide it as a residue of uncertainty to be
handled administratively. Instead we must emphasize the need to set and

implement priorities (Reilly, 1991) as part of a strategy of proactive risk *management* (see Salzwedel, 1991; Ladeur, 1985; 1986). The perverse result of the existing practice of co-ordination between administration and science, is that it tends to undermine confidence in scientific expertise which has to shoulder the responsibility which administration rejects. The main responsibility in risk management should, on the contrary, belong to administration. Risk strategies must be able to rely on a certain confidence, in the same way that the traditional control strategies based on experience were able to. They are not designed according to a normal deductive logic, but to one of plausibility and for this reason they must contain a potential for self-observation and revision (Dubucs, 1987). Accordingly, they should be reinforced by a procedural rationality which allows the planning and modelling of a variety of operations and options. By stressing the experimental character of scientific modelling as *one* element of a complex operational strategy, it is possible to lay open the provisional character of its knowledge base as the best *available* knowledge, as well as the need to formulate stop rules which accept the time dependency and the limited nature of decisions made under conditions of undecidability (Dupuy, 1988). In a heterogeneous society differentiated into subsystems which each follow their own *eigenvalues* there is no alternative to the procedural compatibilization of different functional values, unless one accepts the risk of blocking the autonomy of the subsystems.

As a final example of the role of confidence as the basis for regulation (as limited by argumentative constraints and procedures of self-observation), let us take the issue of the release of genetically engineered organisms. If a strategy were to set limits to biotechnology which were contrary to the opinion of the majority of respected scientists (Tiedje *et al.*, 1989), this would shatter the "ecological field" formed by the network of institutions involved and would create risks for its reproduction. If environmental risks are not to be ignored altogether, it should be considered that the most devastating catastrophes to date have been produced by the risks generated by society when liberal democratic institutions and, thus, the potential for a plural, social self-observation and self-description have been damaged. This is why a conception of complex second-order proceduralization could be a productive alternative to a substantial "deep ecology" (see Böhler, 1991) that neglected the cultural limits of institutional self-modification. Any possibility of finding a substantial social consensus on environmental issues has been destroyed by the process of differentiation and individualization.

NOTE

1 The English version of this paper has been extensively revised and rewritten by Lindsay Farmer.

REFERENCES

Aalders, Marius V.C. (1991) Regulation and Internal Company Environmental Management in the Netherlands. Paper delivered at the Annual Meeting of the Law and Society Association. 26–29 June 1991, Amsterdam.

Albert, Michel (1991) *Capitalisme contra capitalisme*. Paris: Seuil.

Allen, Peter M. (1988) Dynamic Models of Evolving Systems. *System Dynamics Review*, **14**, 109–130.

Amendola, Mario & Gaffard, Jean-Luc (1986) La technologie comme expression de l'environnement: quelques suggestions pour une interprétation du processus d'innovation. *Economie Appliquée*, **39**, 473–492.

Amendola, Mario & Gaffard, Jean-Luc (1988) *La dynamique économique de l'innovation*. Paris: Economica.

Applegate, John S. (1991) The Perils of Unreasonable Risk: Information, Regulatory Policy, and Toxic Substances Control. *Columbia Law Journal*, **91**, 261–333.

Arthur, Brian W. (1990) Positive Rückkopplung in der Wirtschaft. *Spektrum der Wissenschaft* No. 4, 122–129.

Ashford, Nicholas A. *et al.* (1985) Using Regulation to Change the Market for Innovation. *Harvard Environmental Law Review*, **9**, 419–466.

Ayres, Robert U. (1987) Optimal Growth Path with Exhaustible Sources: An Information-Based Model. *IIASA Laxenburg* (Austria).

Ayres, Robert U. (1988) Technology: The Wealth of Nations. *Technological Forecasting and Social Change*, **33**, 189–201.

Baecker, Dirk (1991) *Womit handeln Banken? Eine Untersuchung zur Risikoverarbeitung in der Wirtschaft*. Frankfurt: Suhrkamp.

Baker, Brian P. (1988) Pest Control in the Public Interest: Crop Protection in California. *University of California Los Angeles Journal of Environmental Law and Policy*, **8**, 31–71.

Bauer, Siegfried (1990) Landwirtschaft und Umweltpolitik—Überlegungen aus ökonomischer Sicht. *Zeitschrift für angewandte Umweltforschung*, **3**, 133–147.

Beckmann, Gotthard (1991) Risiko als Kategorie der Gesellschaftstheorie. *Kritische Vierteljahresschrift für Gesetzgebung und Rechtswissenschaft*, **74**, 212–240.

Bienaymé, Alain (1988) Technologie et nature de la firme. *Revue d'économie politique*, **98**, 823–849.

Böhler, Dietrich (1991) Mensch und Natur: Verstehen, Konstruieren, Verantworten. *Deutsche Zeitschrift für Philosophie*, **39**, 999–1019.

Borries, Dietrich F.W. von (1990) Informationssysteme als Instrumente der Bodenpolitik. *Zeitschrift für angewandte Umweltforschung*, **5**, 25–35.

Boynton, Andrew C. & Victor, Bart (1991) Beyond Flexibility: Building and Managing the Dynamically Stable Organization. *California Management Review*, **34**, Fall, 53–66.

Bratman, Michael E. (1992) Practical Reasoning and Acceptance in a Context. *Mind*, **101**, 1–15.

Breuer, Rüdiger (1976) Direkte und indirekte Rezeption technischer Regeln durch die Rechtsordnung. *Archiv des öffentlichen Rechts*, **101**, 46–88.

Breuer, Rüdiger (1989) Die internationale Orientierung von Umwelt- und Technikstandards im deutschen und europäischen Recht. *Jahrbuch des Umwelt- und Technikrechts*, **9**, 43–116.

Bundesamt für Landwirtschaft (Ed) (1990) *Expertenkommission 'Direktzahlungen in der Schweizerischen Agrarpolitik'*. Swiss Government Document, Bern.

Bunge, Mario (1974) Towards a Philosophy of Technology. In Michalos, A.T. (Ed). *Philosophical Problems of Science Technology*. Boston: Allyn and Bacon.

Buxbaum, Richard, M. (1990) Institutional Ownership and the Restructuring of Corporations. In Baur, J.F. *et al.* (Eds). *Festschrift Steindorff.* Berlin/New York: de Gruyter.

Child, James W. (1990) The Moral Foundations of Intangible Property. *The Monist,* 73, 578–600.

Coffee, John C. (1991) Liquidity versus Control: The Institutional Investor as Corporate Monitor. *Columbia Law Review,* 91, 1277–1368.

Coodley, Cheryl (1984) Risk in the 1980s: New Perspectives on Managing Chemical Hazards. *San Diego Law Review,* 21, 1015–44.

Coval, S. *et al.* (1986) The Foundations of Property and Property Law. *Cambridge Law Journal,* 45, 457–475.

Csanyi, Vilmos & Kampis, György (1987) Modelling Society: Dynamic Replicative Systems. *Cybernetics and Systems,* 17, 233–247.

Day, Richard H. (1987) The Evolving Economy. *European Journal of Operational Research,* 30, 251–7.

Demsetz, Harold (1979) Ethics and Efficiency in Property Rights Systems. In Rizzo, M. (Ed). *Time, Uncertainty and Disequilibrium.* Lexington: Lexington Books.

Dierkes, Meinolf (1985) Corporate Social Reporting and Auditing: Theory and Practice. In Hopt, K. & Teubner, G. (Eds). *Corporate Governance and Directors' Liabilities.* Berlin/New York: de Gruyter.

Dimento, Joseph F. (1986) Der Consensus Workshop: Ein geeignetes Forum für Grenzwertfestsetzung? In Winter, G. (Ed). *Grenzwerte.* Düsseldorf: Werner.

Dosi, Giovanni (1982) Technological Paradigms and Technological Trajectories: A Suggested Integration of the Determinants and Directions of Technological Change. *Research Policy,* 11, 142–162.

Dubucs, Jacques-Paul (1987) Sur la logique des arguments plausibles. *Philosophie,* No. 14, 15–31.

Dupuy, Jean-Pierre (1988) On the Supposed Closure of Normative Systems. In Teubner, G. (Ed) *Autopoietic Law: A New Approach to Law and Society.* Berlin/New York: de Gruyter.

Fabio, Udo di (1991) Entscheidungsprobleme der Risikoverwaltung. *Natur und Recht,* 13, 353–359.

Favereau, O. (1989a) Marchés internes, marchés externes. *Revue Economique,* 40, 273–328.

Favereau, O. (1989b) Valeur d'option et flexibilité: De la rationalité substantielle à la rationalité procédurale. In Cohendet, P. & Llerena, D. (Eds). *Flexibilité, information et décision.* Paris: Economica.

Fehl, Ulrich (1983) Die Theorie dissipativer Strukturen als Ansatzpunkt für die Analyse von Innovationsproblemen in alternativen Wirtschaftsordnungen. In Schüller, A. *et al.* (Eds). *Innovationsprobleme in Ost und West.* Stuttgart: Fischer.

Fischman, Robert L. (1992) Biodiversity and Ecological Management. *Environmental Law,* 22, 435–502.

Fitzroy, Felix & Kay, Neil (1991) The Corporation in an Uncertain World: Competition Efficiency and Governance. In Sugarman, D. & Teubner, G. (Eds). *Regulating Corporate Groups in Europe.* Baden-Baden: Nomos.

Florida, Richard (1991) The New Industrial Revolution. *Futures,* 23, 559–576.

Flournoy, Alyson C. (1991) Legislating Inaction: Asking the Wrong Questions in Protective Environmental Decision Making. *Harvard Environmental Law Review,* 15, 327–386.

Foray, Daniel (1991) Repères pour une économie des organisations de recherche-développement. *Revue d'économie Politique,* 779–808.

Gaines, Sanford E. (1990) Science, Policy, and the Management of Toxic Risks through Law. *Jurimetrics*, 30, 271–321.

Garber, Ellen J. (1987) Federal Common Law of Contribution under the 1986 CERCLA Amendments. *Ecology Law Quarterly*, 14, 365–388.

Gauchet, Marcel (1979) De L'avénement d'individu à la découverte de la société. *Annales—Economies, Sociétés, Civilisations*, 34, 451–463.

Godard, Oliver (1990) Environnement, mode de coordination et systemes de légitimité. *Revue Economique*, 41, 215–241.

Harter, Philip, J. (1982) Negotiating Regulations: A Cure for Malaise. *Georgetown Law Journal*, 71, 1–118.

Hayek, Friedrich A. von (1973) *Law, Legislation and Liberty*, Vol. 1. London: Routledge & Kegan Paul.

Herman, Robert *et al.* (1990) Dematerialization. *Technological Forecasting and Social Change*, 38, 333–347.

Hoppe, Ernst & Beckmann, Martin (1989) *Umweltrecht*. München: Beck.

Huber, Peter (1985) Safety and the Second Best: The Hazards of Public Risk Management in the Courts. *Columbia Law Review*, 85, 277–337.

Imai, Kenrich & Itami, Hiroyuki (1984) Interpenetration of Organization and Market. *International Journal of Industrial Organization*, 2, 285–310.

Jarillo, J. Carlos (1988) On Strategic Networks. *Strategic Management Review*, 9, 31–49.

Kirat, Thierry (1991) Pourquoi une théorie évolutionniste du changement technologique? *Economie Appliquée*, 44, 29–58.

Klein, Art (1991) What Does it Mean to be Green? *Harvard Business Review*, 69, (July/August), 38–47.

Klein, Eckart (1989) Grundrechtliche Schutzpflicht. *Neue Juristische Wochenschrift*, 42, 1633–1640.

Kloepfer, Michael (1989) *Umweltrecht*. München: Beck.

Kunz, W. (1985) *Marktsystem und Information*. Tübingen: Mohr.

Ladeur, Karl-Heinz (1985) Zum planerischen Charakter der technischen Norm im Umweltrecht. *Umwelt- und Planungsrecht*, 253–260.

Ladeur, Karl-Heinz (1986) Alternativen zum Konzept der Grenzwerte im Umweltrecht. In Winter, G. (Ed). *Grenzwerte. Interdisziplinäre Untersuchungen zu einer Rechtsfigur des Umwelt-, Arbeits- und Lebensmittelschutzes*. Düsseldorf: Werner.

Ladeur, Karl-Heinz (1987) Jenseits von Regulierung und Ökonomisierung der Umwelt. *Zeitschrift für Umweltpolitik und Umweltrecht*, 10, 1–22.

Ladeur, Karl-Heinz (1988) Umweltrecht und technologische Innovation. *Jahrbuch des Umwelt- und Technikrechts*, 5, 305–333.

Ladeur, Karl-Heinz (1989) Zu einer Grundrechtstheorie der Selbstorganisation des Unternehmens. In Faber & Stein (Eds). *Auf einem dritten Weg, Festschrift Helmut Ridder*. Neuwied: Luchterhand.

Latin, Howard A. (1982) Environmental Deregulation and Consumer Decision Making under Uncertainty. *Harvard Environmental Law Review*, 6, 187–239.

Latin, Howard A. (1983) The Significance of Toxic Health Risk: An Essay on Legal Decision Making under Uncertainty. *Ecology Law Quarterly*, 10, 339–395.

Lemoigne, Jean-Louis (1987) Systémographie de l'entreprise. *Revue internationale de Systémique*, 1, 499–511.

Lewis, H.W. (1980) The Safety of Fission Reactors. *Scientific American*, 242, No. 3, 33–35.

Luhmann, Niklas (1988) Closure and Openness: On Reality in the World of Law. In Teubner, G. (Ed). *Autopoietic Law: A New Approach to Law and Society*. Berlin/New York: de Gruyter.

Luhmann, Niklas (1991a) *Soziologie des Risikos.* Berlin/New York: de Gruyter.

Luhmann, Niklas (1991b) Der Ursprung des Eigentums und seine Legitimation. Ein historischer Bericht. In Krawietz, W. *et al.* (Eds). *Technischer Imperativ und Legitimationskrise des Rechts.* Rechtstheorie: Beiheft.

Lyndon, Mary L. (1989) Information Economics and Chemical Toxicity: Designing Laws to Produce and Use Data. *Michigan Law Review*, 87, 1795–1861.

Mackaay, Ejan (1986) Les biens informationels. *Cahiers Science—Technologie—Société*, No. 12, 134–151.

Magat, Wesley A. (Ed) (1982) *Reform of Environmental Regulation.* Cambridge, Mass.: Balinger.

Masten, Scott E. (1988) A Legal Basis for the Firm. *Journal of Law, Economics and Organization*, 4, 181–198.

Mazmanian, Daniel A. & Morell, David L. (1991) EPA: Coping with the New Political Economic Order. *Environmental Law*, 21, 1477–1491.

Miles, Raymond, E. & Snow, Charles C. (1986) Organizations: New Concepts for New Forms. *California Management Review*, 28, Spring, 62–73.

Mitroff, Ian I. (1990) The Idea of the Corporation as an Idea System. *Technological Forecasting and Social Change*, 38, 1–14.

Müller, von (1930) Möglichkeit und Wahrscheinlichkeit im Polizeirecht. *Preußisches Verwaltungsblatt*, 51, 92–94.

Müller, Johannes (1989) Bericht über die Funktionen von Hecken als Beispiel einer ganzheitlichen Ökosystem-Betrachtung. *Jahrbuch der Akademie der Wissenschaften zu Berlin*, 44–53.

Nelson, Richard R. & Winter, Sidney G. (1977) In Search of a Useful Theory of Innovation. *Research Policy*, 6, 36–76.

Nonaka, Ikujiro (1988a) Toward Middle-Up-Down-Management. *Sloan Management Review*, 29, No. 3, 9–18.

Nonaka, Ikujiro (1988b) Creating Organizational Order out of Chaos: Self-Renewal in Japanese Firms. *California Management Review*, 30, Spring, 57–73.

Nonaka, Ikujiro (1990) Redundant, Overlapping Organization. A Japanese Approach to Managing the Innovation Process. *California Management Review*, 32, Spring, 27–38.

Nonaka, Ikujiro (1991) The Knowledge Creating Company. *Harvard Business Review*, 69, Nov./Dec., 96–104.

O'Leary, Rosemary (1991) Judicial Oversight of EPA Decision-Making. Paper presented at the Annual Meeting of the Law and Society Association, 26–29 June 1991, Amsterdam.

Office of Technology Assessment, US-Congress (1990) *Beneath the Bottom Line. Agricultural Approaches to Reduce Agrichemical Contamination of Groundwater, Summary.* Washington D.C.

Orléan, André (1985) Incertitude et paradoxe. *Economie Appliquée*, 38, 133–153.

Pejovich, Steve (1987) Freedom, Property Rights and Innovation in Socialism. *Kyklos*, 40, 461–475.

Pernice, Ingolf (1992) Rechtlicher Rahmen der europäischen Unternehmens Kooperation im Umweltbereich unter besonderer Berücksichtigung von Art. 85 EWGV. *Europäische Zeitschrift für Wirtschaftsrecht*, 3, 139–143.

Polanyi, Michael (1962) *Personal Knowledge.* Chicago: Chicago University Press.

Priddat, B.B. (1989) Transformation der ökonomischen Vernunft. Über P. Ulrichs Versuch zur "Moralisierung der Ökonomie". In Seifert, E.K. & Pfriem, R. (Eds). *Wirtschaftsethik und ökologische Wirtschaftsforschung.* Bern/Stuttgart: Haupt.

Probst, Gilbert J.B. (1987) *Selbstorganisation. Ordnungsprozesse in sozialen Systemen aus ganzheitlicher Sicht.* Berlin/Hamburg: Parey.

Püttner, Günter (1991) Der informale Rechtsstaat. *Kritische Vierteljahresschrift für Gesetzgebung und Rechtswissenschaft*, **74**, 63–73.

Rasmussen, Jens (1991) Event Analysis and the Problem of Causality. In Brehmer B. & Leptat, J. (Eds). *Distributed Decision Making: Cognitive Models of Cooperative Work*. Chichester: Wiley.

Rehbinder, Eckard (1989) Andere Organe der Unternehmensverfassung. *Zeitschrift für Unternehmens- und Gesellschaftsrecht*, **18**, 305–368.

Rehbinder, Eckard (1990) Ein Umweltschutzdirektor in der Geschäftsführung der Großunternehmen. In Baur, J.S. *et al.* (Eds). *Festschrift für Steindorff*. Berlin/New York: de Gruyter.

Rehbinder, Eckard (1991) Reflexive Law and Practice: The Corporate Officer for Environmental Protection as an Example. In Febbrajo, A. & Teubner, G. (Eds). *State, Law, Economy as Autopoietic Systems*. Milano: Giuffrè.

Reilly, William (1991) Taking Course Toward 2000: Rethinking the Nation's Environmental Agenda. *Environmental Law*, **21**, 1359–1374.

Renn, Ortwin *et al.* (1990) Technische Risikoanalyse und unternehmerisches Handeln. In Schütz, M. (Ed). *Risiko und Wagnis*. Pfullingen: Leske.

Roe, David (1988) Barking up the Right Tree: Recent Progress in Focusing the Toxics Issues. *Columbia Journal of Environmental Law*, **13**, 275–283.

Rosegger, Gerhard (1991) Diffusion through Interfirm Cooperation. *Technological Forecasting and Social Change*, **39**, 81–101.

Salzwedel, Jürgen (1991) Umweltrecht und Umweltstandards. In Organisationsforum Wirtschaftskongreß (Ed). *Umweltmanagement im Spannungsfeld zwischen Ökologie und Ökonomie*. Wiesbaden: Gabler.

Sater, Rachel (1990) EPA's Pesticide in Groundwater Strategy: Agency Action in the Face of Congressional Inaction. *Ecology Law Quarterly*, **17**, 143–177.

Schmidtz, David (1991) *The Limits of Government*. Boulder: Westview.

Schroeder, Christopher H. (1986) Rights against Risks. *Columbia Law Review*, **86**, 495–562.

Schumpeter, Joseph A. (1934) *The Theory of Economic Development*. Cambridge, Mass.: Harvard University Press.

Senge, Peter M. (1990) The Leader's New Work: Building Learning Organizations. *Sloan Management Review*, **31**, Fall, 7–23.

Shapiro, Sidney A. (1990) Biotechnology and the Design of Regulation. *Ecology Law Quarterly*, **17**, 1–70.

Sharples, Frances E. (1987) Regulation of Products from Biotechnology. *Science*, **235**, 1329.

Silver, Larry D. (1986) The Common Law of Environmental Risk and some Recent Applications. *Harvard Environmental Law Review*, **10**, 61–98.

Silverberg, Gerald (1991) Adoption and Diffusion of Technology as a Collective Evolutionary Process. *Technological Forecasting and Social Change*, **39**, 67–80.

Stachowiak, Herbert (1983) Einleitung. In Stachowiak, H. (Ed). *Modelle—Konstruktion der Wirklichkeit*. München: Fink.

Staehelin-Witt, E. & Blöchliger H. (1990) Die Umweltbank als Instrument langfristiger Umweltpolitik. *Neue Züricher Zeitung*, No. 268, Nov. 17/18, 33–34.

Steger, Ulrich (Ed) (1988) *Umweltmanagement: Erfahrungen und Instrumente einer umweltorientierten Unternehmensstrategie*. Frankfurt/Wiesbaden: FAZ/Gabler.

Steger, Ulrich *et al.* (1991) Unternehmensstrategie und Risikomanagement. In Steger, U. (Ed). *Umwelt-Auditing. Ein neues Instrument der Risikovorsorge*. Frankfurt: FAZ.

Stenzel, Paulette, L. (1987) The Need for a National Risk Assessment Communication Policy. *Harvard Environmental Law Review*, **11**, 381–413.

Stewart, Richard B. (1981) Regulation, Innovation, and Administrative Law: A Conceptual Framework. *California Law Review*, **69**, 1259–1377.

Stewart, Richard B. (1985) Economics, Environment and the Limits of Legal Control. *Harvard Environmental Law Review*, **9**, 1–22.

Stewart, Richard B. (1988) Regulation and the Crisis of Legalization in the United States. In Daintith, T. (Ed). *Law and Economic Policy: Comparative and Critical Perspectives*. Berlin/New York: de Gruyter.

Susskind, Larry E. (1989) Four Important Changes in the American Approach to Environmental Regulation. In Archibugi, F. & Nijkamp, P. (Eds). *Economy and Ecology: Towards Sustainable Development*. Dordrecht: Kluwer.

Susskind, Larry E. & Madigan, Denise (1984) New Approaches to Resolving Disputes in the Public Sector. *The Justice System Journal*, **9**, 179–203.

Taylor, Lauren P. (1987) Management: Agent of Human Cultural Evolution. *Futures*, **19**, 513–527.

Teubner, Gunther (1987) Unternehmenskorporatismus. New Industrial Policy und das Wesen der juristischen Person. *Kritische Vierteljahresschrift für Gesetzgebung und Rechtswissenschaft*, **2**, 61–85.

Teubner, Gunther (1989) *Recht als autopoietisches System*. Frankfurt: Suhrkamp.

Teubner, Gunther (1990a) *Beyond Contract and Organization? The External Liability of Franchising Systems in German Law*. Manuscript, December 1990.

Teubner, Gunther (1990b) "Verbund", "Verband" oder "Verkehr"? Zur Außenhaftung von Franchising-Systemen. *Zeitschrift für das gesamte Handelsrecht*, **154**, 295–324.

Teubner, Gunther (1991) Unitas Multiplex: Corporate Governance in Group Enterprises. In Sugarman, D. & Teubner, G. (Eds). *Regulating Corporate Groups in Europe*. Baden-Baden: Nomos.

Thévenot, Laurent (1989) Equilibre et rationalité dans un univers complexe. *Revue Economique*, **40**, 147–197.

Tiedje, James M. (1989) The Planned Introduction of Genetically Engineered Organisms: Ecological Considerations and Recommendations. *Ecology*, **70**, 298.

Trauberman, Jeffrey (1983) Statutory Reform of "Toxic Torts": Relieving Legal, Scientific and Economic Burden on the Chemical Victim. *Harvard Environmental Law Review*, **7**, 177–296.

Ulrich, Hans & Probst, J.B. (1988) *Anleitung zum ganzheitlichen Denken und Handeln*. Bern/Stuttgart: Haupt.

Ulrich, Peter (1989) Können sich Ökonomie und Ökologie wirtschaftsethisch versöhnen? In Seifert, E.K. & Pfriem, R. (Eds). *Wirtschaftsethik und ökologische Wirtschaftsforschung*. Bern/Stuttgart: Haupt.

Ulrich, Peter (1991) Betriebswirtschaftliche Rationalisierungskonzepte im Umbruch— Neue Chancen ethikbewußter Organisationsgestaltung. *Die Unternehmung*, **45**, 146–166.

Van Cleve, Georg (1991) The Changing Intersection of Environmental Auditing, Environmental Law, and Enforcement Policy. *Cardozo Law Review*, **12**, 1215–1240.

Vardaro, Gaetano (1991) Before and Beyond the Legal Person: Group Enterprises, Trade Unions, and Industrial Relations. In Sugarman, D. & Teubner, G. (Eds). *Regulating Corporate Groups in Europe*. Baden-Baden: Nomos.

Weise, Peter D. & Brandes, Wolfgang (1990) A Synergetic View of Institutions. *Theory and Decision*, **28**, 173–187.

Wildavsky, Aaron (1988) *Searching for Safety*. New Brunswick/Oxford: Transaction.

Willard, Charles A. (1991) L'argumentation et les fondements sociaux de la connaissance. In Lempereur, A. (Ed) *L'argumentation*. Liège: Magara.

Williamson, Oliver E. (1991a) Comparative Economic Organizations: The Analysis of Discrete Structural Alternatives. *Administrative Science Quarterly*, 269–296.

Williamson, Oliver E. (1991b) Economic Institutions: Spontaneous and Intentional Governance. *Journal of Law, Economics and Organization*, 7, 159–187.

Winter, Gerd (1983) *Alternativkommentar-BGB*, Vol. 4. Neuwied/Darmstadt: Luchterhand, § 906.

Winter, Gerd (Ed) (1990) *Informationsrechte in Europa*. Baden-Baden: Nomos.

Wolf, Rainer (1987) *Der Stand der Technik—Geschichte, Strukturelemente und Funktion technischer Risiken am Beispiel des Immissionsschutzes*. Opladen: Westdeutscher Verlag.

Zervos, C. (1989) The Anatomy of Safety Evaluations. In Bonin, J.J. & Stevenson, D.E. (Eds). *Risk Assessment in Setting National Priorities*. New York: Plenum.

CHAPTER 14

A Game without Rules? The Ecological Self-Organization of Firms

François Ost[1]
Brussels

The issue of the self-responsibility of firms in the area of the environment is currently enjoying a success that is coupled with that of the movement for deregulation. This movement has railed against the dysfunctionalities induced by the interventionism of the welfare state, while supporting initiatives adopted by firms with a view to accounting better for the negative consequences of their actions for ecological balance. The theoreticians of autopoiesis have seen this as an exemplary confirmation of their analyses, which have expressed scepticism regarding imperative, binding (heterogenic) ways of guiding behaviour for a long time.

This brief study will seek to give an initial evaluation of a few institutions of positive law inspired by these new axioms. We shall review the phenomena of the contractualization of environmental law, the procedure for environmental impact assessment, the institution of an eco-audit and the mechanism for assigning an eco-label. An assessment of these laws can, however, only be made on the basis of definite premises. To the extent that those adopted here depart significantly from those supplied by the autopoietic paradigm, the first section of this chapter will be devoted to their presentation.

AN ETHICAL AND THEORETICAL FRAMEWORK

Effectiveness of means or value of ends?

To deal with the topic of the self-organization of firms in the environmental sphere while retaining the maximum of critical lucidity, it is best to start by explaining the ethical and political position one intends to adopt, and the conceptual instruments to be employed.

Environmental Law and Ecological Responsibility: The Concept and Practice of Ecological Self-Organization
Edited by G. Teubner, L. Farmer and D. Murphy
© 1994 John Wiley & Sons Ltd

On the first point, the aim is to substitute an ethical question for a functional question. All too often the topic of the self-organization of firms is treated from an exclusively functionalist perspective. Starting from the perpetually repeated finding of the failure of the welfare state, its counter-productiveness and perverse effects, we are told that the hour of deregulation has struck and that if firms are made self-responsible then more satisfactory results will be brought about. But results in relation to what objective? Better consideration for ecological equilibria, it is said, and consequently better protection for the environment. With the goal thus vaguely alluded to, as if it were something obvious and a matter of common agreement, attention is concentrated on the means by which it can be attained. It is clear, however, that it is not enough to set a goal in order to define an ethical attitude and determine a policy: other competing or opposed goals may in fact redefine it or cancel it out. It is, there-fore, best once a new goal is identified (in this case protection of the environ-ment) to ensure a philosophical foundation for it and to determine its place in a hierarchy of values, so as to identify a clear order of priorities for decision-makers. That firms will consider the effects of their actions on the ecosystem is not enough to satisfy us as to the place this new aim will be accorded in rela-tion to other factors such as the search for profitability or the maintenance of employment.

The truly relevant question is, therefore, whether responsibility towards the environment has become a priority. Without being able to go too far into this here (see Ost, 1991), it seems that the ecological imbalances threatened today are so extensive as to genuinely pose the question of survival, and of meaningful survival, for future generations. In these circumstances, responsi-bility, if not towards nature then at least towards future generations, emerges as a new categorical imperative (Jonas, 1990) or, translated into political lan-guage, as the first priority.

Obviously, this position on ends does not free us from considering the appropriate means by which it can be reached. As a secondary or derivative question, therefore, the functionalist approach regains legitimacy. Our ques-tions will accordingly be directed towards assessing the relevance of a model of self-organization of the firm in relation to protection of the environment, conceived of as a genuine priority. Setting environmental protection as a pri-ority clearly does not imply an under-estimation of the other objectives tradi-tionally attributed to firms, like the search for a return on capital invested or a guarantee of employment. Many accommodations will accordingly be neces-sary with an eye to reconciling these differing interests as far as possible. The lessons that Max Weber drew specifically from an "ethics of responsibility", as distinct from the "ethics of conviction" apply here as elsewhere, to the extent that concern for principle does not prevent the seeking of provisional and sectoral compromises (Weber, 1959: 171–3).

All this leads us to conceive of the new image of the self-organization of firms in the area of dealing with ecological problems as an interesting emergent feature of post-modernity, always on the condition that the determination of the goals to be attained should remain the outcome of a public debate where all interests involved (which, it should be stressed at the outset, do not necessarily converge), can be given a hearing. This must also include the interest of future generations, whose "lobbying" power is, as we know, extremely weak. The notion of the responsibility of private actors cannot, then, be conceived of outside a framework determined by the law, in its concern for the general interest. Equally, its implementation implies the maintenance of vigilant judicial control.

Autopoiesis or game?

In studying the theme of the self-responsibility of firms, it is not intended simply to subordinate the functionalist question of the effectiveness of the means to the ethical question of the desirability of the end; the game paradigm must also be substituted for the paradigm of autopoiesis[2]. This change in conceptual tools quite clearly entails a change in the viewpoint adopted. The model of autopoiesis is to some extent located within the firm, and the current discourse of self-responsibility legitimates the firm by offering a rationalization of its actions. The game model by contrast takes into consideration the totality of social interactions in which the firm is an important, though not exclusive, actor. By taking these other actors and competing interests into account, the ideas of external regulation and conflict are rehabilitated, just as the discourses and practices of self-responsibility are relativized.

Without going back over critical discussions of autopoiesis on the basis of the game paradigm (see Ost 1986; van de Kerchove & Ost, 1988: 150–9), we can nevertheless note a few arguments that justify our preference for the game model.

Both models share the notion that post-modern society and law quite clearly require the application of complex models. The paradigm of simplicity developed by Descartes has in fact become definitively obsolete, as have the associated ideas of linear causality, ontological dualism (thought-matter, observer-observed) and of apodictic certainty. By contrast we need notions of multiple and circular causality, the interweaving of elements and involvement of the observer, and the idea of uncertainty. Order has henceforth to make compromises with disorder. These ideas lead us to take seriously the fecundity of paradox, something to which both autopoiesis theory (Luhmann, 1988) and game theory (Ost & van de Kerchove, 1992) attribute decisive importance.

What game theorists accuse autopoiesis theorists of is not yet taking paradoxes sufficiently seriously, thereby yielding to what Luhmann himself calls the temptation to "deparadoxification". By insisting on the closure of the system (even if accompanied by cognitive openness to the environment), on the mechanical binarity of its code, and on the self-programming of its programme, by conceiving the relations between subsystems as dialogues between "black boxes" (a modern version of the dialogue of the deaf, Willke, 1986: 208) the autopoiesis model, we feel, relapses into fascination with the simple. And this is so even if it is indeed obliged to bring in a little complexity and paradox on the margins of the system in order to allow the problems of genesis, becoming and that of insertion into a larger system to be pointed to. To give absolute privilege to the *autos* is already to simplify, since we know that "I" is also, and always already, "another", and since we have known, since at least Heraclitus, that "the same" and "the other" entertain relationships that absolutely cannot be disentangled. The thought of autopoiesis is solipsistic thought, a thought of the *solus ipse* that cannot manage satisfactorily to pose the question of relationships, being attached to such an autarkic initial axiom. The thought of game theory is by contrast resolutely dialectical, favouring relationships over elements or, more exactly, affirming that one cannot even *posit* the one without the other. For the dialectics of game theory, there is necessarily a presence of the one *in* the other; no "identity" can be conceived of except as shot through by the play of a difference which is, as it were, the echo within it of the work of the "other". Complex recursive relationships of reciprocal interaction and engendering are set up between one and the other (Morin, 1977: 186). It is a dialectical game, from which emergent properties arise, absolutely irreducible to the code and programme of each of the elements involved, according to the essential rule of complexity that everything is at the same time more and less than the sum of the parts making it up (Morin, 1990: 114).

This conception of a dialectical game allows us to transcend the conventional representations of system, as its elements and its environment. According to the traditional image, a system is a homogeneous whole, coherent, integrated and differentiated from its environment. It "contains" simple elements that it assembles according to its specific structures, and it "detaches" itself from the environment, from which assignable boundaries separate it. By contrast with this conception, more imaginary than real, it should instead be maintained that a system is a fragile, unstable configuration, only partially integrated and not entirely differentiated from the systems surrounding it. A system always contains "non-system": the order that it brings is never total, so that disorder continues to play within its order (Barel, 1979: 48). Conversely, the system always overflows its limits. If, then, it is still possible to speak of the boundaries of a system, they are something like a Möbius strip: both internal and external, they return upon themselves. It has to be understood that a system is at the same time several systems and one only, and one system in many. It can in no way be a (meta-)level exempt from the paradoxical game of the distinction and confusion of levels.

This further entails that the element or subsystem does not exist as an autarkic identity: it is at the same time "total part", an image of the system as a whole, and virtual part, never fully actualized. In reality, the whole and the part, the system and its environment, are co-determined by the very inter-play of their relations, the *field* that their interactions sketch out. The field, explains Barel, is that "set of fluxes that return upon themselves, curving and crossing" (1979: 165), so that we are no longer really in the presence of systems and elements, but more of forms, nodes, "patterns" or phases of the game (like the "deal" in cards, at each sitting, and the set of "openings" it permits) which are more or less stable. There is, Morin clarifies:

> no individual on the one hand and society on the other, species on the one hand and individuals on the other, firm with its organigramme, production programme and marketing studies on the one hand, and on the other its human relations problems, its staff, its public relations. The two processes are inseparable and interdependent (1990: 116).

This model, adequate to the complexity of the phenomena it seeks to give an account of, finds expression in the concept of game, this concept being brought down to its most fundamental, if not most apparent, meaning: that of play in the physical sense, defined as "movement within a framework": it is an interval, a gap, a space left free within a given framework, which is the condition for the possibility of ludic movement. For there is no game except where there is *play* in this primary sense: too much play and everything dislocates, the match becomes meaningless; not enough play and every-thing becomes blocked, meaning becomes fixed in sterile repetition (Henriot, 1989: 88).

From this initial sense of game there derive the manifold pairs of features associated with it, emerging two by two from the dialectical linkage which they bring about; we shall come back to these. But over and above this dialec-tical fecundity, the game model presents one other advantage by comparison with the concept of autopoiesis. While the latter is borrowed from the lexicon of biology, and the conceptual field associated with it comes from cybernetics. the notion of game is by contrast taken from the basics of the most recurrent and the most universal anthropological experiences. While in the first case the concept presents only remote relationships, always debatable, with the social and human phenomena to be explained, the idea of the game by contrast deploys a conceptual space and a field of metaphors taken directly from social reality. One must in fact see that the notion of game is associated with a vast set of concepts whose explanatory power in the social sciences proves particu-larly fecund when it comes to portraying the field of a real set of problems. Speaking of the game entails asking about the *player*, about the *rules* and about the *stakes*. The *objects of play* must be identified, *a meaning* picked out, and the *boundaries between in and out of play* traced. This picks out a concep-tual network able to do justice to the complexity of reality.

Finally, the interest of the game concept is that being itself dialectical in nature, it generates a whole series of conceptual pairs that are particularly welcome when it comes to throwing light on the complexity of social, and particularly legal, reality. One can list at least five, the tension between which well defines the nature of the game, which is nothing but the gap that separates them and at the same time holds them together. These are strategy and representation, co-operation and conflict, reality and fiction, regulation and indeterminacy, and internality and externality (van de Kerchove & Ost, 1992). While it may be shown that certain games undoubtedly present features that bring them closer to one pole than another, it remains the case that within each of the pairs taken there is not one that can be defined by a "pure" or "simple" criterion that excludes its opposite, unless their nature as a game is lost. *Between* reality and fiction, co-operation and conflict, strategy and representation and so on, there is in each case a discernible third way that is not reducible to either of its components. This is not the middle way of compromise, halfway between one and the other, but the emergent property of their history and dialectical interactions.

It thus appears that playing is at the same time struggling for something and representing something (Huizinga, 1951: 35); the two dimensions, combat and fiction, overlap, since the players simultaneously pursue a strategic objective that takes part in calculations of interest, and a symbolic objective that derives from the desire to make sense of or to reaffirm values.

In the same way, it is apparent that one plays simultaneously with and against something: a partner, chance or oneself. The game, even the most competitive or cruel one, retains a minimum of rules, and to that extent presupposes a minimal dose of consensus. In any game there is both participation and resistance; interests are simultaneously (to a variable extent from one game to another, of course) convergent and divergent.

Is the game part of the order of reality or of fiction? Both, undoubtedly; or more exactly, it shifts both categories for its own benefit. The game deprives everyday conventions of reality, though not thereby being unreal. The game creates or institutes its reality which is, as it were, a sur-reality, or second reality. In law, the performativity of legal language, of which legal "fictions" constitute the extreme archetype, institutes a sur-real universe of this type, characterized by a type of il-lusion (from *in-ludere*, to enter the game) which is creative.

Is the game free, spontaneous, creative, or does it respond to rigid rules? Here again we shall say that the game is simultaneously regulated and indeterminate, convention and invention. While in English *play* has more to do with the player's improvization and *game* with the pre-existence of rules, it is clear that these two axes interact. Even an unruly game such as the upheaval of carnival still presupposes a few rules (if only limits of time and space), just as a game as formalized as, say, chess cannot be conceived of without a margin for play left to the player's initiative. One can guess the prizes that can be won in the legal sphere from this dialectic of freedom and rule, which in particular allows us to theorize the question of legal change, conceived sometimes as the changing of the strokes authorized by the rules, and sometimes as changing the rule itself (playing *on* the rules and playing *by* the rules).

Finally, can the game be defined as cut off from the world or immersed in it? While it is clear that every game presupposes a minimum differentiation from its environment, this does not imply insularity of the game. Just as gestalt psychology has shown that the figure sticks out from the background though remaining joined to it, so the game presents movable, reversible boundaries in relation to the context from which it is drawn. Like Alice's looking-glass, the "as if" space of the game authorizes two-way transgression.

It is not possible within the limited framework of this chapter to mention all the implications of the foregoing analyses, for the understanding of law in general or the self-organization of firms in particular. It will, nonetheless, be agreed that the detour through both the ethical and the theoretical questions leads to the same conclusion: the issue of the self-responsibility of enterprises in environmental matters ought to be relocated within a wider framework. That is to say, an ethical framework that subordinates assessment of the relevance of this discourse to the solid priority represented by care for a future that can be meaningful for future generations, and a theoretical framework that reintegrates the new "stroke" represented by the practices of ecological self-organization into the "overall game" that the firm plays along with various partners (the authorities, public opinion, environmental defence associations, and competing firms). This is a complex game with numerous actors and diverse stakes. It cannot be conceived without a minimum of external regulation, nor without a certain number of conflicts.

This reminder is all the more necessary because one of the possible digressions of the discourse of the self-responsibility of firms consists, we feel, in effacing the very idea of the conflict of competing interests, as well as underestimating the need for public regulation of the game. Everything would then happen as if the various actors involved, communicating in a very vague ecological consensus, ceased being adversaries to act as partners, now rivals in imagination and goodwill for the common objective. In these circumstances there would be a real risk that external constraints would weaken, that the *show* of action would replace action itself, that *representation* of ecological concern would mask the reality of strategies of market occupation. At the same time, the need for legal intervention would be questioned, in favour of strategies of persuading the public that have more to do with conditioning to publicity and political marketing.

The concept of "sustainable development" introduced in the 1987 "Brundtland" report (*Our Common Future*) represents a good example of the ambiguity of some contemporary ecological discourse. As Hajer makes very clear (1991: 3–8), this concept has proved to be able to mobilize diverse economic and political forces, though on the basis of a false reconciliation of environment and development, of ecology and economy, of North and South. The inevitable oppositions of interest and ideological struggles seem to have been transcended by the representation of a future of reconciliation, thanks to the progress of science and the reign of rational planning. The apocalyptic discourse calling for a cessation to growth is replaced by the mobilizing discourse of responsible, shared progress. Where even yesterday there was reasoning in terms of a *zero sum game*, where the players appeared as adversaries, and one's gain was necessarily the other's loss, it is now supposed to be possible to play a *positive sum game*, with the actors becoming collaborators in the growth of the common good.

While one should certainly welcome the involvement of a greater number of actors in the concern for environmental protection, that does not mean abandoning all critical vigilance or sacrificing over-hastily the legal mechanisms that age-old knowledge of the ways and byways of human and social games have made it possible to develop.

Some authors, carried away by the enthusiasm aroused by the metaphor of "positive sum games", do not always seem to us to retain adequate lucidity in relation to its possible diversion. Koppen, for instance, sees an excellent illustration of the application of autopoietic theory to ecological issues in the institution of "environmental" mediation (1991). Starting from the need to make environmental policy and market forces compatible, she seeks new legal instruments able to produce "consensual steering". Environmental mediation is supposed to be one of them, with an eye for both the development of public policies and the resolution of conflicts. The mediator, external to the groups involved, would seek to convince the protagonists to move beyond their rigid positions and adopt the viewpoint of partners in a positive sum game. This would, in short, be an intervention of a sort of pedagogical nature, teaching a "win-win approach" to negotiation, also termed a "mutual gains approach" and "integrative bargaining" (Koppen, 1991: 6). Better still, the mediator is termed a "policy therapist", opening up communication pathways and inducing people to engage in dialogue. In the language of autopoiesis, this means managing to bring into harmony the codes specific to the various subsystems involved. She ends by herself noting that the mediator ought to be a "supra-natural being" (1991: 8), capable of combining a large number of mutually exclusive requirements. Is this not intervention by a *deus ex machina* a confession of the more imaginary than real nature of this institution? In the case of juvenile delinquency or family conflicts, where the mechanism of mediation has long been known, involvement of a mediator may certainly, in very precise situations, restart a blocked social dynamic. The efficacy of the mechanism, however, presupposes a clear affirmation of the normative reference framework, as well as the maintenance of the possibility of recourse to a judge given the power, where necessary, to impose the fair solution required by a concern for the common good. Failing such guarantees, the law of mediation might as well be "the law of the strongest", a law all the more dangerous for being portrayed as consensual.

The game theory of law, which does not cease to insist on the necessary paradoxical and dialectical nature of legal phenomena, ought to forearm us against such dangers. Using the language of co-operation, it recalls the existence of at least virtual conflict, irrespective of the inventiveness of the actors, highlights the presence of a normative framework behind the artifices of show, and points to strategic manoeuvres. We should not lose sight of the fact that the social role of law does not consist only in *resolving* conflicts, *co-ordinating* behaviour, and producing *consensus*. At the risk of being provocative, we shall maintain that it is also its task to visualize differences, offer alternatives to action and forms for the expression of dissent. Located at an equal distance between the myth of a functionalist society where optimal social integration seems to be located beyond legal intervention, and the opposing myth of a Marxist society where class struggle goes on before that level, the pluralist conception of society that we defend dialectically articulates the formation of compromises on certain values and procedural rules on the one hand, and the existence of numerous areas of dissent on the other. In such a society the role of law is essential, consisting as much in rendering social divisions visible as in finding ways for their compromise. As Freund points out, it is "because law nourishes conflict that it is also capable of terminating it" (Freund, 1983: 328). This is as true at the level of making law as at that of settling disputes before the courts. While it is clear that the rule of law is a "game rule" capable of producing a certain dose of social harmony and consensus, it is nonetheless certain that it simultaneously appears as a reflection of conflict that preceded its adoption, and as the source for and the stake in new conflicts that follow and that may have as their object its interpretation, its modification or its abrogation. Any rule of any game produces this effect of simultaneously stabilizing the game and getting it going again. No rule, whatever its classifying effect, can escape from conflicts "by" the rule, concerning the admissibility of the "strokes" it authorizes, nor from second-order conflicts "on" the rule, concerning the advisability of retaining it.

For the judicial settlement of differences, this appears as "regulated struggle", the merit of which is at least as much that of allowing (peaceful) expression of real antagonisms running through the body politic as of finding an (at least) provisional settlement. In formalizing conflicts about "rights" in this way, supported by "rules", legality does not seek at all costs to transform protagonists into "partners". If, following their confrontation, conciliation wins the day and they agree on a "positive sum game", so much the better. But that is not the essential object of the law, which is more concerned that the game should not take place between "enemies", for whom all strokes would be permissible, rather than between "adversaries" whose divergent interests do not exclude forms of agreement and co-operation.

Law thus introduces *play* into social discourse and practice, while preserving both the facilities of the polemical language of difference and exclusion and the attractive merging involved in identity and consensus. It does this by ceaselessly recalling the truth that the best-assured identities are always shot through with differences, just as there is no difference that cannot be grasped on a basis of identity, play, variety, externality; or once again, the third party. This is the *tertium* that classical logic always sought to exclude, because it favoured the paradigm of simplicity—a simplicity of identification and exclusion ("them-us", "partners-enemies"). The paradigm of complexity, which interweaves identities and differences, and respects the possibility of dissent at the heart of consensual formations, calls for the intervention of the third party: a judge in a position of real externality, a law that no-one is able to appropriate, a public space that appears as the empty space calling for debate, rather than as the publicity screen saturated with conditioning images or propaganda messages.

A CRITICAL DISCUSSION OF A FEW MECHANISMS OF THE SELF-RESPONSIBILITY OF FIRMS

Contractualization of environmental law: responsible co-management or conventional evasion?

We have known that for some years now the practice of environmental contracts between the authorities on the one hand and firms, groups of firms, professional organizations or whole industries on the other has been developing. These agreements are aimed at involving the private sector directly in the setting and attainment of certain environmental policy objectives that government intends to pursue. They are concerned, as is the case in Belgium, with use of CFCs, batteries and phosphates, with regulating the use of a substance or the composition of a given product. They may also be about the rehabilitation of a site or take the form of agreements to do particular research and communicate information.

These new legal instruments emerge at the point where two social phenomena meet: deregulation on the one hand and the rebirth of social corporatism on the other. The highlighting of certain counter-productive effects of the interventionist state brought with it, from the mid-1970s onward, renewed questioning of the coercive, binding form that public policy was taking. At the same time as the rigidities of unilateral regulation were denounced, the merits of concertation, decentralization or self-regulation were praised. Moreover, these movements led to the recognition and the subsidizing of powerful private groups, representing whole sectors of social life and of economic activity. In exchange for their participation in public policies, these organizations were granted a sort of monopoly of representation of their sector (Morand, 1991: 181). At the junction of these two movements, the contractual form emerges quite naturally as the tool for collaboration between public and private "partners".

These contracts, however, raise a number of questions. Are there real reciprocal commitments involving mutual obligations? This seems to be the case in France with the "branch contracts", where the authorities undertake to give signatory firms technical and financial assistance, and the state sometimes abandons coercive action and negotiates the application of previous legislation (Lascoumes, 1991: 224). Or is it more a case of "declarations of intent" reflecting a unilateral commitment by firms? And, if this is any more than a sort of moral commitment, devoid of any legal binding force, is it a commitment to results or to means? And in these circumstances, what means are available to the authorities to verify and sanction non-respect for these agreements? Finally, is this really a legal instrument or is it more a case of a slightly formalized mode of political action?

Despite these questions, the supporters of these agreements stress their manifold advantages. Whether political leaders or representatives of big industries, the analysis is the same (Smet, 1991: 136 (the Belgian Secretary of State for the Environment); Costa, 1991: 129). These contracts are seen as being capable of ensuring that firms become responsible, and are henceforth motivated to ensure implementation of the new norms of their "own" production, because of their direct involvement in defining them. These contracts are further to guarantee speed of execution of the objectives, by contrast with the traditional slowness in generating regulatory texts, at the same time as imbuing them with the necessary flexibility, since corrections suggested by experience can be incorporated at any moment into the undertakings.

Outside observers are, however, rather more cautious, and have asked a number of critical questions. Ginjaar (1991: 100), for instance, stresses the ambiguity of the environmental covenants (*milieucouvenanten*) in the Netherlands, since some see their role as to prepare or implement a public regulation, while for others (the firms in this case) the object is to replace the normative framework. This is certainly the fundamental question raised by environmental contracts: does government ultimately need to divest itself of its regulatory power and seek to secure through negotiation some of the objectives it no longer thinks it can attain by traditional police actions? Or is the point more to anticipate future regulation by experimenting, in collaboration with firms, with the "feasibility" of objectives considered, while spreading over time the investments needed to cope with them? Or, is the point to secure the execution of the existing regulations by associating the addressees with their implementation?

Were it to prove that environmental contracts were now replacing binding regulations that were no longer wanted, objections to the practice would be very serious. In purely legal terms, first of all it is clear, as the French Council of State has pointed out, that government cannot contractually divest itself of its police powers. Second, in terms of legal policy, it should be noted that the technique of contractual negotiation conducted in the absence of a normative framework would present a number of risks:

(a) the risk of a breakdown in equality between firms. Each firm would be dependent on their own "bargaining power", and so there would be real danger that the most powerful among them would, through contract, secure privileges ("private law" in the fullest sense of the term) that they could not secure through the law (Lascoumes, 1991: 231). Without even considering the extreme case of complete official deregulation, one might think that, even in the case of maintenance of legislation, the existence of an environmental contract would encourage the authorities to display an increasing flexibility in verification, and tolerance in the attitude they would adopt towards the signatories of these contracts, rather than to others. And to the risk of discrimination one must add the danger of unofficial, disguised deregulation (Prieur, 1987: 328);

(b) the risk of the authorities being "captured" by the firms that they are supposed to control and deregulate. Without even going so far as to explicitly divest the administration of regulatory power, it is clear that environmental contracts might bring about collusion between public decision makers and private entrepreneurs so as to place the former under the influence of the latter. In other words, the "norm" protecting natural environments would never go beyond the concessions allowed by industrial circles. In that case one would be far from making the defence of nature into the priority we mentioned in the first section of this chapter. Instead of the much-vaunted "flexible" law, we would only be facing a "soft" law which would, under the pretext of pragmatism, put up with only what polluters had consented to bear. This sort of "conventional administration" would have the effect of fixing the situation and preventing the periodical raising of the threshold of environment protection requirements (Untermayer, 1980: 101);

(c) the risk of reducing the democratic nature of public action by "privatization". There are several indications of the reality of this danger. First of all, there is the very low internal democracy of the powerful interest groups that negotiate with the state; second, for some of them, there is the question of their representativeness within the sector concerned. More broadly, the difficulty of identifying the requirements of the general interest in terms of a process of summation of group interests has been pointed to (Morand, 1991: 206). There is a real risk of distorting policies in favour of the short-term and most powerful interests; but as we know, protecting the interests of future generations requires advocates of the long-term. In this connection, the processes of "conventional administration" are once again marked by the opacity of their application. Everything happens away from the spotlight of democratic checks and in the absence of adversary debate. The question of setting environment protection norms becomes depoliticized and seems to be reduced to technical parameters and economic ratios (Lascoumes, 1991: 235). Once adopted, environmental contracts also remain sheltered from external questioning, since the effect of the agreements is to block action by

third parties, such as the environmental defence associat
Oevelen, 1991: 54).

We can see that the practice of environmental contracts is liable to become involved with two very different types of game. In the first the point is to anticipate or perfect a rule of the game which has been publicly and democratically defined in terms of the regulated exchange of all of the arguments involved. The "strokes" authorized by such a rule are always, moreover, capable of being discussed before an arbiter, a legally neutral and impartial third party. In the second the point is to replace a rule which has been found wanting, perhaps because of the absence of a real will to apply it, by *ad hoc* sectoral arrangements. One category of players, the most powerful firms in the sector, will then arrogate to themselves the power of control over the rules of the game, modulated to their interests. Furthermore, the arbiter is put "in touch" to the extent that the rule of the game that he is supposed to interpret loses its position of externality to the parties.

The advantages of rapidity, flexibility and responsibilization brought by environmental contracts are today feeding on a very favourable climate. This enthusiasm must not, however, be allowed to conceal the critical questions we have just put, without which the game they lead us into might well prove to be a "fool's game". The favourable discourse bound up with them may be seen to result more from conditioning through advertising than from public space understood as the playing field of democratic debate.

Assessment of environmental impacts: objective expertise or publicity showcase?

Of all the instruments of environmental policy, environmental impact studies are undoubtedly amongst those that best reflect its specific features. As a preventive tool, impact studies ought, at least theoretically, to make a reality of the basic principle of environmental law that "prevention is better than cure". To this end, the EC Directive of 27 June 1985 lays down the rule that "the authorization of public and private projects likely to have notable impact on the environment ought not to be given before assessment of the major effects these projects are likely to have on the environment" (Preamble, Directive 85/337/EEC). Before embarking on more detailed consideration of the mechanism thus set up, we should stress its *ratio legis*, for fear of distorting its function and failing to perceive what is really at stake: if a verification obligation is imposed before an administrative authority takes a decision authorizing industrial or infrastructural projects, this is because these projects represent a *risk* to the environment. A risk means a potential danger, that is to say a damage or detriment to those living close at hand, to the community, to future generations. Of course, all social activity involves risks and there can be no question of banning this or that type of conduct simply because it might prove risky. At the very least the authorities are invested with the duty of measuring these risks, after the most objective assessment possible. It is necessary to weigh up interests, which, in addition to objectivity in information, calls for the balanced representation of the viewpoints involved. Can one measure the progress achieved here by the environmentalist cause? Where even yesterday the "progress" held necessarily to result from all investment (public or private, from this point of view, is largely irrelevant) in industry or infrastructure acted as an undisputable source of legitimacy, today the same projects are required to present a positive balance sheet, considering the inevitable environmental costs they entail. In any case, it is the least costly of the alternatives that must be adopted in this connection. The legal conclusion that clearly results is that the process of administrative decision for authorization of major industrial projects takes place in a context of the, at least virtual, opposition of interests. The game that is embarked on here is neither solitaire nor a game of pure co-operation. In some aspects at least, it proves to be conflictual. It is, therefore, important for the rule to be fixed clearly and for the arbiter to enjoy genuine independence.

Bearing this in mind, what exactly is meant by "environmental impact assessments"? From the EC Directive, one may pick out the following indicators: "assessment" is intended to be a genuine procedure (to assess is to weigh, to measure) of balancing the interests of the various people or groups of people involved; the "impacts" are the project's effects of every kind, both negative and positive, both short-term and those that might arise in the long term, both direct and indirect; the "environment" is given a very broad definition, in line with the systems paradigm that characterizes the ecological approach (the Directive talks about "direct and indirect effects of a project on man, animals, plants, the soil, water, air, climate and landscape; interactions among these factors; property and cultural heritage"). Taking the Directive seriously therefore implies that an objective, scientific study has to be done, which cannot be reduced to the project's "publicity showcase" (Jadot, 1992: 197ff.).

A similar conclusion emerges from a consideration of the manifold functions that an impact study worthy of the name is expected to perform in the complex (co-operative/conflictual) game we have described. The study is first of all a conceptual tool, that should inspire the project's director. If done as early as possible, it could offer him various alternatives, including those least unfavourable to the environment, whereas if done too far downstream from the decision its role risks being reduced to that of justifying the option taken. The impact study is thus a tool for informing and involving the public. In accordance with Art. 6 of the Directive, requests for authorization and information gathered shall be kept available to the public, which is to be placed in a position through appropriate procedures to express its opinion before the project starts. Finally, the impact study is a decision-making tool for the administrative authorities called upon to decide on the project's admissibility.

Given these different tasks, the study is quite clearly a complex institution. It is intended to supply scientific illumination for a decision of a political nature. This implies bringing about a dialogue between two largely heterogeneous discourses—each presenting, for instance, their own spatio-temporal framework. But it is also asked to bring out questions and objections of a "popular" nature during an administrative procedure. And still more fundamentally, it is supposed to allow the expression of a democratic logic (broad information, public enquiry, adversary debate) in the course of a process which is more of a private, technocratic nature. Being over-determined in this way, it is hardly surprising that the impact study should have been the object of such diverse applications, so much so that the EC Directive, which has gaps in many respects, remains compatible with often highly divergent national legislation. On the one hand there is a logic of "publicity", with the accent on the public's "right to information" and the independence of the author of the study whose task is conceived in terms of a "public service". On the other, a "private" logic prevails with the emphasis on the promoter's freedom of enterprise. In this case the procedure is simplified to the extreme and the task of the author takes the form of an enterprise contract. The impact study will then appear as either an additional legal formality to be completed by the entrepreneur or promoter along the obstacle course leading him to secure all the necessary administrative organization for his project, or else more or less as publicity, intended to calm the broader public's apprehensions and attract the favours of political decision makers. In the first case, by contrast, the study, whose impartiality will be guaranteed, ought to permit both substantive improvements to the project in terms of protection of the environment, or its banning should the balance sheet turn out too negative for ecosystems.

Two questions seem particularly important in deciding which of these two logics is the inspiration of a particular piece of legislation: the legal description of the mission conferred on the author of the study, and the guarantees offered by the procedural framework in which the study takes place.

As far as the first is concerned (the Directive being silent on this point) all systems seem possible, from the involvement of a panel of independent, sworn experts appointed by the authorities, to the French system which accepts studies done by the project director himself (Huglo, 1992: 87). In this connection the ambiguities of the system introduced by the Walloon Decree of 11 September 1985 point to the difficulty legislators often find themselves in over clearly defining the game that they want to see played. As Déom points out very clearly (1992: 181), the status of the author is dominated by paradox:

> Chosen by the project director and paid by him, he must nevertheless display a certain independence and supply the public authorities with reliable information. Accepted by the executive (according to criteria that are, however, very vague) and subject in this connection to a certain control, he must nonetheless operate alongside the project director, and to some degree represent his interests.

She goes on to talk about this as a "double allegiance". Is this really the case since the author's relationships with the project director are contractual in nature and leave open the possibility to the author of breaking off the contract should the progress of the study fail to satisfy him? Chosen and remunerated by the promoter, the author is obviously drawn into the private logic of the enterprise contract. The ambiguities nonetheless remain, since the text virtuously provides in Art. 18(2) that persons responsible for impact studies shall be treated in the same way as persons "charged with a public service" as far as application of the penal provisions against corruption is concerned. One would be willing to bet that this can scarcely be applied since, logically, the person who would be most inclined to make such an attempt at corruption is the very one who is already paying the author. The author's independence is, therefore, both affirmed and virtually compromised. While the mission ought, if it is to fulfil the various functions expected of it, to appear as an expert report in the public service, the Walloon legislation shapes it in the form of a private enterprise contract. Is this sort of ambiguity acceptable when one knows that the study is part of a contract which, without being of an adversary nature as a legal expert report is, can nonetheless be termed "semi-adversary", since we have accepted that every industrial project or major piece of infrastructure offers real and virtual risks to the environment?

In terms of the procedural framework, either the study is a limited stage, with restricted publicity, in a technocratic-type decisional process where business secrecy is backed up by administrative secrecy. Or else the point is to set up a new administrative logic, on the model of judicial procedures, marked by the need for the broadest possible publicity and the concern to ensure respect for the adversary principle. If some useful effect is to be attributed to the 1985 Directive, would it not be better to favour the second option, the only one in a position to guarantee that the balancing of interests performed by the administrative authorities will be done using the balance of justice, and not the clay pot and iron pot of the fable?

In the complex game set up between private interests and public institutions in the framework of our advanced capitalist states, the impact study ought to appear as a phase of relative scientific objectivity, guaranteed by some political impartiality. If the study is the outcome of the responsibility of firms alone, we cannot be sure that this aspect of relative externality has yet been guaranteed.

Eco-audit and eco-label: right to know or power to seduce?

Two regulations of the Council of the European Communities, still under discussion[3], one for a "Community Environmental Audit Scheme" (eco-audit), the other for a "Community System of Ecological Labelling" (eco-labelling), further illustrate the political will to increase firms self-responsibility in the area of environmental protection. These two texts are, moreover, part of a line of preventive action aimed at reducing nuisances at source, by rational planning of production and the promotion of clean technologies and products. In both cases, the point is to make the policy pursued by the firm in these areas more transparent, make the public aware of the efforts being made and to award the most efficient firms with a sort of good conduct certificate that they would be authorized to use in their publicity campaigns. The public's supposed preference for products and firms with that distinction would then, through the flexible play of the law of supply and demand on the market, lead to the repayment of investments made and to the launching of a general campaign in favour of the most environment-friendly technologies.

Positive as these prospects are, we propose, however, to put a few critical questions in connection with them, inspired by the same concerns as have guided us up to now. Should one, for instance, favour a system of deliberate self-commitment by certain firms in sectors they have chosen, or instead one of respect by *all* firms for *legal* obligations? Are the procedures for awarding the eco-label and doing the eco-audit of a sufficiently adversary nature? Do they give guarantees of independence and impartiality? Is the publicity that seems to be the real objective of both schemes inspired by a genuine concern for public information, or does it derive more from a media message? In other words, is the game one where the cards are on the table, or is it instead related to a game like poker where all moves, including bluffs and cheating, are permitted?

To answer these questions, let us consider each of the regulatory texts. First the eco-audit. This is a tool for the firm's environmental administration, consisting in the systematic, periodic, objective evaluation of its policy, programme and equipment in that sphere. In addition to better information for the public and authorities, it is expected to bring a steady improvement in the performance of firms. Among these aims there is also that of respect for the various legal regulations (Art. 7; currently Art. 2h); the Preamble takes care to underline that the audit system "must be regarded as a complement to, rather than a substitute for, respect for environmental standards and regulations" (the clarification is no longer included in the present Preamble).

This system will not, however, be of general application. It is planned to reserve it to firms that deliberately commit themselves to it. Article 3 provides that enterprises "may" ask for entry to the audit scheme, for one or several of their industrial sites, or for one or several categories of their products (the new version no longer mentions products). They will in this case conform to the rules of the audit, including in particular the obligation to "establish, in accordance with their own needs and choices, an internal system for environment protection", and, of course, the obligation to have the audit done regularly, have it evaluated, and keep it available to the public and the authorities (Art. 4, now Art. 3).

The text further provides for a system of certification by outside auditors of the reliability of the audit reports. This system requires that a variety of national and supra-national authorities be set up. It is further provided that "Member States may engage in action to promote the eco-audit scheme, in particular by introducing appropriate incentives to encourage participation in the eco-audit scheme (reduction in charges for pollution emitted, loans, priorities in public contracts etc.)" (Art. 17; now Art. 12, with the second part of the phrase deleted).

What is one to think of this system? It will first of all be noted that, following the logic of self-responsibility, it has a purely voluntary foundation. Firms conform to it if they want to, and for the sites and products where they feel that it might be advantageous. They are to develop an internal system of environment protection "in accordance with their own needs and choices". It will be agreed that this is to push the logic of "self" very far. One would, however, have no objections if it were established that this sort of scheme is, as the Preamble says, a complement to and not a substitute for the general legal obligations. If it means doing more than the law and better, then so much the better; if on the contrary the point is the *ad hoc* negotiation of the application of obligatory texts, with the state sometimes "buying" conformity in the form of reductions in charges or of loans, it is clear that this sort of scheme would be unacceptable. This conclusion is similar to that already drawn about the scheme for contractualizing environment law that we discussed above. Is our criticism misplaced because the text explicitly rules out this hypothesis? Certainly the most flagrant abuses are unlikely; it is not unrealistic, by contrast, to fear that this scheme of self-responsibilization will gradually, in a climate of euphoric deregulation, replace the normative framework that is both binding and general. In that case, would we not see a growing differentiation between a few very efficient enterprises, at the cutting edge of research in terms of clean technology, and the bulk of the industrial sector whose polluting activities would scarcely be censured any longer? Moreover, how is one to guarantee that the bound measures by "clean" firms would be maintained once they had acquired a dominant market share or ecology has gone out of fashion—especially since in the meantime the present binding mechanism would have been weakened? Here too, then, it is important for progress in the direction of self-responsibility to be accompanied by corresponding progress in the direction of defining general, binding normative objectives. Here again, the game model is instructive: the players' inventiveness has meaning only against the background of the rules of the game.

Our second question relates to the nature of the publicity attached to the audit. In connection with this, the text studied contains a number of very informative passages. The Preamble (old version) says that there is a "clear need to *improve public acceptance* of certain activities (of firms) through greater transparency"; further on there is talk of developing a "*partnership*" with the public (here again we almost explicitly come up against the model of the positive sum co-operative game). Several provisions echo this concern for information/seduction of the public: Article 8 (amended into Art. 5) provides for an "environment statement" to be drawn up to inform the "wider public". Written in "summary, non-technical form", it should however "fully and objectively" reflect the results of the audit and likewise be certified by outside verification. Article 9 (now Art. 11) provides that firms entered in the eco-audit scheme should have the right to use the eco-audit *logo* for sites and/or production activities covered. Finally, Art. 17 (now Art. 12) provides that states may also join this great promotion campaign by "publicizing the eco-audit scheme and informing the public about it".

Many precautions have certainly been taken to guarantee the objectivity of the information contained in the audit report and the outside auditors have the specific mission of ensuring the reliability of these documents. One cannot, however, avoid the impression that the real *raison d'être* of the text is the concern to reinforce the firm's image. In these circumstances, this "licence to seduce" ought not to be a substitute for a much more fundamental legal institution, namely the citizen's "right to know", since every attack on the environment can be analysed as a detriment to the "common goods" that the Belgian and French Civil Codes (Art. 714) claim belong to no-one and are for the benefit of all (Jadot, 1992b: 14). Another EC Directive, dated 7 June 1990, on "the freedom of access to environment-related information" might serve as a reference here. It makes access to information (through the authorities, it is true) a genuine right enjoyed by "every physical or legal person asking for it, without any need to show an interest" (Art. 3). Refusal to communicate the information requested would have to be for stated reasons, and the person turned down would have a remedy (Art. 4). The scope of this principle is further extended by Art. 6, which makes it a duty on states to ensure that "information relating to the environment held by agencies with public responsibilities related to the environment and controlled by the public authorities should be made available on the same terms". One can see the potential field of application of this principle, even if at present it remains encumbered by numerous exceptions (Art. 3). Starting from the idea that the environmental game is at least as much conflictual as co-operative, we admit our preference for this logic of "cards on the table" information over the publicity system of the audit. The regulation on ecological labelling leads us to a similar conclusion.

Inspired by the German "Blue Angel" system, the EC machinery provides for the setting up of a uniform Community scheme of ecological labelling aimed at singling out the least polluting products. The assessment is made according to the comprehensiveness principle, "from cradle to grave", that is to say throughout the product's lifetime from manufacture to disposal, through transport, treatment and use. This system too is optional. It is based on a request for the award of the label coming from the manufacturer or importer (Art. 12, now Art. 10). Once awarded, the label can be used only by means of a contract between the firm concerned and the competent authorities (Art. 15, now Art. 12).

The system provided for by the proposed regulation posits two stages. The first is the establishment of categories of "labelable" products and relevant criteria. This decision is taken by the EC Commission on the advice of a consultative committee (Arts. 8 and 9, now Arts. 6 and 7, much amended). Thereafter, in a second stage, the ecological label could be awarded to individual products. The decision would be taken by a jury made up of "one representative per Member State plus one representative of the following interest groups: industry, commerce, consumer organization, ecological organization, trade union, media" (Art. 13, now Art. 10, much amended).

This proposal calls for a number of comments. The first are very similar to those made on the eco-audit project where we noted, firstly, the optional nature of the system in accordance with the logic of self-responsibility of firms. All very well, we might repeat, as long as the binding norm, applicable to all firms, is not thereby eliminated. One will also have noted (this is the second comment) the very "promotional" nature of the instrument employed; here too a logo is used to distinguish the clean product. Personally, we are not at all sure that giving this or that product good marks in this way is the best way to guarantee objective information for the citizen. Would it not be preferable for each product to be accompanied with an explanatory note of the type existing in the area of pharmaceuticals, or at least to develop a system of labelling by stars according to the harmfulness of the product (Jadot, 1992c)? One should never lose sight of the fact that no product, no technology, is truly "clean" or without effect on the environment.

Our third comment concerns the proposed procedure for both the general definition of categories or products and the relevant criteria for awarding the label to a specific product. Once again two logics clash, while the Economic and Social Committee (ECS), set up by the European Community, resolutely advocates "as broad as possible privatization of this procedure" (the point being to associate the various economic and social interest groups with the decision from the stage of determining categories of product and criteria to be met) (ECS Opinion 610/91, 12.9.1991), authors more aware of the principles of public law, by contrast, rightly point out that a public authority cannot delegate or share the decisional power with which it has been invested. Thus, while it is perfectly logical for the Commission to have itself informed by technical advice from an *ad hoc* body, it is not acceptable for that body to be given the last word (Jadot, 1992c). By adopting a mixed system, borrowing from both logics in turn, the text of the regulation once more reflects the difficulty of choosing between the private logic of contract and the logic of public service. If the procedure planned were at least more adversary in nature (we would point out in this connection that the Quebec eco-logo system provides for a procedure of public enquiry and a possibility of counter-reports), some of the concerns expressed here might be somewhat appeased.

"Right to know" or "power to seduce", was the question that we asked in relation to the eco-audit and eco-labelling projects. The texts studied certainly allow both interpretations. At least one ought to know that the statement "eco" is no longer necessarily a synonym for environment protection. The recent Belgian Law of 14 July 1991 on trade practices has for the first time, as far as we know, banned misleading publicity in particular relation to "effects on the environment" (Art. 23). Does this not amount to a recognition that the colour green, which has become of considerable economic importance in the great economic game, is no longer sheltered from opposing fashions?

CONCLUSIONS

The strategy of self-responsibilization of firms in the ecological sphere consists in "internalizing externalities produced", as economists say, while preserving their autonomy (Rehbinder, 1991: 598). In the concern to improve their image before the broad public by taking account of the negative effects of their actions on natural environments while controlling from within the normative process for regulating that action, firms are today adopting the discourse of self-responsibility.

It is understandable that the theorists of autopoiesis are tempted to see this phenomenon as empirical application of their models. Had they not shown, some time ago now, that the capacity for one subsystem to interfere with another was very limited, and in any case random, since every subsystem ultimately had access only to its own environment, that is, to its own internal image of the world outside? If "eco" (we mean the external environment) has no other substance, on this view, than what the "auto" can confer upon it, there is indeed no other solution than self-regulation.

We have endeavoured to discuss the theories of self-organization, in both the philosophical sphere, by opposing them with a dialectical mode of thought (for which "auto" is not even conceivable apart from "eco"), and in legal terms by replacing them with a game model of social interactions. A consideration of four contemporary institutions of environmental law, contractualization, impact studies, the eco-audit and eco-labelling, has led us to argue systematically for the reinsertion of machinery for self-responsibilization of firms in a broader framework, where conflicts and rules (this time in the sense of external regulation) are not absent. By insisting on the reintroduction in each case of a solid normative framework, a genuine adversary debate, a publicity (in the sense of transparency of procedures) worthy of the name, and a neutral and impartial judge, we have sought to rehabilitate *tertium*—by which we have in mind distance, externality, the *play* essential to the movement of ideas and actions. The tertium, the third party, is the social "third dimension". This is what ultimately gives real substance to what is, without it, reduced, as in the "auto" model, to a fresco, certainly ingenious, frequently seductive, but confined to the two dimensions of its selfhood. Undoubtedly, as the history of painting shows, it is not impossible to bring perspective into this flat space; a minimum distance from self to self that suggests the image of depth. But if we wish to go beyond the sphere of the image and *trompe-l'oeil*, we must introduce real distance, the distance brought by the third dimension introduced by a real space where a third party comes in, bringing play, history and movement.

As Rehbinder has demonstrated in connection with the German institution of a person responsible for environmental matters within firms (a rather disappointing experiment, since the power of this "employee" remains extremely limited) it is not enough, in order to modify firms' conduct towards the environment, for them to interiorize this or that constraint, if outside pressure is not kept up at the same time:

> It is, in fact, unlikely for a subsystem whose primary orientation is optimizing performance to consider outside interests, in the event of conflicts of objectives, more than the administrative political system would do (Rehbinder, 1991: 601).

Similar conclusions have been arrived at in our analysis, done on the basis of the theory of the social as a game and of ethics as an argued hierarchicalization of priorities.

Will we then, have convinced the upholders of autopoiesis and of the self-responsibility of firms? In a recent interview given to *Journal M.*, Teubner expressed sympathy for the green movement which, he said, was radically questioning the results of the functional differentiation affecting our societies, and the jargon that accompanied each of these differentiated subsystems. But, he immediately added,

> the tragic aspect of this movement lies in the fact that what they positively desire—reintegration of fragmented society into the public discourse that would make the language games into mere tools in the hands of the public space—is bound to remain an illusion (Teubner, 1991: 40).

Certainly, the idea of a single, supreme language game, a sort of "game of games", simultaneously and functionally integrating all needs and all codes, will prove an illusion, just as did the desperate effort to create a universal language like Esperanto out of all sorts of bits and pieces. Conversely, we regard as equally "illusory" and "tragic" the monologue engaged in by each differentiated subsystem which, juxtaposed or superimposed, can offer only the sorry sight of incommunicability. Babel, as it were, against Esperanto. But is it really true that we are condemned to these alternatives? Are Babel and Esperanto not once again simplistic images: a soliloquy by a differentiated subsystem in the first case, and the great monologue of the integrated social macro-system in the second? Between Babel and Esperanto, is there not room for a third way: the path of complexity which, on the model of the game of dialogue and of translation from one language into another, articulates, in dialectic fashion, through successive approximation and undoubtedly at the cost of many errors, "auto" with "eco", "I" and the "other"? If there is still some room for the law in the post-modern world that is coming into being, it can only consist in blazing that narrow path. Were it, by contrast, to prove that that path was now closed, there would be every reason for believing that media processes will soon replace legal ones.

NOTES

1 Translated from the French by Iain Fraser.
2 There is no room here to discuss in depth the positions supported by the supporters of autopoiesis. What interests us in relation to the topic of the self-responsibility of firms is, in any case, more the way these positions are today being spread and understood, rather than the details of the positions defended by Niklas Luhmann and his disciples.
3 On ecological labelling we have used the *Proposal for a Council Regulation Concerning a Community System of Ecological Labelling* presented by the Commission on 1 February 1991 (OJ C75/5, 20.3.1991). This text was adopted in an amended form by the Environment Council on 2 December 1991, and is now entitled *Draft Council Regulation for a Community System of Ecological Labelling*. For the Eco-Audit, we used the *Draft Proposal for a Council Regulation Laying Down a Community Scheme of Environmental Audits*. This text was adopted by the Commission in an amended form, and is now entitled *Proposal for a Council Regulation allowing Voluntary Accession by Firms in the Industrial Sector to a Community System of Eco-Audits* (January 1992). The analyses are based on the initial versions of the two texts. We shall confine ourselves to indicating the most substantial changes made to the texts between this and the second versions. In general terms it will be noted that some of the criticisms made of the initial forms of the texts have been met in the formulations now being debated.

REFERENCES

Barel, Yves (1979) *Le paradoxe et le système. Essai sur le fantastique social*. Grenoble: Presses universitaires de Grenoble.
Costa, P. (1991) Avantages et inconvénients des conventions sectorielles. Le point de vue de l'industrie. In Bocken, H. & Traest, I. (Eds). *Milieubeleidsovereenkomsten. Conventions sectorielles: instrument de gestion de l'environnement*. Brussels: Story Scientia, p. 127ff.
Déom, Diane (1992) Le statut juridique de l'auteur de l'étude. In CEDRE *Les études d'incidences, un progrès juridique?* Brussels: Publications des Facultés Saint-Louis, p. 181ff.
Freud, Julius (1983) *Sociologie du conflit*. Paris: P.U.F.
Ginjaar, L. (1991) Het gebruik van Milieuconvenanten in Nederland. In Bocken, H. & Traest, I. (Eds). *Milieubeleidsovereenkomsten. Conventions sectorielles: instrument de gestion de l'environnement*. Brussels: Story Scientia, p. 89ff.
Hajer, Maarten (1991) Environmental Performance Review as Instrument of Ecological Modernism. Paper presented at Environmental Performance Review, Oslo.

Henriot, Jacques (1989) *Sous couleur de jouer. La métaphore ludique.* Paris: José Corti.

Huglo, Christian (1992) L'étude d'impact écologique en droit français. In CEDRE *Les études d'incidences, un progrès juridique?* Brussels: Publications des Facultés Saint-Louis, p. 87ff.

Huizinga, Johan (1951). *Homo ludens. Essai sur la fonction sociale du jeu*, 8th edn, translation by Seresi, C. Paris: Gallimard.

Jadot, Benoît (1992a) Des études d'incidences: pour qui, pour quoi? In CEDRE *Les études d'incidence, un progrès juridique?* Brussels: Publications des Facultés universitaires Saint-Louis, p. 195ff.

Jadot, Benoît (1992b) *L'audit de l'environnement et l'information du public* (forthcoming manuscript).

Jadot, Benoît (1992c) Considerations du droit de l'environnement et du droit public sur le label écologique. Forthcoming, in Actes du Colloque organisé par CEDRE sur le thème: *Quelle règles pour l'institution d'un label écologique?* Brussels, 5 November 1991.

Jonas, Hans (1990) *Le principe responsabilité. Une éthique pour la civilisation technologique*, translated by Jean Greisch. Paris: Editions du Cerf.

Koppen, Ida (1991) Environmental Mediation: An Example of Applied Autopoiesis? In In 't Veld, R. T. *et al.* (Eds). *Autopoiesis and Configuration Theory: New Approaches to Societal Steering*, Dordrecht: Kluwer.

Lascoumes, Pierre (1991) Les contrats de branche et d'entreprise en matière de protection de l'environnement en France. Un exemple de droit négocié. In *L'Etat propulsif. Contribution a l'étude des instruments d'action de l'Etat.* Paris: Publisud, p. 221ff.

Luhmann, Niklas (1988) The Third Question: the Creative Use of Paradoxes in Law and Legal History. *Journal of Law and Society*, **15**, 153.

Morand, Charles-Albert (1991) La contractualisation corporatiste de la formation et de la mise en oeuvre du droit. In *L'Etat propulsif. Contribution a l'étude des instruments d'action de l'Etat.* Paris: Publisud, p. 181ff.

Morin, Edgar (1977) *La méthode. I. La nature de la nature.* Paris: Seuil.

Morin, Edgar (1990) *Introduction à la pensée complexe.* Paris: ESF éditeur.

Ost, François (1986) Entre ordre et désordre, le jeu du droit. Discussion du paradigme autopoiétique appliqué au droit. *Archives de philosophie du droit*, **31**, 133.

Ost, François (1991) Faut-il légiférer en matière d'environnement? *Cahiers de l'Ecole des sciences philosophiques et religieuses*, **10**, 63.

Prieur, Michel (1987) La déréglementation en matière d'environnement. *Revue juridique de l'environnement*, 327.

Rehbinder, Eckard (1991) Reflexive Law and Practice. In Febbrajo, A. & Teubner, G. (Eds). *State, Law, Economy as Autopoietic Systems: Regulation and Autonomy in a New Perspective.* Milan: Giuffrè p. 595ff.

Smet, Miet (1991) Milieubeleidsovereenkomsten in België. In Bocken, H. & Traest, L. (Eds). *Milieubeleids-overeenkomsten. Conventions sectorielles: instrument de gestion de l'environnement.* Brussels: Story Scientia, p. 133ff.

Teubner, Gunther (1991) *La théorie des systèmes autopoiétiques. Journal M*, February–March, 36.

Untermayer, Jean (1980) Le droit de l'environnement. Un premier bilan. *Année de l'environnement 1980*, **1**, 1.

Van Oevelen, A. (1991) Privaatrechtelijke aspecten van milieubeleids-overeenkomsten. In Bocken, H. & Traest, L. (Eds). *Milieubeleidsovereenkomsten. Conventions sectorielles: instrument de gestion de l'environnement.* Brussels: Story Scientia, p. 3ff.

van de Kerchove, Michel & Ost, François (1988) *Le système juridique entre ordre et désordre.* Paris: P.U.F.

van de Kerchove, Michel & Ost, François (1992) *Le droit ou les paradoxes du jeu*. Paris: P.U.F.
Weber, Max (1959) *Le savant et le politique*. Paris: Plon.
Willke, Helmut (1986) Diriger la société par le droit? *Archives de philosophie du droit*, **31**, 189.

CHAPTER 15

Ecological Responsibility without Subject: Differences between Japan and the West

Toru Hijikata[1]
Tokyo

"ENLARGING THE KITCHEN AND NARROWING THE GARDEN"

Japan ranks among the great powers of the world. Yet amidst all the hard work, we have barely begun to reflect on our life. Development? And for what? Our development is normally described symbolically as enlarging the kitchen (to eat) and narrowing the garden (for the enjoyment of one's life). And so, as we Japanese discover new conceptions of the "ethics of enterprise" or "ecological responsibility", it is worth examining Japanese religious behaviour, for there is a major difference in the consciousness of responsibility between the West and Japan.

RELIGIOUS COMMUNICATION

Japanese religious behaviour differs from that of the Western world. It is based on the amulet and the oracle paperslip. The amulet or charm is faith, God protects against evil. Yet if it is not effective, this will be understood as an exceptional occurrence. There is no loss of faith or discouragement, for the disappointment is taken as one's own mistake or that of family or friends. The next time the amulet will be used again.

Oracle paperslips foretell the future, God's deeds are true prophesies. If the prophecy is contrary to the wishes, the predicted fate disagreeable, the slip of paper will be attached to the branch of a tree in the temple compound in the hope of avoiding the negative fate. This signifies the reversal of the bad into a good fate. Thus, one possesses two alternatives: either trust one's good fate or reverse the bad.

Environmental Law and Ecological Responsibility: The Concept and Practice of Ecological Self-Organization
Edited by G. Teubner, L. Farmer and D. Murphy
© 1994 John Wiley & Sons Ltd

Is that religious communication? Perhaps you would say no, because it seems to be too simple. You would say that this is not religion but superstition. Should a well known prophet declare that green towels bring bad luck, would this be believed? Surely not, but nevertheless nobody would buy a green towel! This, of course, is not faith but iconoclasm but, as I see it, it is the typical Japanese religious behaviour. It is used in special or difficult situations as well as for general predictions. A remarkable phenomenon in this society.

More remarkable still when we remember the difference between danger and risk. Danger is a future peril attacking from outside. Risk, by contrast, is something that one looks upon in one's own decisions, the consequences of which one must shoulder. We suppose that the orientation of society, looking from the present to the future, has changed from one of danger to one of risk. Yet the amulet and the paperslip refer to danger: Protestantism understands this as risk. Wishes and predictions are the forces impelling the Japanese to turn outwards, to Gods. Not yet speaking of risk, they interrogate religious powers over good or bad events to come. Is such a conviction possible in modern society?

SHORT-LIVED LUCK, NOT HAPPINESS

Another part of Japanese religious background can be seen in the definition of Samsara in Buddhism, the soul's never-ending renewal—life then death then life, like the turning of a wheel. The end of human life returns periodically, as part of a continuous circulation of degeneration and regeneration. Samsara sees the animal and human as transitory, one following the other, yet following the same path. Man in the present becomes animal in the next world. There is no future, but an uninterrupted presence. Thus, rescue and the fulfilment of wishes and deeds in the present time are obligatory, forming the passing luck and happiness of the present. This so-called presentism makes it possible to understand future bad luck as harmful or damaging and something to be endured. This is the religion of defence, maintaining the status quo.

RELIGION AS A GAME

Japanese religious understanding makes no clear differentiation between God and man, as Christianity does. Man might become god in a simple way, by purifying and exercise. A Shinto shrine is dedicated to a real man, a general or the Tenno, dedicated as a god. Families believe that their ancestors are turned into gods in their own homes. The Shinto view that all men bear godliness within them, links God and man, but also means the holiness of World and man. The sin of human beings is their uncleanliness, accumulating on the outside body which can be washed off. The outside can be purified. And because bad luck might be caused by uncleanliness, good luck is achieved by its purification. The soul of a delinquent will be purified after his death;

a corrupt statesman may declare that he is purified after a certain lapse of time.

The differentiation between holiness and unholiness is clouded, even interchangeable. Religion plays the role of circulation, dividing holiness and unholiness, but also to avoid the circulation so that uncleanliness might itself reproduce to produce the next misfortune.

So it is just as well that the Japanese do not have a consciousness of "the original sin". There are major differences between us in the consciousness of risk and sin. Westerners consider environmental disruption as (original) sin. Buddhists teach that the human being is one of the many parts of nature. These sins are not due to one's own actions but to the weight of uncleanliness which is imposed by the outside world—like natural disasters. To fight the outside world, formal power ceremonies are necessary, together with the reliance on other external powers.

This religion is like a game: you rely on the outer forces of purification to be freed from sins, misdeeds and uncleanliness, thus avoiding bad luck. It thus becomes clear that there is a difference between Japan and the Western World. We cannot stand on the same platform for ecological responsibility, for we cannot find the subject with responsibility that is man under the Protestant ethic. And if this is the ethic that has an elective affinity with the development of capitalism, can environmental ethics control the behaviour of enterprises today?

Generally speaking, the Japanese are told that Japan has solved the problem of environmental pollution. Or perhaps it is also true to say that Japanese enterprises have been engaged in solving the problem for a long time. This, at least, must be looked at from the point of view of Japanese enterprise.

THE PROBLEM SHAPED BY THE ENVIRONMENT

Our environmental history has been shaped by Japan's geography. It is an island with fast flowing rivers. Unlike European rivers, most streamed so rapidly to the sea that we were accustomed to throw everything into their waters. The environment was cleansed by the water flowing away. Now the environment of enterprise is shaped by the flow from outside. After centuries of closure, recent Japanese history has witnessed the pursuit of Western civilization and culture. These we take in and develop ourselves. It is said that almost entire cultures and civilizations have been imported. Conscious of our talent in modifying them, we are at the same time conscious of the lack of originality. Nikon copied Contax, Canon copied Leica, Nissan copied Austin and so on. Cultures are consumed and re-exported.

Our limited natural resources have also produced a reliance on the natural resources of other countries. From a global point of view we are open to the charge of being an assailant of the environment. The population of Japan is 2.4% of the world. We consume 28% of the shrimp, over 30% of the tuna and use 25% of the wood, 16% of the oil, 27% of the coal, 30% of the ironstone, and so on. There is more. Ordinarily big enterprises buy these resources from small traders all over the world. And so the price tends to be determined by the Japanese. In this way Japan has grown economically: a product itself of the movement between mass consumption and mass export-ation. And so it is that the mass consumption of foreign resources has caused Japanese environmental disruption.

A further irony. As environmental regulation increased at home, Japanese enterprise and technology moved abroad to exploit the freedom offered by countries with little or no legal regulation. At the same time, following the oil crisis we have developed energy-saving technology which, if exported world-wide could, it is said, save 30% of current energy use. And Japanese enter-prises will then assist under-developed countries to save energy by the export of more technology.

Ecological problems become a commercial item. With their dependence on export, Japanese enterprises must send a message internationally that their products respect the environment. Anticipating the trend of the times, they seek a new business opportunity. This gives the opportunity to create business without criticism.

WHY ECOLOGY IS THE FATE OF ENTERPRISE

Nowadays the consuming public watches the action of enterprises and is quick to blame them for environmental disruption. If an enterprise causes damage, it not only suffers criticism but will also lose business. It is enterprise that carries the burden of establishing that they have not caused pollution. In this situation we can say that the existence of enterprise depends on its ecological behaviour. All try to give the impression that they have considered the environ-ment. If an ecological problem arises the enterprise must react immediately regardless of its actual involvement in order to maintain a good reputation. If they fail, they will be criticized and their business threatened. In this sense, ecology is the very fate of the enterprise.

ECOLOGY AS A RELIGION. (REGULATION WITHOUT REASON ...)

It seems today that ecology functions as a religion. The environmental problem is radical in that there is no exterior. The earth as a whole is under threat and this must underpin all of our social activities. This is to take up the traditional function of religion. But, in this context, is it still possible to rely on some external force such as nature, reason or God? The problem is the same whether we are talking about uncleanliness or sin. And in this sense, ecology can only integrate so far as religion plays an integrative role in our society. Today the province of religion seems more diffuse, taken up by other social systems. Yet insofar as we still rely on religious communication, whether fate or sin, this must be taken into account. It can provide no universal code, but perhaps a universal myth, laying a new code of ecologically good/ecologically bad over the activities of differentiated systems. And this does not operate through reason but for its resonance with societal circumstance.

Certainly Japanese enterprises try to cope with ecological problems in their lives. And their motivations seem to be formed by the economic rationality of enterprise or efficiency, at most activating the new mythical code. However, we are not subject to Enlightenment reason, but sociological enlightenment.

NOTE

1 The English version of this paper has been extensively rewritten by Lindsay Farmer.

CHAPTER 16

Constructing the Responsible Organization: Accounting and Environmental Representation

Michael Power[1]
London

INTRODUCTION

The organizational linkages between our hopes for the natural environment and our bureaucratic and technological capabilities are often underplayed. Indeed, it is characteristic of certain forms of "anti-technocratic" thinking to ignore the administrative implications of reform. From this point of view accounting, as the embodiment of economic rationality, is part of the problem rather than the cure (Gorz, 1989; Power, 1992) and we require a radical abandonment of its "organized irresponsibility" rather than a "green" reconstruction of its mission. And yet, as Sagoff (1988) has argued, unrealizable hopes may be even more damaging than a realizable pragmatism. Weber was pessimistic about the rise of "specialists without vision" but it may be that the "visionaries without speciality" are equally problematic. Accordingly, we need to explore the forms of administrative expertise, such as accounting, through which ecologically relevant transformations of corporate and individual behaviour may be possible.

Popular perceptions of ecological crisis have their basis in complex structures of knowledge and in this respect so-called "green consciousness" is cognitive in part. However, these knowledge systems may also be highly selective and certain types of "risk" can be made publicly visible at the expense of others (Douglas & Wildavsky, 1982)[2]. Theorists such as Luhmann (1989) remind us of the enormity of the reformist task facing ecologists by emphasizing the need to know *how* social systems react to their environments. For Luhmann, this issue is a general one and the "natural environment" is simply a special case.

Environmental Law and Ecological Responsibility: The Concept and Practice of Ecological Self-Organization
Edited by G. Teubner, L. Farmer and D. Murphy
© 1994 John Wiley & Sons Ltd

He also argues, to the irritation of green activists, that we cannot make utopian prescriptions in advance of a deeper knowledge of the functional capabilities of society itself. From Luhmann's point of view, improvements in systems description must *precede* the articulation of responses to the crises registered in social systems. Against romantic aspirations to an unconditional "openness" to the natural environment, Luhmann emphasizes the closed and organized preconditions of any such openness.

Luhmann reworks the hermeneutic boundary between the "tradition" and "horizon" of cognition and dehistoricizes it as the boundary between system and environment, a boundary which is an internal product of the system itself. Such a position is clearly antithetical to the aspirations of radical environmentalists ("deep greens") from whose perspective Luhmann appears to ignore the "reality" of an issue which has arisen precisely because of the corrupt rationalities embodied in social systems. Indeed, on a particularly unsympathetic reading of his position, Luhmann seems to be saying that there is no environmental crisis because the "environment" is itself a systems-relative construct. Against Luhmann, radical environmentalism seeks to regain contact with nature, not the social systems which threaten it. But for Luhmann the selective processes of closure which determine the systemic reception of environmental perturbations are poorly understood. Hence we must

> work out how society reacts to environmental problems, not how it ought to or has to react if it wants to improve its relation with the environment (Luhmann, 1989: 133).

In this sense environmentalism and welfarism face similar forms of difficulty as "political programmes" which presuppose, but cannot ensure, the possibility of their own efficacy (Miller & Rose, 1990). Luhmann's argument that we need to understand modes of system autonomy before we can make interventionist proclamations can be seen in the light of a general theoretical and practical climate of attention to technologies of implementation and administrative control. If there is no intervention without representational media such as accounting, then it follows that the "crisis of the environment" is as much a crisis of these representational technologies as it is of political will.

The argument of this chapter has four components. First, accounting is considered in the light of the idea of autopoietic systems. It is argued that traditional systems categories, such as those used by Luhmann, must be refined to accommodate hybrid bodies of expertise, such as accounting. Second, it is argued that the "resonance" capacity, in Luhmann's sense, of accounting is not simply informational but may also be expressive and symbolic. From this point of view, technically problematic accounting flows may be highly effective in reproducing and "celebrating" a certain form of organizational rationality. Third, proposals for a "greening" of accounting are considered in the light of these general reflections. It is argued that, enthusiasm for experimentation notwithstanding, such proposals often inherit and embody the difficulties of orthodox accounting. The fundamental issue for green accounting is to transform organizational rationality and, by implication, organization boundaries. Fourth, a tentative proposal is made for a form of green accounting which emphasizes mechanisms for re-internalizing enterprise externalities and which may establish new symbolic linkages between organizations and their environments. Overall, the chapter suggests that these *symbolic* connections must be established before more detailed and technical refinements of information flows can be anchored. Emerging forms of green accounting may be over-zealous in their pursuit of a certain technical credibility at the expense of attempting to shift the normative domain of visibilities which constitute organizational boundaries.

ACCOUNTING, AUTOPOIESIS AND EXPERTISE

Accounting is difficult to "allocate" in Luhmann's systems-theoretic terms; as the "language of business", its calculative rationality seems to belong to the economy. From this point of view accounting is the formal code of the economic system and the binary code of payment/non-payment is further embodied in double-entry bookkeeping. But as a technology of social control it appears as an extension of law and the goals of regulation. Here the binary code of legal/illegal determines methods of disclosure which are policed, at least in theory, by auditors. Because of this, the systems purist might argue that accounting is really two subsystems which, in strictly autopoietic terms, cannot "interact". But given the cultural variability of its relation to legal and economic institutions (Napier & Noke, 1992; Nobes & Parker, 1991: Chs. 1 and 2; Miller & Power, 1992), it could also be asked whether accounting is a "system" in its own right, a body of functionally differentiated administrative knowledge with its own autopoietic implications (Robb, 1991). To answer this question we need to consider the nature of Luhmann's systems theory in more detail.

In the context of social theory the concept of autopoiesis provides a new emphasis on modes of organizational autonomy and self-reproduction. However, while it provides a critique of certain (naive) beliefs in regulatory control and interaction, it also raises the spectre of the impossibility of control. The enduring puzzle for systems theory informed by the concept of autopoiesis is that, in formal terms, system "interaction" never happens, binary codings are mutually exclusive, and the problem of regulatory control becomes the sceptical task of "stimulating" modes of self-organization (Kennealy, 1988). There are two ways out of this puzzle. One way is to allocate problems of interaction and of autonomy to different theoretical levels[3]. The second way is to "deconstruct" the puzzle by accepting the possibility of interaction and then determining the implications of this for the theory. This deconstruction has three steps.

First, we accept the possibility of interaction between systems such as the economy and law in some rudimentary sense. Second, we reaffirm the claim that interaction is only internal in order to retain the concept of autopoiesis. Third, it follows that our system categories must be adapted in order to describe the "new" system in which these interactive arrangements occur. In other words, the puzzle of systems interaction can be turned on its head to suggest that the real problem is that of system specification. This implies that the categories of law, science and economy are too crude to characterize the reality of institutional differentiation and interaction. Interaction between these grand systemic configurations occurs because of important hybrid subsystems and practices which have been created out of both. Of course, the more one is reliant upon finding new subsystems (like new Ptolomaic epicycles) to save the insights of autopoiesis, the more one must question whether the systems-theoretic vocabulary can bear the theoretical weight that is placed upon it.

Without doubt Luhmann is correct to argue that we need to know *how* "society structures its capacity for processing environmental information" before we can address the question of improvements. But these structures are not always readily reducible to the categories of law, economics, politics and science. Luhmann is perhaps a social theorist who cannot fully digest classical German preoccupations with the cultural categories of science, law and art. The problem of the possibility of interaction between such broad system categories naturally arises. But perhaps it is our understanding of these "windowless monads" which must change rather than our belief in the possibility of interaction. From this point of view, it is necessary to replace these grand system categories with more nuanced conceptions of institutional configurations of *expertise*. Kant argued in his *Metaphysical Foundations of Natural Science* that if current definitions of matter make (gravitational) action at a distance problematic, then it is the concept of matter which must change to accommodate the *possibility* of the phenomenon whose *actuality* we know. Consequently, he effectively *redefined* matter as being where it interacts. The lesson to be learned from Kant's conceptual shift is that, without giving up on the insights of autopoiesis, we still need more refined descriptions of the "systems" which enable interaction. One dimension of this descriptive refinement involves closer attention to "hybrid" knowledge systems, such as accounting.

Understood as an *a priori* claim, the theory of autopoiesis seems to be "anti-realist" about the natural environment and the latter is simply the systems-theoretic analogue of Kant's thing-in-itself which is necessarily posited by social systems. The idea is that we cannot adopt a viewpoint which is outside the opposition between system and an environment of which we have no unmediated knowledge. Not surprisingly, this claim has been interpreted as defeatist and conservative by Luhmann's ecological critics. However, this pessimistic reading of the pure core of Luhmann's position is a mistake. At this abstract level he is arguing very simply that openness, as a property of information systems, is always relative to processes of closure and transformation by the elements of that system. Knowledge of the environment is always selective and relative to systems based knowledge. This is one of the fundamental insights of autopoiesis; that there is no "pure" knowledge of environment. The latter is known and posited from *within* interactive system elements:

> Information is thus a purely system internal quality. There is no transference of information from the environment into the system ... system boundaries have to be drawn so that the world acquires the possibility of observing itself. Otherwise there would be pure facticity alone (Luhmann, 1989b: 18).

Luhmann's quasi-Hegelian argument also shares a structure with philosophical debates about realism and theory dependent observation. But if there is no theory neutral observation language, this does not mean that we are committed to a *particular* theory. Similarly, even if "natural" environmental knowledge is necessarily autopoietic this does not preclude the possibility of shifting the terms of that autopoiesis. Thus Luhmann's theory becomes *empirically* interesting at this very point when we consider the different modes in which autonomy as "self-reproduction" occurs and, more generally, why system differentiation occurs in one way rather than another.

Before pursuing these empirical issues in more detail it is necessary to clarify the theoretical point just a little further. It can be argued that Luhmann's various "binary codes" are "quasi-transcendental" conditions of the possibility of certain practices of discourse or communication. The binary code specifies a formal structure which governs the validity of any communicative claim to knowledge. Unlike Habermas's communicative theory, this process of "validity redemption" is entirely internal to specifically differentiated system structures rather than broader lifeworld constructs[4]. But at the level of a *formal* code structure the theory is institutionally indeterminate. Indeed there seems to be no *a priori* reason why, for example, legal communication constituted by the legal/illegal code, is necessarily confined to recognizably legal institutions[5].

In this sense binary codings are *portable*, an attribute which has important theoretical implications for questions of regulatory control. At the level of the code, system autopoiesis is impregnable but perhaps too anaemic in institutional terms. More important are the bodies of expertise through which the code(s) are instantiated. Hence, a "governing expertise" perspective (Miller & Power, 1992) can be contrasted with that of "autopoiesis" (Teubner, 1990) in terms of differing emphases upon system interdependence, susceptibility to expert invasion, and autonomy. Expert practice is often constituted at the intersection between binary codes (true/false; right/wrong; legal/illegal). Indeed, bodies of expertise simultaneously comprise a knowledge claim and an authority claim which often *demands* ambiguity in the precise instantiation of codes. In other words, expertise allows the binary code of legitimate/not legitimate to be conducted in the form of the true/false code. Accounting expertise can be regarded as a "hybrid subsystem" in which different formal binary codings are engaged *simultaneously* within the code of resource/claimant (debit/credit).

If binary codes are "portable" then expert practices may represent "sites" of competition either for the pre-eminence of a particular code (economic versus legal) and its institutional counterpart (market versus state) or for control over interpretation of concepts which could be anchored in different institutional ways[6]. Binary codes do not simply describe ready-made configurations of expertise but are fluid in relation to their possible institutionalizations. They sustain the rhetorical basis by which the competition for control of particular practices is conducted. Furthermore, we can restate the problem of system interaction and regulatory intervention in such a way as to retain the autopoietic emphasis upon the autonomy of system elements but which understands this in relation to very specific bodies of expertise which do not always, or automatically, fit into one system category. The problem of systems "resonance" in Luhmann's sense can now be expressed more in terms of *institutional embodiments of different codings by experts which give rise to different communicative potentials and restrictions* and less in terms of an impersonal systems-theoretic logic. With this in mind we must now consider accounting in greater detail.

ACCOUNTING IN ORGANIZATIONAL CONTEXT

The concept of autopoiesis is relevant to the regulatory control of organizations because it emphasizes the elements of self-reproduction which constitute system autonomy. This perspective requires a shift in the theoretical centre of gravity of policy-thinking; from planning to evolution and from control to autonomy. The autopoietic point of view expresses a principle of unity within which system and environment are moments. Hence it reconciles and contains traditional oppositions between "market" and "hierarchy"; the economic and the political; conceptions of the firm as a contractual nexus and richer conceptions of organizational action. In placing autopoiesis beyond attempts to reduce one side of the opposition to the other, they are reconfigured as environment and system, the respective elements of which in organizational contexts are monetary transactions and decisions:

> autopoiesis theory regards organization as a process of the internal differentiation of the economic system into an organized and a spontaneous area (Luhmann, quoted in Teubner, 1990: 75).

On this view organizations are autopoietic systems whose basic elements "are not payments but decisions". Rather than the firm as a nexus of contracts,

contracts are always the environment of the organization as system. This reworking of the relation between market and hierarchy, traditionally regarded as being opposites (Williamson, 1975), has important implications for accounting. It has already been argued that it is difficult to allocate accounting to a particular system in the sense of the economy or law. But the important point is that accounting is not an *economic* medium as such but is rather the medium *between the economic environment and the organization* in Teubner's (1990) sense. Accordingly, institutional variations in the form of accounting reflect and determine different patterns of market activity and organizational arrangements. For example, differences between capital market institutions in the UK and Germany underly the respective differences in their financial reporting regimes.

Accounting is the filter for variable forms of environmental uncertainty and translates "spontaneous" disturbances into the "manageable" for decision making purposes[7]. Meyer (1986) has argued that where organization is strong and hierarchies confront stable environments then accounting will be less necessary than where market uncertainty is high. Teubner (1990) makes much the same point by contrasting market orientation as receptivity to "variety" with hierarchy as "redundancy" in the sense of a "structural restriction of decision making contexts" which concentrates power within the organization. The implication is that strong hierarchies do not require sophisticated modes of accounting because they are confident of their boundaries. Thus, accounting is variably implicated in constituting bureaucratic forms of control *within* organizations. In contrast to the "regulatory trilemma", which is usually framed in terms of the prospects for regulatory instruments which are *external* to the domain to be regulated, accounting is often *internal* and constitutive of the informational reality of the target domain[8]. This invests accounting with a special potential as a regulatory device. It already has an organizational presence which shapes the form of system (organizational) autonomy and which filters environmental disturbance according to the rules of its communicational elements, thereby promoting a certain form of "organizational rationality" (Montagna, 1990).

Recent socio-legal debates about the nature and possibility of legal intervention in the corporate enterprise (Rehbinder, 1991) have explored the possibility of "reflexive" legal instruments which, in contrast to hierarchical regulatory styles, stimulate processes of self-organization[9]. But as Stone (1986) argues, even a "reflexively" conceived law presupposes the possibility of some effective *internal* administrative procedures and requires considerable social knowledge of its own. This regulatory dependence reflects the more general problem of "liberal" modes of government (Miller & Rose, 1990) which seek to enact change "at a distance" from the "target" organization and thereby "intervene" by stimulating processes of self-organization[10]. According to Miller and Rose, the retreat from a certain direct and hierarchical kind of intervention has generated a new role for technologies such as accounting which can be readily internalized by regulatees. While this gives accounting its distinctive regulatory potential, the spectre of the "regulatory trilemma" (Teubner, 1987) remains. Accounting interventions, such as the imposition of green disclosures, always run the risk of being ineffective (decoupled), over-imperialistic (colonizing) or captured (colonized) in particular organizational settings.

The regulatory mission which attempts to act on one component of a system necessarily embodies assumptions about how that component functions in the system as a whole. For example, financial regulation presupposes a knowledge of how accounting functions in the organizational decision-making process. However, accounting does not always function in these contexts as it is "ideally" expected to do so. Burchell *et al.* (1980) argue that accounting will only function in its ideal sense in *uncontested* routine contexts, which are likely to be rare. Elsewhere it may function, not so much to facilitate decision making as to "rationalize" decisions which have already taken place, that is to ceremonialize organizational actions within an accounting model and to reaffirm hierarchical structures, rather than to provide "information". The informational ideal of accounting may, therefore, function as an important institutional "myth" which permits regulating and regulated organizations to be linked and to become, at a certain programmatic level, "isomorphic" (Meyer & Scott, 1992). In this way administrative technologies such as accounting and audit are not simply technical artifacts but represent a powerful "symbolic" resource, and a distinctive regulatory "style" (Hood, 1991) for organizations which may need to legitimate activities to constitu-encies of insiders and outsiders.

On this view, financial reporting may function more as the ideological embodiment of economic rationality (Montagna, 1990) and less to provide unambiguous information to users in, say, the capital markets. This *expressive* dimension of accounting, which may appear to be a defect from a narrow managerial decision perspective, has importance in the context of "environmental" reporting proposals[11]. Accounting provides selective but constitutively significant representations of the corporate entity. The detailed language of assets, liabilities, costs and profits provides not merely a range of corporate imagery but also a vocabulary which is important for the self-reproduction of organizational rationality, at least in the USA and UK if not in continental Europe. As Armstrong (1987) and Espeland and Hirsch (1990) argue, accounting functions to represent the corporation as a *financial* rather than as a productive entity, a process which is partly in retreat as a new "politics of the product" refocuses organizational priorities (Miller & O'Leary, 1993).

Accounting functions as a common "language" which links together, and gives expression to, varied interests. In this sense it could be said to have a communicative role and potential which is wider than concerns with "efficient" decision-making. It is this, rather than any explicit "decision usefulness", which gives accounting its power and, in the UK and USA at least, which gives the accountant a prominent position in the organizational hierarchy. Accounting bears a complex relation to organizational behaviour because it provides the categories through which participants perceive both themselves and the organization. Far from being a merely technical and neutral informational medium, it has constitutive implications for the organization and the individuals within it by virtue of having a "reflexive" relation to the organization itself. Patterns of informational relevance which link organizational participants are simultaneously constructions of actors who must perceive themselves as "calculable" and "governable" selves within the administrative structure (Hopwood, 1987; Miller & O'Leary, 1987).

Accounting is therefore as much, if not more, a *social* technology as an economic one. Against a subsumption of accounting within the "economics of information", it is accounting, as a social epistemology, which provides the conditions of possibility for actions which can be described as "economic". In system-theoretic terms accounting functions to define the operational limits of the enterprise and provides a technology which reinforces very particular conceptions of organizational autonomy and receptivity to environmental disturbance. Indeed, the entity assumptions which drive accounting can be decoupled from legal conceptions of organizational form; accounting may be adapted to define the boundary of the cost centre, the subsidiary, the branch, the partnership, the franchise and even the group. As every student knows, accounting entities are not necessarily legal entities and while lawyers may agonize about legal status accounting has created its own indifference to such concerns.

To the extent that accounting *creates* categories of visibility as the basis for organizational action then it seems to fit Luhmann's description as a technology of the "self-knowledge of systems". Accounting is a technology of selective "openness" in so far as it determines and expresses particular structures of receptivity to environmental disturbance[12]. Its status as an information or communication system, which links and creates the conditions under which broader system configurations can interact, is both a window of hope for new conceptions of corporate control and also a potential for a new "iron cage" of technocratic reason. This ambivalence in the role of accounting arises from its complex relation to informedness. For example, accounting disclosures which come to be accepted as institutionally legitimate may have the consequence that the organization is not subject to critical scrutiny. Strategies of "openness" may, therefore, function as much to deter as to invite enquiry, an important point in the context of green accounting initiatives.

If our general conception of accounting knowledge must be expanded beyond the narrowly instrumental to embody the symbolic and expressive, it must be repeated that the relation between the latter and organizational action is contingent. We cannot be sure of the precise behavioural impacts of shifts in accounting policy. However, a more modest and realistic possibility suggests itself. Changes in accounting, rather than giving rise to detailed regimes of decision making, may nevertheless shift the terms of organizational discourse and from this point of view the accounting "entity" can be regarded as a flexible horizon of cognitive possibility. Stone (1986) argues that "responsible" behaviour, however defined, is in part a cognitive issue. Applied to organizations this suggests that one possibility for "reflexive" intervention is the introduction of new informational categories via accounting technology. There is no guarantee that desired behaviours will ensue but, given the constitutive role played by accounting, it may be possible to shift perceptions of organizational boundaries by creating a new moral vocabulary and thereby a new configuration of the managed and the spontaneous and of the private and public domains (Deetz, 1992).

ACCOUNTING AND ENVIRONMENTAL RESONANCE

While experiments in "green" accounting are by no means only a recent phenomenon, there has lately been an intensification of interest (for example, see Pearce et al., 1989; Gray, 1990a; 1990b; 1991; 1992; 1993)[13]. However, the considerable "enthusiasm" in the UK and other countries (Carey, 1990; Owen, 1992) is not matched by a consensus about the shape of green accounting practice. There are undoubtedly differences in the values and motivational structures which could orientate adaptations of traditional forms of accounting. For example, between programmes for enhancing corporate "environmental" disclosures within the existing corporate report, and a more radical abandonment of financial reporting in favour of different information flows, there are many possibilities. A crude, but nevertheless useful, distinction can be drawn between those initiatives which are based around the corporate annual report in its present form and those which are not. This distinction is reflected in the highly simplified structure of Figure 16.1 below.

On the left-hand side of Figure 16.1 we can locate green reporting initiatives which attempt to work through the annual report and experimentation with voluntary financial and non-financial disclosures has been extensive (see, for example, Roberts, 1992; Gray, 1993: Ch. 11). Underlying these efforts is the questionable assumption that *mainstream* financial reporting is sufficiently understood for its "green" potential to be readily evident. In other words, a particular model of the informational benefits of normal disclosures to users of financial statements is presupposed. More radical critics (e.g. Medawar, 1976) of corporate activity stress the virtues of disclosures not for information purposes but for greater transparency and publicity of corporate activity. From this point of view the "greening of accountancy" is allied to certain democratic aspirations for the transformation of working life. Current trends for voluntary environmental disclosure in the accounts of European companies are merely cosmetic—extensions of corporate public relations activity which

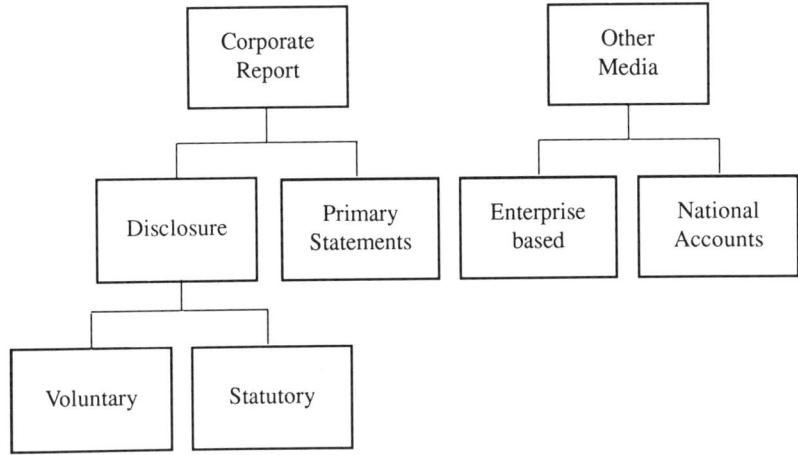

Figure 16.1   The structure of green accounting initiatives.

may be "decoupled" from the mainstream internal information flows within the corporation.

Statutory environmental disclosures in the annual report are currently very limited (Canada, USA and Norway) and UN voluntary proposals seem to have a low level of compliance (see Gray, 1993: 216). As far as more integrated forms of green accounting within the primary statements (profit and loss account, balance sheet) of the corporate report are concerned, there have been some interesting experiments in attempting to cost the environmental effects of corporate activity. Gray (1993: 223) describes some of the Dutch initiatives, notably that of the BSO/Origin company in producing an environmental "value added" statement. In general, continental Europe has been the source of varied attempts to link traditional forms of financial reporting to environmental variables.

Many of these initiatives have a longer history (Dierkes & Preston, 1977; Ullman, 1976) than is readily apparent and shade into the area of Figure 16.1 where forms of green accounting can be differentiated from the primary statements in the corporate report. These forms vary between those which remain tied to the enterprise as the primary accounting "entity" and those which operate at different levels of aggregation, such as the national accounts. Of particular note are the proposals arising out of the EC "Eco-Management and Audit Scheme" which became a Regulation in March 1993. Accounting for external purposes takes the form of an "environmental statement" which may include, among other matters, a summary of figures on pollution emissions, waste production, consumption of raw material, energy and water, and noise. This environmental statement is required to be audited by an "accredited environmental verifier". The scheme applies to "sites" rather than the corporation as a whole and is presently voluntary. It remains to be seen whether the statement will be routinely incorporated into the annual report and how it may interact, if at all, with the financial statements.

Gray (1990a; 1993) occupies a middle position between two extremes of cosmetic adaptation and radical rejection of mainstream financial accounting. As Hajer (1991) has noted, the idea of "sustainable development" has been articulated primarily at the level of a new political orientation. Gray attempts to develop this concept as an adaptation of existing concerns, both practical and theoretical, about "capital maintenance". Capital maintenance is relative to the interests of a particular class of claimants; if the claimant is defined as future generations then the concept of capital itself must change. Borrowing from Pearce *et al.* (1989) Gray differentiates man-made capital from natural capital. The latter is further subdivided into "core" or irreplaceable capital (e.g. the ozone layer) and renewable capital. This sustainable cost approach requires that the profit and loss account be charged with notional reductions in sustainable capital. In promoting the experimental use of notional costs Gray's proposal (1993: 294) suggests that the operational or technical ambiguities of these new concepts of capital, and the problems of costing them "accurately", are less important than their potential in redefining the action horizon of the enterprise.

Luhmann's concept of "resonance" has a bearing upon all these accounting initiatives. The natural environment is always "known selectively" via mechanisms which translate uncertainty into certainty and Luhmann argues for the need to "heighten resonance capacity". This is fundamentally a question of information systems which construct the difference between system and environment internally. Thus, from any system-based perspective we need to understand the particular properties of system resonance and must examine each functional system according to its own "specific resonance capacity", i.e. the particular categories of visibility which construct the universe of "facts" for the system. For Luhmann this implies a reflexive approach which enables the internal observation of systems which are already self-observing. While he does not articulate explicitly the technologies which would enable such self-observation, audit and accounting are clearly distinctive possibilities. However, the concept of autopoiesis does not provide a basis for adjudicating between the green accounting initiatives encompassed by Figure 16.1.

From Luhmann's point of view, we cannot appeal to ideals of "representational faithfulness" as a benchmark because we have no external criterion by which "faithfulness" could be judged. Luhmann also has very little to say on the subject of a "politics of truth" which could address the technocratic dimensions of the various accounting experiments. Benchmarks of environmental "performance" take shape within an interested policy arena (Nelkin, 1975; Douglas & Wildavsy, 1982; Ravetz, 1990) in which scientific concerns for the truth are by no means autonomous from politics. The language of systems differentiation, in which attempts at "de-differentiation" appear romantically naive, itself systematically overlooks the configurations of expertise which are implicated in the creation of "regimes of truth". However, it does not follow from the claim that there is no pure access to the environment that we cannot internally differentiate the systems of environmental attention represented in Figure 16.1. One possibility is to address the "immanent" potential of accounting systems to represent that which they do not (Choudhury, 1988). Another possibility is to consider the internal potentials for changing the character of corporate "personality", a subject which is close to lawyers' hearts (Stokes, 1986; Graham, 1989) and for which the constructive role of accounting is significant.

Teubner (1990) argues for the possibility of an *expressive* potential for a "corporate actor" over and above the "micro-corporatist producers coalition". In other words, the very notion of an impersonal corporate actor embodies a potential for a reconfiguration of the debits and credits which characterize the boundary of the organization, that is,

> the institutional strengthening of the corporate "actor" as the liberation of an impersonal action context which imposes effective constraints on the action of the individual participants (Teubner, 1990: 82).

Given the symbolic function of accounting discussed above, it can be suggested that new modes of accounting can contribute to the identity formation of this corporate actor not so much as the representation of external interests *within* the organization, but, more ambitiously, as the reconstruction of the interests *of* the organization. In this respect one can pass judgement on the various green accounting initiatives embodied in Figure 16.1 in terms of the extent to which the forms of accounting linkage to the natural environment challenge traditional assumptions about the relevant entity. It is to this that we now turn.

ACCOUNTING AND THE RE-INTERNALIZATION OF EXTERNAL EFFECTS

It has already been suggested that the boundary between system and environment is necessarily a knowledge system of its own. Accounting is an organizational "veil" or "membrane" through which environmental disturbances are communicated; systems-theoretic boundary imagery must dispense with ideas of impermeable barriers and absolute conceptions of autonomy (Nedelsky, 1990). But if it is theoretically plausible to model the function of accounting in this way, and to elaborate its potential to construct a moral vocabulary with its centre of gravity in a new corporate identity, it is nevertheless an empirical question as to whether any particular adaptation of accounting will have a desired outcome. Given a pessimism about "unintended effects" then we seem to be no further than traditional concerns with the efficacy of legal intervention. The argument that forms of "green accounting" may provide new categories of organizational visibility does not determine the precise form which information systems may take. Whichever form is chosen is always selective and new capacities for openness to the environment require new forms of closure. The sociology of risk perception and of the policy arena in which environmental risks are articulated (Douglas & Wildavsky, 1982; Ewald, 1990; Ravetz, 1990; Beck, 1992) also points to the social and cultural determinants of particular preoccupations about the environment[14]. All this must be borne in mind when considering the potential for a greening of accounting.

A further danger arises that new categories of organizational disclosure will simply give rise to offsetting innovations in "creative environmental accounting". As McBarnet and Whelan (1991) have shown in mainstream accounting contexts, new "rules" create new opportunities for "creative compliance" in which the spirit of those rules is circumvented. Hence, changes will fail at the level of establishing detailed accounting rules alone. At this level "creative accounting" and "non-compliance compliance" will thrive. The only practical method to avoid this is to address the constitutive processes by which accounting might contribute to the formation of a new corporate actor in Teubner's sense, not so much by particular rules but by promoting a new moral vocabulary and new categories of relevance and visibility. The existing "positivities" of accounting practice do not necessarily exist independently of that practice itself (Hines, 1988). Thus, *it is not new rules which are needed but new facts.*

In this respect the history of the "value added" concept in the UK provides an instructive model and emphasizes the institutional preconditions for any form of accounting. Burchell *et al.* (1985) argue that the technical ambiguity of "value added" accounting statements allowed them to express a potential for a new co-operative conception of the corporate enterprise, linking dissatisfaction with the existing forms of financial accounting, macro-economic concerns with national efficiency and productivity, and preoccupations with labour relations and collective bargaining mechanisms. Value added reports seemed to provide a vehicle to unite these disparate interests within a common programmatic ambition. Far from merely reflecting a historically contingent constellation of interests, value added accounting also represented a common vocabulary and new positivities of "co-operative" corporate life through which such interests could be articulated.

The "value added" experiment in UK financial reporting in the 1970s had a very limited life but there are some interesting lessons for the range of current attempts to create a "green" accounting. Such attempts will fail merely as technical innovations unless they can appear to represent and link wider constituencies of concern. For example, in the UK there are extensive programmes to promote the compatibility of environmentalism and efficiency. New languages of "waste" and "sustainability", albeit subordinated to existing organizational codes via the rubric of efficiency, may effect a new moral environment within the organization. Accordingly, accounting has a "colonizing" potential because of its status as a "symbolic" practice capable of appealing to diverse groups (Broadbent *et al.*, 1991; Power & Laughlin, 1992).

The value added event in the UK also suggests that it is not so much a question of creating new incentive structures for individual and collective behaviour but, more profoundly and problematically, of changing the common sense phenomenology of the organization; a new universe of facts rather than new behaviour in the existing universe of facts. From this point of view, concepts themselves are important technologies for change (for example, the "value for money" concept has had an important influence in the UK public sector) especially where these concepts are articulated through information systems. Indeed, as the UK "value added" statements have shown, technically ambiguous concepts can play an important symbolic role. One could push this point even further and suggest that "crude" forms of green accounting may have a legitimate role in challenging dominant organizational rationalities. The remainder of this chapter makes some tentative suggestions in this direction.

In the context of new conceptions of environmental liability, Teubner (Chapter 2) has argued for the "de-individualization" of environmental risks into appropriate collective "pools" which must then be re-individualized at the level of the enterprise. This is analogically suggestive of a form of what might be described as "risk pool" accounting. The important feature of such a proposal is that we do not account, initially at least, at the level of the enterprise. There is rather a new entity, a risk pool, for which a natural asset *Ökobilanz* is in principle possible. Risk pool accounts are prepared for each relevant unit and the flows of natural "stocks" for these units are recorded, quantified in money terms and allocated to the relevant enterprises which are members of the pool as a notional charge[15]. This charge could be displayed prominently in the profit and loss account as, for example, a deduction from bottom line profit and in the balance sheet as a "negative" asset, the enterprise share of the stock of polluted resource.

This proposal has parallel with that of Gray (1993: 293–4) and there are similar operational difficulties. For example, how do we determine the boundaries and membership of risk pools and how can natural asset flows be priced and equitably allocated? However, in terms of a *symbolic*, rather than an informational, strategy these are not difficulties which are insurmountable and it would be instructive to experiment with numbers known to be arbitrary. The significance of this proposal lies not so much in the allocation of particular numbers to particular enterprises but the *general principal of linkage* between accounting levels of aggregation appropriate to relevant ecosystems and those of enterprises. Rather than starting from enterprise accounting as it is and evolving new disclosures, we need to start from outside the enterprise and allocate "costs" back into it. To put it in Teubner's terms, we need a form of accounting which de-individualizes and then re-individualizes. While there have been numerous experiments in natural resource accounting (Pearce *et al.*, 1989: Ch. 4), the problem of re-internalization or linkage is critical. Figure 16.1 suggests a dichotomy between macro-economic and enterprise levels of accounting, although, as Gray (1993: 222) notes, there have been attempts to link corporate accounts to national accounts through environmental "satellite" accounts. These initiatives, in France and Norway, seem to have been abandoned on grounds of technical workability. The question is whether a principle of linkage can be retained despite the detailed technical difficulties of valuation and allocation.

Risk pool accounting in the manner described would not depend on getting costs "right" but on providing new internal visibilities of external effects in a document which embodies considerable "symbolic capital" for the enterprise—the corporate report. It could also indicate the potential role for accounting as a powerful "coupling" technology linking science and the economy. It is likely that corporate enterprises would object to "unfair" allocations of notional costs known to be crude even though the proposal as it stands has no cash implications. This is because it would nevertheless impose "symbolic" costs on organizations, charges against what Teubner (1991) has described as their "normative surplus value". This normative surplus could well become a normative deficit which one might also describe, following Habermas, as a legitimation crisis for the enterprise. Equally, risk pool accounting could be instrumental in constructing conceptions of new responsibilities in relation to broader accountability constructs. In this sense the difference between current innovations in environmental disclosure and risk pool accounting corresponds to that between new information for old motives and new information as a basis for new motives. The important point about risk pool accounting is that accounting flows must originate outside the enterprise and are then (re-)internalized by them. In this way the direction of information flow is at least as important as the *content*. The direction of information flow embodies conceptions of responsibility and in this way risk pool accounting would function as a moral technology.

Accounting embodies a potential to represent external interests within the organization. This is a role that it already plays, albeit presently in relation to the narrow interests of abstract shareholders and creditors. Risk pool accounting could function as an instrument of "reflexive intervention" by virtue of its internalization of hitherto invisible external effects. While the regulatory trilemma cannot be fully eradicated to leave guarantees of "effective" intervention, accounting, even in the form of soft voluntary disclosures, offers opportunities for instantiating a new moral vocabulary which other regulatory technologies do not possess. In this respect its calculative anonymity can be an advantage.

These speculations about a form of "risk pool" accounting may not be so far fetched given recent EC initiatives (Commission of the European Communities, 1992: 67) in which it is argued that there is a need for a "redefinition of accounting concepts, rules, conventions and methodology so as to ensure that the consumption and use of environmental resources are accounted for as part of the full costs of production". The Commission also calls for instruments to "internalize all environmental costs incurred". At its most fundamental the greening of accounting requires an expansion of the spatial and temporal horizons of the organization and experimentation with multiple "entity" concepts. In particular, forms of linkage are necessary which flow from right to left on Figure 16.1. Luhmann has noted that time has to build itself into the organizational system which must develop its own future/past perspectives and its own temporal urgencies. In this respect there are structural similarities between attempts to make accounting systems more "strategic" and attempts to make them green. Sustainability can provide an important "storyline" (Hajer, 1991) for accounting, a new underlying concept of organizational value, only if it can in some way provide a linkage between ecosystem and enterprise accounting. Like the "value added" statements of the 1970s, risk pool accounting could take the form of a "sustainable capital" statement from companies, which for all its technical crudity, would symbolize new organizational commitments for disparate interest groups.

Dangers of a "technocratic" nature lurk in these tentative proposals (Habermas, 1971). The calculative capacity of accounting to act across organizational space and time may generate risks not so much of capture but of "proceduralization" by a new breed of environmental experts and risk managers (Beck, 1992). Accounting is an embodied practice conducted by experts who can define the problems they face to suit the skills they possess, thereby assuring their position in a market for their services. Weber's problem of "expert vision" may, therefore, be reflected at a broader policy level in terms of "regulatory styles" which determine what is seen as constituting relevant knowledge for regulatory purposes (Ravetz, 1990). Accordingly, Luhmann's concern to move away from populist "obscure formulas of balance" must also confront the competition by experts for control of environmental discourse:

> Social Rationality would naturally require that the ecological difference of society and its external environment is reintroduced within society and used as the main difference. We have to begin from the idea that there can be no privileged place for this, no authoritative organization, consequently no constitution that could transform the ecological difference into binding guidelines for further information processing (Luhmann, 1988: 137).

While the widest possible plurality of environmental resonance capabilities may be desirable, Luhmann does not fully confront the difficulties in establishing appropriate institutional settings for this diversity. He resists the urge to "standardize" via "binding guidelines" but he does not address the dangers of a mobile industry in expert "mediation" which is capable of transforming and standardizing in its own image. Like accounting, even audit is far from being a neutral monitoring technology in this respect. By projecting a public image of being "boring" and routine, audit hides the sense in which it can actively transform organizations to render them auditable according to its expert-specific conceptions of risk (Power, 1991). Given the early experiences of the EC Eco-Management and Audit Regulation, it is likely that there could be no "risk pool" accounting in the sense described above without conferences, courses, texts, state agencies, accreditation procedures and a host of other administrative practices. Accordingly, environmental "resonance" in Luhmann's terms may tell us less about the dangers we "really" face and more about the "risk experts" we happen to trust.

CONCLUSIONS

Adorno and other members of the Frankfurt School recognized that even in the most corrupted practice, the most sentimental representation of the human condition, residues of our deepest hopes and longings are present. Only on the basis of these distorted fragments could their precarious critical project make sense as a transformation "from within" rather than as a set of externally imposed values from a "holy family of critics". Immanent critique is precisely that technique of questioning the existing universe of facts on the basis of residues (resonances) of other possibilities within them. Similarly, the project of green accounting that matters most is not simply the creation of new detailed disclosure requirements which can be "monitored" by new auditing experts, but the creation of an accounting with the potential to change the "factual universe" of the organization. New liabilities can help to initiate such changes in organizational reality by mediating external effects and thereby challenging internal information systems.

A rapidly expanding market for environmental services (Power, 1991; Gray & Collison, 1991) has led to realignments of, and claims to, expertise, particularly in the heterogeneous market for environmental audit. New coalitions of expertise in the form of consultancies have been generated around environmental initiatives such as the EC Eco-Management and Audit Scheme. These experts have also been actively engaged in the market for corporate liability advice. Such a reordering of practical knowledge implicates processes of organizational politics which can be overlooked if the environment is conceived as a merely technical problem. For example, one increasingly visible dimension of the new market for environmental experts created by existing and anticipated European legislation could be a repositioning of scientific expertise within corporate hierarchies. The once marginal "health and safety" officer may find a new organizational prominence. But for corporate cultures such as the UK, where the finance function generally dominates scientific and operational expertise, this could lead to considerable organizational conflict. Eco-Audit may have implications for internal corporate structures and the bodies of knowledge through which they are sustained. The urge to standardize, which is characteristic of a certain style of regulatory intervention, is also the road to expert "capture" by "specialists without vision" in the market for creative compliance.

It has also been tentatively suggested above that new conceptions of knowledge of the enterprise are necessary to avoid "vision without specialism". From this point of view the role of something like a "risk pool" accounting has important but ambivalent potentials. By virtue of its constitutive role within organizations and the manner in which it mediates a variety of system codes, accounting is a powerful technology for organizational change because it may already be a feature of important information flows within and outside organizations; it is already a technology of environmental, if not ecological, linkage. Hopes for a new corporate actor can be constructed through new categories of corporate information which reflect the public, external character of corporate action. Here there may be an important role for concepts such as "sustainability" which are at least as vague as "efficiency", "performance" and "value for money" but which can be instrumentally embodied in linkages between different aggregation levels for organizational activity. Accounting disclosure of, even notionally imposed, environmental costs may create new internal visibilities for external interests. In this manner accounting has a potential for "re-coupling" organizations to their natural environments[16].

NOTES

1 The author is grateful for the helpful comments of Rob Gray, Anthony Hopwood, Richard Laughlin, Peter Miller and Gunther Teubner.

2 Indeed, Ewald (1990) reminds us that the concept of risk itself belongs to very particular technologies, such as insurance, for the treatment and management of events.

3 This is suggested by Luhmann's (1989a: 139) discussion of the conceptualization of system dependence and independence.

4 According to Luhmann (1989a: 145), argumentation is "not a search for convincing rational grounds but ... a way of mastering contingency and as a condensation of the systemic context". But the difference between Habermas and Luhmann on this point may be less great when Habermas's transcendentalism is understood more contextually (see Power, 1993).

5 Luhmann (1989a: 142) seems to fall back upon an institutional definition of system boundaries which reflects a theoretical conservatism about the possible instantiations of the binary code legal/illegal.

6 Space does not permit an elaboration of this point but a discussion of the concept of "true and fair" in UK financial reporting would be a good illustration. There is considerable debate about whether the concept is "purely legal" or not, but semantic deliberations tend to give way to territorial disputes between accountants and lawyers about who is to be final arbiter of the matter. Binary code structures are insufficient to characterize the stakes in these debates.

7 According to Teubner, organization can never be exhausted by "transactional cost"

interpretations of it. Organizational "surplus value" always escapes the explanatory horizon of transactions cost theories which, at the point at which they abstract and refine themselves, are relegated to the "environment" of the very phenomenon which they are attempting to explain.

8 Against this, the external origins of internal accounting practices must also be emphasized. In this sense, accounting can be linked to organizational environments in ways which inhibit and constrain autonomy (Meyer, 1986).

9 Indeed there may be some irony that, as legal theorists are seeking a new and modest realism about the possibility of external intervention in the internal workings of organizations, accounting and organization theorists have been developing an increasing awareness of the *external* pressures upon *internal* organizational arrangements (See Meyer, 1986; Meyer & Scott, 1992) and their very varied possible effects (Laughlin, 1991).

10 Looking in from the outside at the legal intervention debates, it is interesting to see how German and American "sociologies" of law reveal common preoccupations despite differences of intellectual tradition. Thus, the problem of how to intervene in corporate behaviour links German concerns with the "effectiveness" of the welfare state and the "disorganization" of capitalism to American concerns which originate in the liberal-pragmatist need to reconcile appropriate forms of regulation to deeper ideals of freedom. At the level of administrative technology the two traditions share a "problem of control".

11 More generally, the expressive dimension of accounting is significant in the context of mobilizing organizational change (Dent, 1990).

12 Of course, accounting is only one particular strategy of "openness". Rights of access and inspection provide a very different basis for becoming "informed" about organizational activities.

13 Preoccupations with environmental accounting and disclosure have a much longer history, particularly in the USA where they have been part of a more general concern with social accounting (see Gray *et al.*, 1987), and the "accountability" of corporations which has grown and shifted its terms of reference over time (Benston, 1982a; 1982b; Ijiri, 1983; Roberts & Scapens, 1985).

14 It is worth contrasting the positions of Douglas and Wildavsky (1982) and Beck (1992) in terms of their respective "ontologies" of risk. For Beck risks are in some sense real and "out there"; the problem is to disentangle the social barriers to the proper articulation of these dangers. In addition, he argues that the risks we face today are a distinctive internal product of the industrial system. In contrast Douglas and Wildavsky focus on the cultural settings within which ideas of purity, danger and risk are expressed and contested. From their point of view there is nothing distinctive about the environmental issues which developed societies now face.

15 In double-entry terms this would require a debit to profit and loss for notional environmental degradation, offset by any compensating expenditures, and a credit to balance sheet as an environmental claim upon the resources of the enterprise.

16 It is important to appreciate the diverse forms of resistance to any such innovation in enterprise accounting. In the UK a recent private members' bill proposed that companies be forced to disclose in full all breaches of environmentally relevant legislation together with details of fines and programmes for remedial action. The response to this initiative included defensive arguments to the effect that such information is usually already publicly available in the registers of local authorities and hence that disclosure in the corporate report would not be of any further assistance. But if the arguments for risk pool accounting have any merit then it should be clear that the point is not simply one of "information" but of "symbolic capital". Such disclosures in the corporate report, even if replicated elsewhere, would mark a

significant change in the representation of the corporate actor. See "Green Tinge to Company Books", *Financial Times*, 15 January 1992.

REFERENCES

Armstrong, Peter (1987) The Rise of Accounting Controls in British Capitalist Enterprises. *Accounting, Oranizations and Society*, 12, 415.

Beck, Ulrich (1992) From Industrial Society to the Risk Society: Questions of Survival, Social Structure and Ecological Enlightenment. *Theory, Culture and Society*, 9, 97.

Benston, George J. (1982a) An Analysis of the Role of Accounting Standards for Enhancing Corporate Governance and Social Responsibility. *Journal of Accounting and Public Policy*, 1, 5.

Benston, George J. (1982b) Accounting and Corporate Accountability. *Accounting Organizations and Society*, 7, 87.

Broadbent, Jane *et al.* (1991) Recent Financial and Administrative Changes in the NHS: A Critical Theory Analysis. *Critical Perspectives on Accounting*, 2, 1.

Burchell, Stuart *et al.* (1980) The Role of Accounting in Organizations and Society. *Accounting, Organizations and Society*, 5, 5.

Burchell, Stuart, *et al.* (1985) Accounting in its Social Context: Towards a History of Value Added in the UK. *Accounting, Organizations and Society*, 10, 381.

Carey, Anthony (1990) The Environmental Challenge That Won't Go Away. *Accountancy Age*, 11 October.

Choudhury, Nandan (1988) The Seeking of Accounting where it is not: towards a theory of non-accounting in organizational settings. *Accounting, Organizations and Society*, 13, 549.

Commission of the European Communities (1992) *Towards Sustainability*. Brussels: DG X1, 27 March.

Deetz, Stanley A. (1992) *Democracy in an Age of Corporate Colonization*. New York: State University of New York Press.

Dent, Jeremy F. (1990) Strategy, Organization and Control: Some Possibilities for Accounting Research. *Accounting, Organizations and Society*, 15, 3.

Dierkes, Meinolf & Preston, Lee (1977) Corporate Social Accounting and Reporting for the Physical Environment: a Critical Review and Implementation Proposal. *Accounting, Organizations and Society*, 2, 3.

Douglas, Mary & Wildavsky, Aaron (1982) *Risk and Culture*. Los Angeles: University of California Press.

Espeland, Wendy N. & Hirsch, Paul M. (1990) Ownership Changes, Accounting Practice and the Redefinition of the Corporation. *Accounting, Organizations and Society*, 15, 77.

Ewald, François (1990) Norms, Discipline and the Law. *Representations*, 30, 138.

Freedman, Judith & Power, Michael (1991) Law and Accounting: Transition and Transformation. *The Modern Law Review*, 54, 769.

Gorz, André (1989) *Critique of Economic Reason*, translated by G. Handyside and C. Turner. London: Verso.

Graham, Cosmo (1989) Regulating the Company. In Hancher, L & Moran, M. (Eds). *Capitalism, Culture and Economic Regulation*. Oxford: Clarendon Press.

Gray, Rob (1990a) *The Greening of Accountancy: The Profession after Pearce*. London: ACCA.

Gray, Rob (1990b) The Accountant's Task as Friend to the Earth. *Accountancy*, June, 65.

Gray, Rob (1991) The Accountancy Profession and the Environmental Crisis. *University of Dundee, Discussion papers in Accountancy & Business Finance*, May.

Gray, Rob (1992) Accounting and Environmentalism: an Exploration of the Challenge of Gently Accounting for Accountability, Transparency and Sustainability. *Accounting, Organizations and Society*, **17**, 399.

Gray, Rob (1993) *Accounting for the Environment*. London: Paul Chapman Ltd.

Gray, Rob & Collison, David (1991) The Environmental Audit: Green Gauge or Whitewash? *Managerial Auditing Journal*, **6**, 5: 17.

Gray, Rob *et al.* (1987) *Corporate Social Reporting: Accounting and Accountability*. Englewood Cliffs: Prentice-Hall.

Habermas, Jürgen (1971) *Towards a Rational Society*, translated by J.J. Shapiro. London: Heinemann.

Hajer, Maarten (1991) Environmental Performance Review as an Instrument of Ecological Modernism. Paper presented at the *Environmental Performance Review*, Oslo.

Hines, Ruth D. (1988) Financial Accounting: In Communicating Reality We Construct Reality. *Accounting, Organizations and Society*, **13**, 251.

Hood, Christopher (1991) A Public Management for All Seasons. *Public Administration*, **69**, 3.

Hopwood, Anthony G. (1987) The Archaeology of Accounting Systems. *Accounting, Organizations and Society*, **12**, 207.

Ijiri, Yuji (1983) On the Accountability-based Conceptual Framework of Accounting. *Journal of Accounting and Public Policy*, 75.

Jensen, Michael C. and Meckling, William H. (1976) Theory of the Firm: Managerial Behaviour, Agency Costs and Ownership Structure. *Journal of Financial Economics*, 305.

Kennealy, Peter (1988) Talking about Autopoiesis—Order from Noise? In Teubner, G. (Ed). *Autopoetic Law: A New Approach to Law and Society*. Berlin: de Gruyter.

Laughlin, Richard C. (1991) Environmental Disturbances and Organizational Transitions and Transformations: Some Alternative Models. *Organization Studies*, **12**, 209.

Luhmann, Niklas (1989a) Law as Social System. *Northwestern University Law Review*, **83**, 136.

Luhmann, Niklas (1989b) *Ecological Communication*, translated by J. Bednarz. Cambridge: Polity Press.

McBarnet, Doreen & Whelan, Christopher (1991) The Elusive Spirit of the Law: Formalism and the Struggle for Legal Control. *The Modern Law Review*, **54**, 848.

Medawar, Charles (1976) The Social Audit—a Political View. *Accounting, Organizations and Society*, **1**, 389.

Meyer, John W. (1986) Social Environments and Organizational Accounting. *Accounting, Organizations and Society*, **11**, 345.

Meyer, John W. & Scott, W. Richard (1992) *Organizational Environments: Ritual and Rationality*, 2nd edn. London: Sage.

Miller, Peter & O'Leary, Ted (1987) Accounting and the Construction of the Governable Person. *Accounting, Organizations and Society*, **12**, 235.

Miller, Peter & O'Leary, Ted (1993) Accounting Expertise and the Politics of the Product: Economic Citizenship and Modes of Corporate Governance. *Accounting, Organizations and Society*, **18**, 187.

Miller, Peter & Power, Michael (1992) Accounting, Law and Economic Calculation. In Bromwich, M. & Hopwood, A.G. (Eds). *Accounting and the Law*. Hemel Hempstead: Prentice Hall/ICAEW.

Miller, Peter & Rose, Nikolas (1990) Governing Economic Life. *Economy and Society*, **19**, 1.

Mingers, John (1989) An Introduction to Autopoiesis—Implications and Applications. *Systems Practice*, **2**, 159.

Montagna, Paul (1990) Accounting Rationality and Financial Legitimation. In Zukin, S. & Dimaggio, P. (Eds). *Structures of Capital: The Social Organization of the Economy*. Cambridge: Cambridge University Press.

Nedelsky, Jennifer (1990) Law, Boundaries and the Bounded Self. *Representations*, No 30, 162.

Napier, Christopher & Noke, Christopher (1992) Accounting and Law: An Historical Overview of an Uneasy Relationship. In Bromwich, M. & Hopwood. A.G. (Eds). *Accounting and the Law*. Hemel Hempstead: Prentice Hall/ICAEW.

Nelkin, Dorothy (1975) The Political Impact of Technical Expertise. *Social Studies in Science*, **5**, 35.

Nobes, Christopher & Parker, Robert (Eds) (1991) *Comparative International Accounting*, 3rd edn. Hemel Hempstead: Prentice Hall.

Owen, Dave (Ed) (1992) *Green Reporting: The Challenge of the Nineties*. London: Chapman Hall.

Pearce, David *et al.* (1989) *Blueprint for a Green Economy*. London: Earthscan.

Power, Michael (1991) Auditing and Environmental Expertise: Between Protest and Professionalization. *Accounting, Auditing and Accountability Journal*, **4**, 3: 30.

Power, Michael (1992) After Calculation? Reflections on André Gorz' "Critique of Economic Reason". *Accounting, Organizations and Society*, **17**, 477.

Power, Michael (1993) Habermas and Transcendental Arguments. *Philosophy of the Social Sciences*, **23**, 26.

Power, Michael & Laughlin, Richard (1992) Critical Theory and Accounting. In Alvesson, M. & Willmott, H. (Eds). *Critical Management Studies*. London: Sage.

Ravetz, Jerry R. (1990) Risks and their Regulation. In Ravetz J.R. *The Merger of Knowledge with Power: Essays in Critical Science*. London: Mansell.

Rehbinder, Eckard (1991) Reflexive law and Practice: The Corporate Officer for Environmental Protection as an Example. In Febbrajo, A. & Teubner, G. (Eds). *State, Law, Economy as Autopoetic Systems: Regulation and Autonomy in a New Perspective*. Milan: Giuffrè.

Robb, Fenton (1991) Accounting—A Virtual Autopoietic System? *Systems Practice*, **4**, 215.

Roberts, Clare (1992) Environmental Disclosures in Corporate Annual Reports in Western Europe. In Owen, D. (Ed). *Green Reporting: The Challenge of the Nineties*. London: Chapman Hall.

Roberts, John & Scapens, Robert (1985) Accounting Systems and Systems of Accountability. *Accounting, Organizations and Society*, **10**, 443.

Sagoff, Mark (1988) *The Economy of the Earth*. Cambridge: Cambridge University Press.

Stokes, Mary (1986) Company Law and Legal Theory. In Twining, W. (Ed). *Legal Theory and Common Law*. Oxford: Basil Blackwell.

Stone, Christopher (1986) Corporate Social Responsibility: What it Might Mean, if it were Really to Matter. *Iowa Law Review*, **71**, 557.

Teubner, Gunther (Ed) (1987) *Juridification of Social Spheres: A Comparative Analysis in the Areas of Labour, Corporate Antitrust and Social Welfare Law*. Berlin: de Gruyter.

Teubner, Gunther (1990) Unitas Multiplex: Corporate Governance in Group Enterprises. In Sugarman, D. & Teubner, G. (Eds). *Regulating Corporate Groups in Europe*. Baden-Baden: Nomos.

Teubner, Gunther (1991) Autopoiesis and Steering: How Politics Profit from the

Normative Surplus of Capital. In In 't Veld, R.J. *et al.* (Eds). *Autopoiesis and Configuration Theory: New Approaches to Societal Steering*. Dordrecht: Kluwer Academic Publishers.

Ullman, Arieh (1976) The Corporate Environmental Accounting System: A Management Tool for Fighting Environmental Degradation. *Accounting, Organizations and Society*, **1**, 71.

Williamson, Oliver E. (1975) *Markets and Hierarchies: Analysis and Anti-Trust Implications*. New York: Free Press.

SECTION V

CONCLUSION

CHAPTER 17

Self-Regulation and the Theory of Institutions

Philip Selznick
Berkeley

I welcome the opportunity to prepare this conclusion, which enables me to present some general remarks on the themes developed, together with some added thoughts on the theory of institutions. I have been greatly impressed by the quality of the discussion which has shown how policy and theory can be fruitfully combined; and I have much appreciated the diversity of the views expressed, especially given what appears to be a largely shared commitment to the arcane language of "systems theory".

I was relieved to discover that some of my fears about "autopoiesis" as applied to the theory and practice of self-regulation turned out to be groundless. Some formulations of the theory, in emphasizing the "closure" and "reflexivity" of systems would lead one to think that a radical autonomy of systems is necessary and inevitable, and that the attempt to govern them by external regulation is doomed to failure. But this is not so. Rather, the perspective developed here, in common with others, argues a *prima facie* case for self-regulation, and the desirability of maximum feasible self-regulation. However, the quest for appropriate *external* regulation is not abandoned. On the contrary, the question is what form it should take and what its limits should be. At that point we return to what Max Weber called, in another context, "the slow boring of hard boards". The emphasis shifts from very general conceptual frameworks to more specific problems and their solution. The latter cannot be found without close attention to variability and contexts, that is, without examining the conditions that determine when self-regulation is necessary, desirable and feasible.

It is important to distinguish "systems" theory from theories of systems. The former is mainly heuristic and metatheoretical: a way of identifying and emphasizing certain abstract aspects of all self-regulating unities. The payoff

Environmental Law and Ecological Responsibility: The Concept and Practice of Ecological Self-Organization
Edited by G. Teubner, L. Farmer and D. Murphy
© 1994 John Wiley & Sons Ltd

(or gamble) is that fresh insights and better hypotheses will be developed. The postulates and nuances of systems theory are no substitute, however, for specific theories about particular systems or types of systems. The former may help in producing the latter, but it is the theory of the system that we really want. Only that will lead to informed decision; only that will take account of variability and context.

In intellectual life we often discover that philosophical or analytical frameworks are only loosely coupled to empirical conclusions. One can accept many aspects of Marxist thought without embracing dialectical materialism, and the same may be said of other systems of thought, many of which gather merciful dust on library shelves. The empirical conclusions stand on their own feet. They do not depend on the validity or fruitfulness of the framework. Indeed, the latter may be discarded when its work is done and when its limits are revealed. This is not to deny the worth of "general theory" in social science as a potential source of sensitivity and insight. Moreover, in such matters, I believe, each generation should be its own master. I wish only to remind you that heuristic standpoints are not self-justifying. Nor are they necessarily innocent. They may burden our minds even as they liberate.

The burden may take the form of tunnel vision. Although systems theory is very general, it tends to focus on technical limits and possibilities. The study of values, of moral outcomes, as they impinge on organization, and as they are affected by organization, is not in the foreground of autopoietic consciousness. Even autonomy and self-regulation are treated more as technical necessities than as precarious and variable ideals.

A different perspective is suggested by the theory of institutions. The latter is by no means incompatible with systems theory, and might even be restated in its terms. Before that is tried, however, we should take stock of what the theory of institutions has to say. Most important, I think, is the idea that institutions embody values. They are moral agents as well as rational systems; they are objects of moral concern as well as instruments for achieving personal and collective goals.

The theory of institutions touches closely on many issues of moral theory. We have to consider, for example, how the creation of institutions affects the meaning of "survival", "self-preservation", and "responsibility". And we ask what costs are entailed when we accept a commitment to *maximize* discrete, selected goals or outcomes. These and many other issues are brought into focus when we become sensitive to the moral significance of institution building. From the standpoint of the theme of self-regulation, the advantage is that we are compelled to ask: self-regulation for what? What values are at stake? How are they affected by the play of organizational imperatives? By the needs and dynamics of systems?

I think we can agree that there are two central problems. These are (a) the nature of the corporate actor, and (b) given that nature, how to make self-regulation effective. In what follows I shall indicate, very briefly, how the theory of institutions deals with these problems[1].

When we think of the enterprise "as an institution" we have in mind, to begin with, the transformation of a *formal* system of organization into an *operative* system. The latter includes all the groups, connections and patterns of decision that have arisen in the course of adaptation and growth. The master process is institutionalization, which I have defined as "the emergence of orderly, stable *socially integrating* patterns out of unstable, loosely organized, or narrowly technical activities". The underlying reality, the basic source of stability and integration, is the creation of social entanglements or commitments. The latter lead to vulnerabilities as well as strengths. Thus a "pure" organization is a special-purpose tool, a system of consciously co-ordinated activities. An institution, on the other hand, is a complex product of social adaptation, a result of multiple and converging interests and forces.

In the transformation of organizations into institutions we may discern a two-step process. The first is foundational and formal. The act of association is itself a quest for "institutional" solutions to problems of economy and co-ordination. Instead of relying on spontaneous interaction or on markets and contracts, a need (or advantage) is perceived for authority and discipline. The formal structure of the organization is a first long step toward institutionaliza-tion. Explicit goals and rules; a chain of command; channels of communi-cation; these *designed* modes of social integration overcome the looseness, instability, and limited rationality of *ad hoc* or contractual arrangements.

But this is only the beginning. Beyond lies what we may call "thick" institu-tionalization, which takes place in many different ways: by sanctifying or otherwise hardening rules and procedures; by establishing strongly differen-tiated organizational units, which then develop vested interests and become centres of power; by creating administrative rituals, symbols and ideologies; by intensifying "purposiveness", that is commitment to unifying objectives; and by embedding the organization in a social environment. On the one hand, these and other institutionalizing forces tend to stabilize expectations, conduct and belief. They create a distinctive, and more or less integrated social reality. On the other hand, each is a potential source of distraction and incoherence. If purposes are to be achieved and values realized, the course of instutionaliza-tion must be monitored and controlled—and this must be done in institutional-ized ways.

As applied to the corporate actor, institutionalization helps make sense of the idea of a corporate conscience. According to classical free market doctrine, no more should be expected of a business, even a large business, than that it operate within the law, and that the law should protect the autonomy and rationality of the enterprise. After all, it is argued, a firm is the very model of instrumental rationality. However, the social costs of moral indifference (dis-torted priorities, defrauded consumers, degraded environments, deformed babies) have created an irrepressible demand for enhanced accountability, more external regulation and a stronger sense of social responsibility. In response to this call for a corporate conscience, theories of the corporation have taken an institutional turn. The enterprise has come to be perceived as infused with value and thickened by commitment.

In recent years there has been a growing understanding that if moral competence is to be meaningful, it must be built into the social structure of the enterprise. Thus understood, the corporate conscience is not elusive or indescribable, nor is it a mainly psychological phenomenon. A corporate conscience consists of specific arrangements for making accountability an integral part of corporate decision making.

The main strategies are institutionalization and the concomitant creation of a corporate culture. Fairness in industrial discipline; affirmative action to overcome discrimination and its effects; environmental protection; quality control; occupational safety; these and other aspirations cannot be achieved through rhetoric or a grudging conformity to external rules. They usually require specialized units capable of determining policy, monitoring practices, and establishing appropriate procedures.

Here the benign face of bureaucracy appears. Bureaucratic hierarchy is more than a formal chain of command. It is an arrangement of parts that have an integrity and continuity of their own. These parts are, in effect, internal interest groups. As sources of energy and centres of competence they may either subvert the enterprise or lend it life and strength. They serve the ends of policy insofar as they are effective guardians of particular standards or objectives. In this way values and policies are anchored in the group structure of the enterprise; in this way the corporate conscience is built up and made good.

In this quest for moral competence, which includes the capacity for self-regulation, the enterprise faces conflicting demands for openness and closure. We usually associate institutionalization with closures. As Mary Douglas put it, institutions "fix processes that are essentially dynamic" (1986: 92). Indeed institution building and culture creation are necessarily stabilizing and integrating forces. Nevertheless, institutionalization can take place in the service of flexibility and openness, for example when procedures for consultation and learning are established and reinforced. This suggests that closure *versus* openness is not a helpful dichotomy. We need *regularized forms of openness*, within a particular system and in the relation between the system and its environment.

Openness looks inward as well as outward. Internally it requires policies and strategies that encourage initiative and commitment; that cut across bureaucratically defined jurisdictions; and that take account of fragile incentives, multiple interests, and the dynamics of co-operation and conflict. In this post-bureaucratic model, communication is less firmly channelled, authority is more diffuse, consultation and participation are more strongly encouraged.

All this takes place within, not outside, a system of order. At stake is the kind of order—not order itself.

Externally, openness strains towards a model of *responsive* administration. The challenge is to maintain institutional integrity while dealing constructively with new problems, new forces in the environment, new demands and expectations. Responsiveness is often wrongly identified with uncontrolled adaptation and capitulation to pressure. But that confuses responsiveness with drift or opportunism—the pursuit of short-term advantages without effective restraint by principle or ends. If an institution is too weak or inept to defend its integrity, we should call it opportunistic rather than responsive. A responsive institution avoids insularity without embracing opportunism.

Thus organizations are open systems, as is often noted, but they are closed as well. The dialectic of openness and closure is the source of many problems of leadership, including how to manage the *blurring of boundaries*. Blurred boundaries are a challenge to institutional integrity, but they may also be essential aspects of institutional life. A school may need to be a good parent, especially where parents are weak or under stress; parents need to be good teachers as well. The blurred boundary is a source of strength for both institutions. It also demands sensitive leadership and institutional design.

This approach is, of course, very different from one that insists, as Luhmann does, that institutional autonomy (or better, integrity) requires a sharp differentiation of system attributes. I argued long ago that the sociology of law must attend to the "distinctively legal", that is, to what is special about legal authority in contrast to other claims to (or sources of) deference and obedience. It does not follow, however, that the administration of justice must be wholly rule-centred or that there cannot be blurred boundaries, in the interests of the integrity of the legal decision itself, between law and economics, law and education, law and psychology, law and politics, law and morality and so on[2].

The perception of blurred boundaries is an important part of the sociological, and the moral perspective. Sociological interpretation is not comfortable with clear lines drawn between one human activity and another. Every practice and every institution is seen as "in society" fatefully conditioned by larger contexts of culture and social organization. Sociology looks to the continuities of life to show how things fit together and are interdependent, and it finds in these continuities the primordial sources of obligation and responsibility. This does not deny the reality of conflict or the importance of autonomy for technical competence and for the moral integrity of specialized institutions. It does say that without attention to fundamental continuities, conflicts are exacerbated and autonomy degenerates into perverse and self-defeating isolation.

Blurred boundaries and overlapping functions are the natural if troublesome offspring of moral decision. Concern for whole persons or for the comprehensive well-being of a group or community must take account of multiple values, which may be competing or complementary. In either case, the moral outcome is not likely to be unidimensional or single-minded. On the contrary, responsibility strains towards the creation of larger unities; towards extension, not restriction; toward sharing, not exclusiveness.

At stake in the quest for institutional integrity, and of special relevance to the question of self regulation, is the creation of *internal moralities*. An internal morality is the set of standards that must be honoured if the distinctive mission of an institution or practice is to be achieved. Thus, the internal morality of legislation, as described by Fuller (1969) requires generality, publicity, intelligibility, and constancy. The internal morality of adjudication includes impartiality and the opportunity for each party to offer proofs and arguments. The internal morality of family life includes trust and shared commitment. The internal morality of bureaucracy includes fidelity to assigned responsibilities and accountability to the institution and its sponsors.

To identify an internal morality (to know what integrity requires) we need a theory of the institution. What counts as integrity and what affects integrity will be different for a research university and a liberal arts college; for a constitutional court and a lower court; for a regulatory agency and a highway department. Each institution, or type of institution, has special functions and values; each has a distinctive set of unifying principles. When an institution is charged with lack of integrity, the charge depends on an implicit conception of what the institution is or should be.

The significance of the internal morality for regulation is not hard to find. If an organization has a well-developed internal morality (driven by the quest for excellence and sustained by the interplay of means and ends) the community's strategy may well shift from *external* to *internal* control. Instead of demanding conformity to standards imposed by legislation and regulation, we may place greater reliance on moral development. In this way, the internal morality of an institution becomes a resource for public policy. If a practice or institution has an effective internal morality we may grant it, and its efficacy may require, broad powers of self-regulation.

We should recognize, however, that some internal moralities are more definite or more effective than others. Most families can be relied on to care for their children without external interference or goading. An internal morality is sustained by parental love and by feelings of solidarity among close relatives. The internal morality of business arises from the demands of rationality, including the utility of good relations among employees, customers and suppliers. However, impoverished families and marginal firms are likely to have weak internal moralities. Thus, the ideal of self-regulation depends on an underlying reality. Deference is unjustified if the institution in question has little or no potential for self regulation.

Ideally, the responsible enterprise is *self*-created, not constructed by putting in place units or procedures mandated by government. This autopoietic ideal, which applies to persons as well as institutions, establishes a presumption and a policy preference. It does not follow, however, that mandated changes are necessarily counter-productive, that they inevitably subvert the integrity of the system, or that they cannot be based on a sensitive awareness of distinctive institutional logics.

The ideal we seek is responsive regulation. As described by Bardach and Kagan (1982: Ch. 5), for example, the "good inspector" tries to limit pollution or enhance occupational safety by techniques of dialogue, diagnosis, and institutional design. Instead of "going by the book", with its emphasis on fixed rules, violations and fines, responsive regulators take specific circumstances into account. They participate with business and other institutions in a co-coperative effort to make law effective. This may involve forbearance to mitigate unreasonableness, or technical consultation to show how improvements may be made without undue cost, such as by bringing the experience of other firms to bear (Rees, 1988). Thus, responsive regulation is more problem than rule-centred, and it tries to bring about the maximum feasible self-regulation.

Responsive regulation is not the same as deregulation. Regulatory agencies must have effective authority to serve the public interest. However, the worth of what they do depends on how that authority is used. Rule-centred law tends to expend authority on securing conformity to rules, not on solving problems. Responsive regulation is less interested in rule-compliance for its own sake than in the mobilizing of energies for the achievement of public purposes. Therefore, the enterprise must be perceived as, and must become, a resource for public policy. Although this requires respect for and deference to its special needs and purposes, it also requires a reconstruction of the enterprise to enhance its capacity for responsible conduct.

The larger objective is to increase the institutional competence of the regulators as well as the regulated. To this end, regulatory agencies must develop strategies of *generalization* which look to principles, purposes and outcomes rather than to detailed rules or prescriptive standards; and strategies of *mediation*, which encourage the formation of "private" regulatory institutions within industries or within sectors of industries. These and related strategies vindicate and even enlarge the authority of the regulators. At the same time, they create frameworks within which self regulation can go forward and can be made to serve public as well as private interests.

NOTES

1 My comments draw in part on the discussion of institutions in Selznick, 1992.
2 Thus Luhmann, recognizing that political forces and economic interests impinge on law, says: "The law's autonomy is in danger only when the code itself is in danger—for instance when decisions are taken in the legal system itself increasingly according to the difference between beneficial and harmful rather than the difference between legal and illegal" (1988: 347). However, the boundary between what is legal/illegal and beneficial/harmful is often blurred, not only in, for example, a legal determination of "the best interests of the child" in divorce or adoption

PHILIP SELZNICK

proceedings, but also in determining what claims of right will be given legal recognition.

REFERENCES

Bardach, Eugene & Kagan, Robert (1982) *Going by the Book: the Problem of Regulatory Unreasonableness*. Philadelphia: Temple University Press.
Douglas, Mary (1986) *How Institutions Think*. Syracuse: Syracuse University Press.
Fuller, Lon L. (1969) *The Morality of Law*. New Haven: Yale University Press.
Luhmann, Niklas (1988) Closure and Openness: On Reality in the World of Law. In Teubner, G. (Ed). *Autopoietic Law: A New Approach to Law and Society*. Berlin: de Gruyter.
Rees, Joseph K. (1988) *Reforming the Workplace: A Study of Self-Regulation in Occupational Safety*. Philadelphia: University of Pennsylvania Press.
Selznick, Philip (1992) *The Moral Commonwealth: Social Theory and the Promise of Community*. Berkeley: University of California Press.

Index

Index compiled by Geoffrey C. Jones